AF361784

Nanometre-Scale Biosensors for Applications in Medicine/ Environment/Food/Biomedical Research

Nanometre-Scale Biosensors for Applications in Medicine/ Environment/Food/Biomedical Research

Editor

Pik Kwan Peggy LO

Basel • Beijing • Wuhan • Barcelona • Belgrade • Novi Sad • Cluj • Manchester

Editor
Pik Kwan Peggy LO
City University of Hong Kong
Hong Kong, China

Editorial Office
MDPI
St. Alban-Anlage 66
4052 Basel, Switzerland

This is a reprint of articles from the Special Issue published online in the open access journal *Biosensors* (ISSN 2079-6374) (available at: https://www.mdpi.com/si/biosensors/Nanometre_Biosens).

For citation purposes, cite each article independently as indicated on the article page online and as indicated below:

Lastname, A.A.; Lastname, B.B. Article Title. *Journal Name* **Year**, *Volume Number*, Page Range.

ISBN 978-3-0365-9696-9 (Hbk)
ISBN 978-3-0365-9697-6 (PDF)
doi.org/10.3390/books978-3-0365-9697-6

Contents

About the Editor

Pik Kwan Peggy LO

Pik Kwan Peggy LO currently works as an associate professor at the Department of Chemistry at City University of Hong Kong. She studied for her Ph.D. in the Department of Chemistry at McGill University in Canada. Following her doctoral studies, she pursued postdoctoral research at the Department of Chemistry and Chemical Biology at Harvard University in the United States. Dr. Lo's research group is dedicated to the development of a versatile platform that comprehensively addresses the understanding, detection, and treatment of non-communicable diseases. Her group adopts an interdisciplinary approach, integrating synthetic organic chemistry, DNA chemistry, nanotechnology, and molecular biology. Her objectives encompass the synthesis of novel (macro)molecules, the construction of nanoscale structures, and the refinement of their photophysical, photochemical, and biochemical properties. These endeavors have culminated in the creation of intelligent and biocompatible materials tailored for biomedical and technological applications.

 *biosensors*

Editorial

Nanometre-Scale Biosensors Revolutionizing Applications in Biomedical and Environmental Research

Pik Kwan Lo [1,2]

[1] Department of Chemistry and State Key Laboratory of Marine Pollution, City University of Hong Kong, Tat Chee Avenue, Kowloon Tong, Hong Kong SAR, China; peggylo@cityu.edu.hk

[2] Biotech and Health Care, Shenzhen Research Institute of City University of Hong Kong, Shenzhen 518057, China

Citation: Lo, P.K. Nanometre-Scale Biosensors Revolutionizing Applications in Biomedical and Environmental Research. *Biosensors* **2023**, *13*, 969. https://doi.org/10.3390/bios13110969

Received: 31 October 2023

Accepted: 2 November 2023

Published: 6 November 2023

Driven by the convergence of nanotechnology, biotechnology, and materials science, the field of biosensors has witnessed remarkable advancements in recent years. These miniature devices can detect specific targets by leveraging recognition receptors, and subsequently translate biomolecular interactions into measurable signals. The advent of nanometre-scale biosensors has opened new avenues for the application of these devices in various domains, including medicine, environment, food, and biomedical research. One of the key advantages of nanometre-scale biosensors lies in the availability of a wide array of recognition elements, such as antibodies, enzymes, nucleic acids, peptides, and carbohydrate-binding proteins. These elements can be easily synthesized and integrated into biosensor platforms, enabling the transformation of molecular binding events into various physical signals, including electronic, optical, magnetic, and mass changes. This versatility allows for qualitative and quantitative analysis of a broad range of biomolecules, proteins, DNAs/RNAs, biomarkers, metal ions, microorganisms, and toxin pollutants.

In this Special Issue, Moabelo et al. (contribution 1) reported the development of a rapid and specific biosensor for the detection of retinol-binding protein 4 (RBP4), a potential biomarker for the early diagnosis of type 2 diabetes mellitus (T2DM). The biosensor utilizes gold nanoparticles (AuNPs) as the sensing platform and employs a colorimetric detection approach. The retinol-binding protein aptamer (RBP-A) is immobilized on the surface of the AuNPs, enabling nanoparticle stabilization. RBP4 binds to the RBP-A, causing detachment from the AuNPs and resulting in the aggregation of the nanoparticles upon the addition of sodium chloride (NaCl). This aggregation leads to a visible color change in the AuNP solution. The developed assay provides a test result within 5 min and demonstrates a limit of detection of 90.76 ± 2.81 nM. The study highlights the advantages of using aptamers in biosensing applications, such as their high specificity, selectivity, low molecular weight, and ease of production. The colorimetric approach based on AuNPs offers simplicity, rapid response, and high sensitivity, making it suitable for use in point-of-care testing (PoCT) and resource-limited settings. This study provides a promising approach for the detection of RBP4, which can aid in the early diagnosis of T2DM.

In their research, Cui et al. (contribution 2) applied a microfluidic live-cell immunoassay, integrated with a microtopographic environment, in order to investigate intercellular interactions in different tumor microenvironments. The platform allows the coculturing of immune cells and cancer cells on tunable substrates and the simultaneous detection of different cytokines. It also enables the investigation of migration behaviors of mono- and co-cultured cells on flat and grating platforms, revealing topography-induced intercellular and cytokine responses. The study employs a microbead-based sandwich assay for on-chip cytokine monitoring and achieves precise quantification with a low sample volume and short assay time. The authors validate the biocompatibility of the co-culture strategy and compare the immunological states of different cell types on different substrates. The integrated microfluidic platform offers an efficient and precise approach for on-chip cytokine detection, eliminating manual sampling procedures and enabling continuous cytokine monitoring. The findings have implications for early immune diagnosis, personalized immunotherapy, and precision medicine. This study presents a novel microfluidic platform that enhances our understanding of intercellular interactions and cytokine secretion, providing valuable insights for the development of optimized diagnosis and treatment strategies in various fields.

In this Special Issue, Lau et al. (contribution 3) demonstrated the design and development of small-molecule fluorescent probes, named DB-TPE, for the selective binding and detection of RNA sequences containing CAG repeats. These repeats are associated with neurodegenerative diseases such as Huntington's disease. The researchers modified a bis(amidinium)-based binder, known for its selective targeting of CAG RNA, and incorporated an environment-sensitive fluorophore to create a turn-on fluorescent probe. The probe exhibited a significant 19-fold increase in fluorescence upon binding to a short CAG RNA sequence. The binding and fluorescence response were found to be specific to the RNA secondary structure with A·A mismatches. This study suggests that the DB-TPE fluorescent probe shows promise for application pathological studies, disease monitoring, and the diagnosis of neurodegenerative diseases linked to expanded CAG RNA repeats.

In this Special Issue, Monteiro et al. (contribution 4) utilized a sensitive and selective label-free photoelectrochemical (PEC) immunosensor for the detection of cardiac troponin I (cTnI), a biomarker of myocardial infarction. The immunosensor platform is based on a fluorine-doped tin oxide (FTO)-coated glass photoelectrode modified with bismuth vanadate ($BiVO_4$) and sensitized by an electrodeposited bismuth sulfide (Bi_2S_3) film. The PEC response of the platform is enhanced in the presence of Bi_2S_3. The cTnI antibodies (anti-cTnI) are immobilized on the platform surface to create an anti-cTnI/Bi_2S_3/$BiVO_4$/FTO immunosensor. When the immunosensor is incubated in cTnI solution, it inhibits the photocurrent generated by ascorbic acid, enabling the detection of cTnI. The immunosensor exhibits a linear relationship between the photocurrent and the logarithm of cTnI concentration within the range of 1 pg mL^{-1} to 1000 ng mL^{-1} The performance of the immunosensor is validated using artificial blood plasma samples, showcasing its successful application for cTnI detection with high recovery values. This study shows the potential of the developed PEC immunosensor for the sensitive and selective detection of cTnI, which can aid in the early diagnosis of myocardial infarction.

In this Special Issue, Zhang et al. (contribution 5) presented a new approach for the detection of silver ions (Ag$^+$) using a minidumbbell DNA-based sensor (M-DNA). Silver ions are common heavy metal pollutants with potential health risks. The M-DNA sensor is designed with an 8-nucleotide minidumbbell structure containing a unique reverse wobble C·C mispair, which serves as the binding site for Ag$^+$. The sensor demonstrates high accuracy in detecting Ag$^+$ in real environmental samples, with a detection limit of 2.1 nM. The M-DNA sensor offers advantages such as fast kinetics and easy operation due to the use of an ultrashort oligonucleotide. This new DNA structural motif provides a promising platform for the development of on-site environmental Ag$^+$ detection devices. The study contributes to the field of nanometre-scale biosensors and their applications in environmental monitoring, particularly in the field of heavy metal detection.

This Special Issue contains several review articles highlighting different aspects of biosensors and their applications in various fields. These articles provide valuable insights into the development and utilization of nanometre-scale biosensors for diverse purposes. Chen et al. (contribution 6) emphasize the significant potential use of electrical impedance biosensors, integrated with handheld and bench-top microfluidic technologies, in various biological sensing applications. These biosensors enable easy operation by personnel without requiring specialized training. The use of microfluidics enhances sensitivity and analytical capabilities, reduces reagent consumption and analysis time, and improves reliability and efficiency through automation. Impedance biosensors offer extensive information about cell properties, eliminating the need for fluorescent labeling or other treatments, and serving as a cell fingerprint for identification and characterization. Lam et al. (contribution 7) summarize the strategies employed in Capture-SELEX, a modified SELEX method, for detecting and characterizing small-molecule-aptamer interactions. They discuss the development of aptamer-based biosensors for diverse applications. The advantages of aptamers over antibodies are highlighted, including thermal stability, low cost, chemical modification capabilities, and ease of generation. Aptamers exhibit selective and tight binding with a wide range of targets, including proteins, cells, microorganisms, and small-molecule contaminants. The article also addresses challenges associated with the Capture-SELEX platform and biosensor development, such as false-positive results and conformational changes of target molecules. Wen et al. (contribution 8) highlight the versatility of metal–organic frameworks (MOFs) in constructing single- or multi-emission signals for analytical purposes. They discuss the engineering and tailoring of photonic units within MOFs, originating from metal nodes, linkers, or guest molecules, to achieve diverse luminescence signals. MOFs show promise when applied to surface-enhanced Raman spectroscopy (SERS) methods, acting as substrates for signal enhancement or forming composite substrates by encapsulating metallic nanomaterials. The tunable properties, high surface area, controllable pore size, and sieving effect of MOFs contribute to improved detection selectivity, sensitivity, stability, and reproducibility. The potential of MOFs for multiplexed detection, integrating multiple signals into a single optical nanoprobe for ratiometric and multimodality measurements, is also discussed. Koyappayil et al. (contribution 9) highlight the recent trends, advances, prospects, advantages, and limitations of using metal nanoparticle-decorated 2D materials in electrochemical biomarker detection. Techniques such as electrochemical impedance spectroscopy (EIS), cyclic voltammetry (CV), differential pulse voltammetry (DPV), and square wave voltammetry (SWV), which possesses uses in biomarker detection systems, are discussed. The role of gold nanoparticles (AuNPs) on 2D platforms in enhancing signals and anchoring antibodies for biomarker detection is also described. Deng et al. (contribution 10) discuss strategies for the preparation and modification of semiconducting polymer dots (Pdots). Pdots have received significant attention due to their favorable photophysical properties and versatility, including large absorption cross-section, high brightness, tunable fluorescence emission, excellent photostability, biocompatibility, and ease of modification. They present recent advancements in the use of Pdots as optical probes for analytical detection and clinical applications in biosensing and disease diagnosis.

This Special Issue aims to showcase cutting-edge developments and innovative approaches that are helping to shape the field of nanometre-scale biosensors. By sharing knowledge and fostering collaboration, our goal is to accelerate progress and pave the way for practical applications that benefit society as a whole.

Conflicts of Interest: The authors declare no conflict of interest.

List of Contributions

1. Moabelo, K.L.; Lerga, T.M.; Jauset-Rubio, M.; Sibuyi, N.R.S.; O'Sullivan, C.K.; Meyer, M.; Madiehe, A.M. A Label-Free Gold Nanoparticles-Based Optical Aptasensor for the Detection of Retinol Binding Protein 4. *Biosensors* **2022**, *12*, 1061. [CrossRef]
2. Cui, X.; Liu, L.; Li, J.; Liu, Y.; Liu, Y.; Hu, D.; Zhang, R.; Huang, S.; Jiang, Z.; Wang, Y.; et al. A Microfluidic Platform Revealing Interactions between Leukocytes and Cancer Cells on Topographic Micropatterns. *Biosensors* **2022**, *12*, 963. [CrossRef]
3. Lau, M.H.Y.; Wong, C.H.; Chan, H.Y.E.; Au-Yeung, H.Y. Development of Fluorescent Turn-On Probes for CAG-RNA Repeats. *Biosensors* **2022**, *12*, 1080. [CrossRef]
4. Monteiro, T.O.; Neto, A.G. dos S.; de Menezes, A.S.; Damos, F.S.; Luz, R. de C.S.; Fatibello-Filho, O. Photoelectrochemical Determination of Cardiac Troponin I as a Biomarker of Myocardial Infarction Using a Bi_2S_3 Film Electrodeposited on a $BiVO_4$-Coated Fluorine-Doped Tin Oxide Electrode. *Biosensors* **2023**, *13*, 379. [CrossRef]
5. Zhang, J.; Liu, Y.; Yan, Z.; Wang, Y.; Guo, P. A Novel Minidumbbell DNA-Based Sensor for Silver Ion Detection. *Biosensors* **2023**, *13*, 358. [CrossRef]
6. Chen, Y.S.; Huang, C.H.; Pai, P.C.; Seo, J.; Lei, K.F. A Review on Microfluidics-Based Impedance Biosensors. *Biosensors* **2023**, *13*, 83. [CrossRef]

7.   Lam, S.Y.; Lau, H.L.; Kwok, C.K. Capture-SELEX: Selection Strategy, Aptamer Identification, and Biosensing Application. *Biosensors* **2022**, *12*, 1142. [CrossRef]
8.   Wen, C.; Li, R.; Chang, X.; Li, N. Metal-Organic Frameworks-Based Optical Nanosensors for Analytical and Bioanalytical Applications. *Biosensors* **2023**, *13*, 128. [CrossRef]
9.   Koyappayil, A.; Yagati, A.K.; Lee, M.H. Recent Trends in Metal Nanoparticles Decorated 2D Materials for Electrochemical Biomarker Detection. *Biosensors* **2023**, *13*, 91.
10. Deng, S.; Li, L.; Zhang, J.; Wang, Y.; Huang, Z.; Chen, H. Semiconducting Polymer Dots for Point-of-Care Biosensing and In Vivo Bioimaging: A Concise Review. *Biosensors* **2023**, *13*, 137. [CrossRef]

Article

Photoelectrochemical Determination of Cardiac Troponin I as a Biomarker of Myocardial Infarction Using a Bi_2S_3 Film Electrodeposited on a $BiVO_4$-Coated Fluorine-Doped Tin Oxide Electrode

Thatyara Oliveira Monteiro [1], Antônio Gomes dos Santos Neto [2], Alan Silva de Menezes [3], Flávio Santos Damos [2], Rita de Cássia Silva Luz [2,*] and Orlando Fatibello-Filho [1,*]

[1] Department of Chemistry, Federal University of São Carlos, São Carlos 13565-905, SP, Brazil
[2] Department of Chemistry, Federal University of Maranhão, São Luís 65080-805, MA, Brazil
[3] Department of Physics, Federal University of Maranhão, São Luís 65080-805, MA, Brazil
* Correspondence: rita.luz@ufma.br (R.d.C.S.L.); bello@ufscar.br (O.F.-F.)

Abstract: A sensitive and selective label-free photoelectrochemical (PEC) immunosensor was designed for the detection of cardiac troponin I (cTnI). The platform was based on a fluorine-doped tin oxide (FTO)-coated glass photoelectrode modified with bismuth vanadate ($BiVO_4$) and sensitized by an electrodeposited bismuth sulfide (Bi_2S_3) film. The PEC response of the Bi_2S_3/$BiVO_4$/FTO platform for the ascorbic acid (AA) donor molecule was approximately 1.6-fold higher than the response observed in the absence of Bi_2S_3. The cTnI antibodies (anti-cTnI) were immobilized on the Bi_2S_3/$BiVO_4$/FTO platform surface to produce the anti-cTnI/Bi_2S_3/$BiVO_4$/FTO immunosensor, which was incubated in cTnI solution to inhibit the AA photocurrent. The photocurrent obtained by the proposed immunosensor presented a linear relationship with the logarithm of the cTnI concentration, ranging from 1 pg mL^{-1} to 1000 ng mL^{-1}. The immunosensor was successfully employed in artificial blood plasma samples for the detection of cTnI, with recovery values ranging from 98.0% to 98.5%.

Keywords: cTnI; PEC immunosensor; bismuth vanadate; bismuth sulfide

Citation: Monteiro, T.O.; Neto, A.G.d.S.; de Menezes, A.S.; Damos, F.S.; Luz, R.d.C.S.; Fatibello-Filho, O. Photoelectrochemical Determination of Cardiac Troponin I as a Biomarker of Myocardial Infarction Using a Bi_2S_3 Film Electrodeposited on a $BiVO_4$-Coated Fluorine-Doped Tin Oxide Electrode. *Biosensors* **2023**, *13*, 379. https://doi.org/10.3390/bios13030379

Received: 11 February 2023
Revised: 6 March 2023
Accepted: 9 March 2023
Published: 13 March 2023

1. Introduction

Cardiovascular diseases, especially those that present with acute morbidity, are considered one of the leading causes of mortality worldwide. Acute myocardial infarction (AMI) is the main concern, defined by pathology as myocardial necrosis due to a prolonged reduction of blood supply to the heart. It is estimated that by 2030, approximately 23.3 million people globally will die annually from AMI [1,2]. In this sense, AMI requires an early, rapid, and effective diagnosis to improve the survival rate and ensure the health quality of patients. Among the known biochemical markers of early AMI, cardiac troponin I (cTnI) is considered the golden standard in medical diagnosis [3]. In addition to being a protein specifically related to myocardial damage, it can also remain in the myocardial tissue for a long time and is released from the cells in levels of a very low concentration within 3–4 h after the onset of AMI symptoms [4,5]. For these reasons, a sensitive method for detecting cTnI is reasonable. Several reliable, sensitive, and robust methods for detecting cTnI have been proposed, including methods based on fluorescence microscopy [6], 2D-chromatography [7], colorimetry [8], surface plasmon resonance (SPR) [9,10], liquid chromatography–tandem mass spectrometry [11], and SERS-based immunoassays [12]. However, some previously proposed methods may be time-consuming, exploit labeled probes, require trained personnel, or require expensive facilities to implement, making their use in point-of-care testing difficult [6,8–12].

Photoelectrochemical (PEC) immunoassay has attracted growing attention in recent years [13–15]. This approach exploits a combination of the high sensitivity of the PEC bioanalysis and the affinity of the antigen/antibody molecules [14]. In principle, a PEC immunosensor can be easily designed using a photoelectroactive material such as a semiconductor as the signal-producing transducer and an antibody as the biological recognition element [14]. PEC measurements can be performed by employing the signal-off strategy in which the photocurrent is decreased while the formation of antibody–antigen conjugates occurs, blocking the transport of redox probes to the sensing surface (such as a label-free immunosensor) [3]. Thus, the high photoresponsive sensitivity of the semiconductor materials on the electrode is an essential aspect for the successful application of these devices.

Some photoactive semiconductor materials present a good biocompatibility, rapid reactivity, and the rapid generation and separation of electron–hole pairs [15]. Among these semiconductors, bismuth vanadate ($BiVO_4$) is a promising material. Bismuth vanadate is an n-type semiconductor that presents a commonly monoclinic crystalline structure with good photocatalytic activity and a bandgap energy of 2.4 eV. It is appropriate to production of charge carriers under visible light irradiation [16]. In order to improve its photoelectrochemical efficiency and reduce the recombination processes of $BiVO_4$, heterojunctions commonly based on semiconductors with a narrower bandgap could be employed, such as bismuth sulfide (Bi_2S_3). Bismuth sulfide also is an n-type semiconductor and has a bandgap of 1.3 eV. Bi_2S_3 presents a reasonable efficiency of photocurrent conversion under visible light [17] and has become attractive for many PEC applications [18–21]. In this paper, based on the properties of the $BiVO_4$ and Bi_2S_3 materials described herein, we report a label-free PEC immunosensor designed with a junction of these two semiconductors to determine the cTnI biomarker in clinical samples in real time, exploiting the effects of the immunoreaction upon the response of the PEC platform to the ascorbic acid (AA) donor molecule.

2. Materials and Methods

2.1. Reagents and Chemicals

Human cardiac troponin I (cTnI), monoclonal cTnI antibody (anti-cTnI), N'-ethylcarbo diimide hydrochloride (EDC), N-hydroxysuccinimide (NHS), bovine serum albumin (BSA), bismuth nitrate ($Bi(NO_3)_3$), ammonium metavanadate (NH_4VO_3), ethylene glycol, and thioglycolic acid were purchased from Sigma-Aldrich (St. Louis, MO, USA). Sodium thiosulfate, ethylenediaminetetraacetic acid disodium salt dihydrate ($C_{10}H_{14}N_2Na_2O_8 \cdot 2H_2O$), sodium hydroxide, sodium dihydrogen phosphate, disodium hydrogen phosphate, ascorbic acid, citric acid, acetic acid, boric acid, phosphoric acid, sodium chloride, potassium chloride, calcium chloride, ammonium chloride, disodium sulfate, potassium dihydrogen phosphate, and urea were acquired from ISOFAR (Duque de Caxias, RJ, Brazil). All aqueous solutions were prepared with water purified in a OS10LXE Gehaka osmose system (São Paulo, SP, Brazil).

2.2. Experimental Apparatus

Photoelectrochemical experiments were performed using an Autolab potentiostat/galvanostat model PGSTAT 128N (Metrohm Autolab B. V., Netherlands) equipped with a Frequency Response Analyzer module, controlled by NOVA software, and coupled to a three-electrode electrochemical cell confined in a box to control the illumination on the photoelectrodes. A commercial 36 W LED lamp was used as a visible light source. A FTO glass photoelectrode (5 cm length × 1 cm width, with a modified area of 0.7×1.0 cm^2), modified with $Bi_2S_3/BiVO_4$, was used as the working electrode. Ag/AgCl/KCl$_{sat}$) was used as the reference electrode, and a Pt wire was used as the counter electrode. Electrochemical impedance spectroscopy experiments were carried out in a 0.1 mol L^{-1} KCl solution containing 5 mmol L^{-1} K$_3$[Fe(CN)$_6$] in the frequency range of 10^{-1}–10^5 Hz under an AC amplitude of 10 mV.

2.3. Synthesis of BiVO$_4$

This procedure was performed according to method proposed by [22]. For this step, 0.4850 g of Bi(NO$_3$)$_3$.5H$_2$O was transferred into a falcon tube containing 5 mL of ethylene glycol and sonicated for 40 min. A mass of 0.1175 g of NH$_4$VO$_3$ was transferred into a falcon tube containing 5 mL of water and heated in a water bath until complete dissolution. The latter solution was then dropped slowly into the former solution under constant stirring, causing the formation of a yellow BiVO$_4$ suspension.

2.4. Construction of Bi$_2$S$_3$/BiVO$_4$/FTO Photoelectrochemical Sensor

Prior to modification, the FTO electrode was washed with water and ethanol to remove any adsorbed species on the surface. The FTO plates were then sustained until the temperature equilibrated at 100 °C to the deposition of the BiVO$_4$ material. A volume of 35 µL of BiVO$_4$ suspension was dropped into 0.7 × 1.0 cm^2 of the FTO heated plate and a BiVO$_4$ film was formed after few minutes. This procedure was repeated three times to completely cover the FTO electrode surface. The BiVO$_4$/FTO platform was annealed at 500 °C for 1 h using an oven with a heating rate of 10 °C min^{-1}. After annealing, the BiVO$_4$/FTO electrode was allowed to cool to room temperature. Next, a Bi$_2$S$_3$ film was electrodeposited onto the BiVO$_4$/FTO platform, according to the method proposed by [23] with few adaptations. An amount of 0.04 mol L^{-1} Na$_2$S$_2$O$_3$ was diluted in a solution containing 4.5 mL purified water and 0.5 mL HCl. After the formation of colloidal sulfur, the solution was kept still at room temperature for 24 h to allow the sulfur particles to precipitate at the bottom of the solution. The supernatant solution was then used for the preparation of another solution containing 0.006 mol L^{-1} Bi(NO$_3$)$_3$ and 0.006 mol L^{-1} EDTA. The Bi$_2$S$_3$ electrodeposition onto the BiVO$_4$/FTO electrode surface was then performed by amperometry, applying a potential of −0.9 V vs. Ag/AgCl/KCl$_{sat}$ for 300 s at room temperature. The Bi$_2$S$_3$/BiVO$_4$/FTO platform was annealed using a hot plate at 200 °C for 30 min.

2.5. Optimization of the Platform Response for the AA before and after the Immobilization of the Biological Materials

Some experimental parameters were optimized to achieve a higher sensitivity of the Bi$_2$S$_3$/BiVO$_4$/FTO PEC platform to the AA donor molecule. Thus, the effect of the applied potential on the sensor response was initially investigated by evaluating the photocurrents in 0.03 mol L^{-1} AA in a 0.1 mol L^{-1} phosphate buffer, at pH 7.4 under different potentials (from −0.2 V to 0.2 V vs. Ag/AgCl/KCl$_{sat}$). The influence of the nature of the buffer solution on the response of the platform was then evaluated by monitoring the photocurrents obtained from 0.03 mol L^{-1} AA under an applied potential of 0 V vs. Ag/AgCl/KCl$_{sat}$ in three different buffer solutions: phosphate, McIlvaine, and Britton–Robinson. The study of the effects of the buffer solutions was performed by maintaining the buffer concentration at 0.1 mol L^{-1} and pH 7.4. The effect of the concentration of the AA donor molecule on the response of the platform was also investigated for AA concentrations from 0.01 to 0.06 mol L^{-1}.

After the optimization of the Bi$_2$S$_3$/BiVO$_4$/FTO PEC platform response for the concentration of the AA molecule, the platforms were modified with the antibody of troponin (anti-cTnI). Initially, the electrode was incubated in an aqueous solution of 0.003 mol L^{-1} thioglycolic acid (TGA) at room temperature for 15 min to introduce carboxylic groups at the surface of the Bi$_2$S$_3$/BiVO$_4$/FTO PEC platform. Following this, the excess TGA was removed with water and 15 µL of an EDC/NHS mixed solution (0.15 mol L^{-1} EDC and 0.10 mol L^{-1} NHS) was added and incubated for 1 h on the functionalized surface to activate the -COOH groups. The activated platform was then incubated with 10 µL of different concentrations of anti-cTnI (from 1 to 5 µg mL^{-1}) at room temperature for a determinate time. The platform modified with the antibody was gently washed to remove the weakly adsorbed anti-cTnI. Furthermore, the EDC/NHS excess was removed from surface with purified water. Next, 10 µL of 1 % (*m/v*) BSA solution was added to block nonspecific binding sites. Posteriorly, the effect of the interaction time between the antibody and the antigen on the response of AA (10, 15, 20, 25, and 30 min) was studied using a concentration of 1 ng mL^{-1} cTnI. After evaluating all the experimental parameters, the conditions that provided the highest photocurrent value were fixed to finally obtain the analytical curve for the determination of cTnI. The analytical curve was obtained after the incubation of the anti-cTnI/Bi_2S_3/$BiVO_4$/FTO PEC immunosensor platform with different cTnI concentrations (from 1 pg mL^{-1} to 1000 ng mL^{-1} cTnI).

2.6. Preparation and Analysis of Artificial Blood Plasma Samples

The performance of the anti-cTnI/Bi_2S_3/$BiVO_4$/FTO PEC immunosensor was evaluated by the determination of cTnI in artificial blood plasma samples using an external calibration method. The artificial samples were composed of 0.138 mol L^{-1} NaCl, 0.004 mol L^{-1} $NaHCO_3$, 0.003 mol L^{-1} KCl, 0.001 mol L^{-1} $Na_2HPO_4 \cdot 3H_2O$, 0.002 mol L^{-1} $MgCl_2 \cdot 6H_2O$, 0.003 mol L^{-1} $CaCl_2$, and 0.507 mmol L^{-1} Na_2SO_4 [24]. The samples were spiked with different concentrations of cTnI (0.05, 2.0 and 50 ng mL^{-1}), and aliquots of 10 µL of each sample were added directly onto the immunosensor surface at the optimal conditions and analyzed in presence of the AA molecule using the developed sensor in three replicates.

2.7. X-ray Diffraction and Scanning Electron Microscopy

X-ray Diffraction (XRD) measurements were obtained with a D8 Advance diffractometer (Bruker) equipped with the LynxEye linear detector, using Cu Ka radiation (k = 1.5418 Å) and operating at 40 kV/40 mA. The diffraction patterns were collected from 15 to 80 ° with a step size of 0.02° and a counting time of 0.6 s. The sample surface was imaged on an EVO HD electron microscope (Zeiss) at 10 kV after being fixed onto stubs using a carbon film.

3. Results and Discussion

3.1. Characterization of the Materials by X-Ray Diffraction and Scanning Electron Microscopy

The materials utilized for the construction of the photoelectrochemical platform were evaluated by X-ray diffraction (XRD). In Figure 1A (red line), the characteristic peaks of the $BiVO_4$/FTO can be observed. The 2θ values of $BiVO_4$ at 18.7, 28.7, 30.4, 34.5, 35.2, 39.7, 42.2, 46.7, 50.3, 53.3, and 58.4 correspond to the (110)(011), (−130)(−121)(121), (040), (200), (002), (211), (051), (240), (202), (−161)(161), and (−321)(321) planes of the monoclinic $BiVO_4$ scheelite (JCPDS 014-0688) [16]. A small amount of the $BiVO_4$ tetragonal phase (JCPDS 014-0133) was also detected, and its most intense peak (200) can be seen at 24.4°. Furthermore, additional diffraction peaks appear after the Bi_2S_3 electrodeposition, as can be seen in Figure 1A (black diffractogram). In this figure, the XRD spectrum of the $BiVO_4$/Bi_2S_3/FTO photoelectrochemical platform shows the characteristic diffraction peaks of the $BiVO_4$/FTO platform. In addition, one can see a peak at 27.2° from the (012) plane of the Bi rhombohedral phase and the (220) and (130) peaks (at 22.3 and 24.8, respectively) of the orthorhombic Bi_2S_3 phase (JCPDS 017-0320), clearly demonstrating that the modification process was effective and that all samples display a high crystallinity. In addition, to show the morphology of the Bi_2S_3/$BiVO_4$/FTO platform, a scanning electron microscopy (SEM) image was obtained. Figure 1B shows smaller particles covering an agglomerated material. These smaller particles resulted from the electrodeposition of Bi_2S_3 on the $BiVO_4$/FTO platform surface after the preparation procedure of the Bi_2S_3/$BiVO_4$/FTO electrochemical platform, which was described in the experimental section.

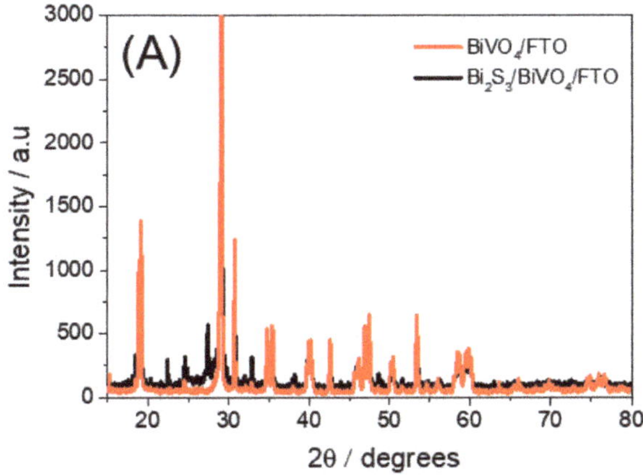
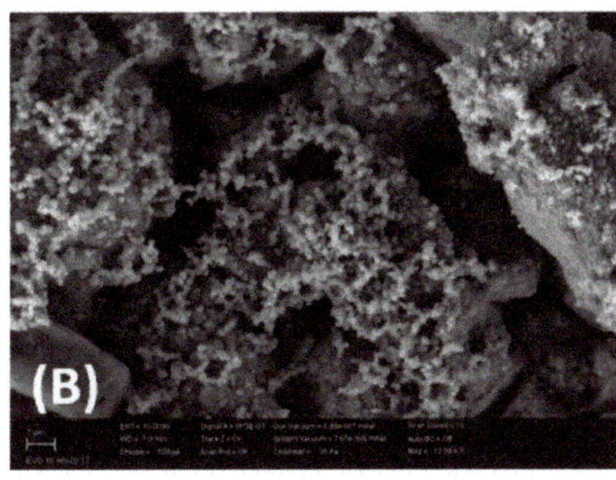

Figure 1. (**A**) X-Ray diffraction patterns of $BiVO_4$/FTO and $BiVO_4$/Bi_2S_3/FTO. (**B**) SEM image of $BiVO_4$/Bi_2S_3/FTO. Scale bar = 1 μm.

3.2. EIS and Photoelectrochemical Characterization of Bi_2S_3/$BiVO_4$/FTO Platform

Electrochemical impedance spectroscopy (EIS) is a powerful tool for monitoring the charge transfer resistance, which can be used to demonstrate the successful fabrication of a PEC-sensing platform [25]. EIS measurements were taken to evaluate the charge transfer characteristics of the semiconductor materials. EIS measurements were also performed to evaluate the electrocatalytic effect of the semiconductor materials and the increase in the electroactive area of the working electrodes.

Figure 2A shows the Nyquist plots of the FTO, $BiVO_4$/FTO, and Bi_2S_3/$BiVO_4$/FTO electrodes, recorded at the open circuit potential and modeled using an adapted Randles equivalent circuit (inset in Figure 2A) consisting of a cell resistance (R_s) in series with a parallel combination of a constant phase element (CPE), considered a non-ideal capacitance, and a charge transfer resistance (R_{ct}) with a Warburg impedance (Z_w). Following the fitting of the parameters presented in Figure 2A. The $BiVO_4$/FTO electrode presented a charge transfer resistance value of 515 Ω, while the Bi_2S_3/$BiVO_4$/FTO electrode exhibited an R_{ct} value of only 93 Ω. These results suggest that the sensitization of the platform by Bi_2S_3 enhanced the electron transfer in the electrode–solution interface, contributing to the improved electrocatalytic properties of the sensing platform. Furthermore, the capacitance values of the $BiVO_4$/FTO and Bi_2S_3/$BiVO_4$/FTO electrodes were determined to be 0.0629 mF and 3.08 mF, respectively. These results indicate an increase in the electroactive area of the surface due to increase in the capacitances.

The effects of the LED light on the electron transfer processes and the lifetime of the electron were evaluated for the different sensing platforms to investigate the nature of their PEC responses. Figure 2B presents the Nyquist plots of the Bi_2S_3/$BiVO_4$/FTO electrode obtained in 0.1 mol L^{-1} Na_2SO_4 solution in the absence and in the presence of visible LED light. As can be seen, there was a decrease in the semicircle diameter of spectra in the presence of LED light, indicating that the R_{ct} value decreased. These results suggest that light irradiation enhances the electrocatalytic potential of the sensing platform since it enables the formation of electron–hole pairs in the Bi_2S_3/$BiVO_4$ composite material. Bode phase plots were obtained for the $BiVO_4$/FTO and Bi_2S_3/$BiVO_4$/FTO electrodes in order to estimate the lifetime of the electron through the following equation [26]:

$$\tau_e = \frac{1}{(2\pi f_{max})} \tag{1}$$

where τ_e is the lifetime of the electron and f_{max} is the maximum frequency in the Bode phase diagram.

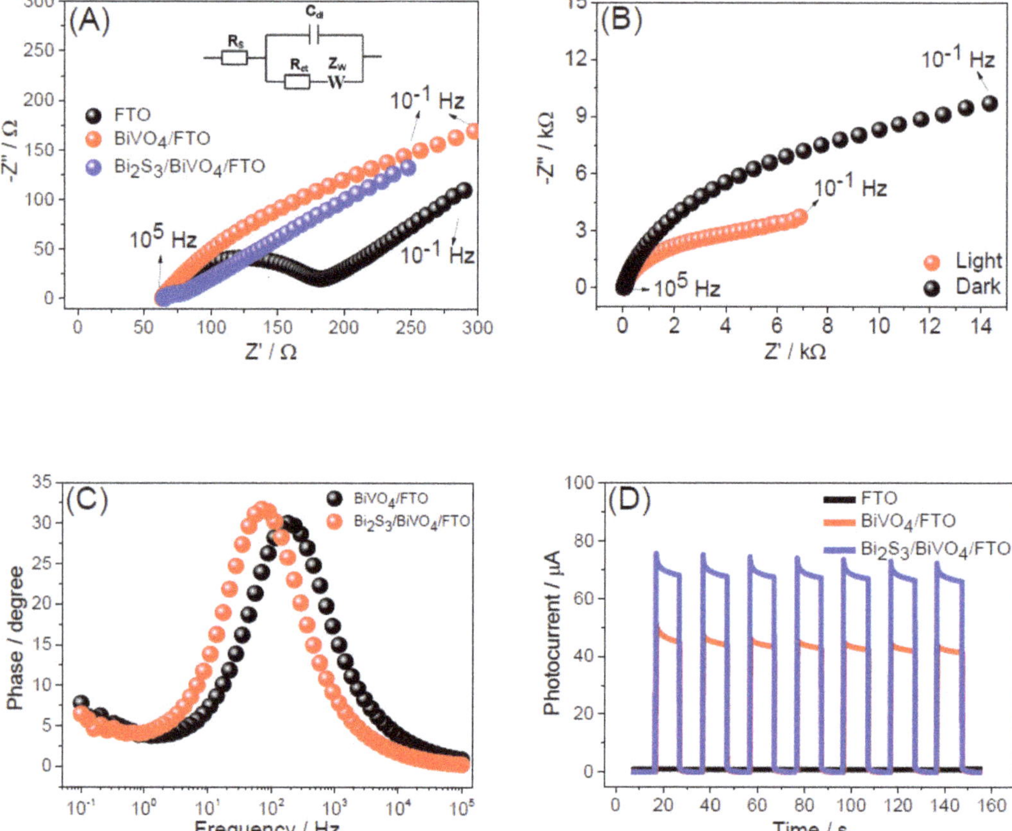

Figure 2. (**A**) Nyquist plots for the FTO (black), BiVO$_4$/FTO (red), and Bi$_2$S$_3$/BiVO$_4$/FTO (blue) electrodes, recorded in 0.1 mol L^{-1} KCl solution containing 5 mmol L^{-1} Fe[(CN$_6$)]$^{3-/4-}$. (**B**) Nyquist plots for the Bi$_2$S$_3$/BiVO$_4$/FTO photosensor in aqueous 0.1 mol L^{-1} Na$_2$SO$_4$ solution, recorded at an open-circuit potential in the dark (black) and under the visible LED light irradiation (red). (**C**) Bode phase plots for BiVO$_4$/FTO (black) and Bi$_2$S$_3$/BiVO$_4$/FTO (red) electrodes in 0.1 mol L^{-1} KCl solution containing 0.03 mol L^{-1} AA donor molecules. (**D**) Photocurrent response of FTO (black line), BiVO$_4$/FTO (red line), and Bi$_2$S$_3$/BiVO$_4$/FTO (blue line) photoelectrodes. Amperometric measurements were performed in a 0.1 mol L^{-1} phosphate buffer (pH 7.4) containing 0.03 mol L^{-1} AA, E$_{appl}$ = 0 V vs. Ag/AgCl/KCl$_{sat}$.

Figure 2C shows that the f_{max} in the Bode phase plots of the BiVO$_4$ platform decreased from 179.6 Hz to 70.38 Hz after sensitization by Bi$_2$S$_3$, reflecting the increase in the lifetime of the electron from 0.886 ms to 2.26 ms. The enhancement of the lifetime of the electron suggests that the formation of the Bi$_2$S$_3$/BiVO$_4$ heterojunction enabled a slower recombination of the electron–hole pairs, which also explains the more favorable electron transfer between composite material and FTO electrode, as shown in the study of Figure 2A.

Additionally, the photocurrent intensity obtained for each individual component of the developed PEC platform was evaluated in the presence of an AA donor molecule (Figure 2D). As can be seen, there was an increase in the photocurrent response for the BiVO$_4$/FTO platform after sensitization by Bi$_2$S$_3$ film, confirming an improvement in the photo–current conversion efficiency due to a greater absorption of visible light, an increase in the electroactive area, and an improvement in the electrochemical properties of the platform. Although BiVO$_4$ is a semiconductor with good photocatalytic activity, its photoelectrochemical efficiency may not reach a desirable level due to the occurrence of charge recombination [27], which is minimized after the conjunction of energy bands in the $Bi_2S_3/BiVO_4$ composite. Furthermore, Bi_2S_3 is a material that presents a high surface activity, further improving the catalytic activity of the PEC platform [28].

3.3. Evaluation of Experimental Parameters on the $Bi_2S_3/BiVO_4/FTO$ PEC Platform Response

Initially, the effect of the applied potential, buffer type, and AA (electron donor molecule) concentration on the response of the $Bi_2S_3/BiVO_4/FTO$ PEC platform were evaluated. The applied potential is an important parameter that can directly influence the analytical performance of a sensor. Thus, the effect of the applied potential on the response of the $Bi_2S_2/BiVO_4/FTO$ PEC platform was investigated, and the results are presented in Figure S1. According to this figure, it can be observed that the photocurrent of the $Bi_2S_2/BiVO_4/FTO$ PEC platform in the presence of 0.03 mol L^{-1} AA increased when the applied potential changed from -0.2 V to 0 V vs. $Ag/AgCl/KCl_{sat}$, and it remained almost constant from 0 to 0.1 V vs. $Ag/AgCl/KCl_{sat}$. These results suggest that using a potential of 0 V is enough to obtain a high photocurrent value. Under these conditions, a higher sensitivity to the system was achieved while maintaining low biasing conditions, making it possible to determine the analyte even at very low concentrations that consume a minimum of energy. In addition, it is possible to significantly reduce or eliminate the possible influence of interfering species on the photoelectrochemical processes. At potentials above 0.1 V, it can be observed that the photocurrent tended to decrease. This can be related to a lower stability of the $Bi_2S_2/BiVO_4$ film, causing a lower efficiency for AA oxidation. In this sense, an applied potential of 0 V was chosen to construct the analytical curve for the determination of the antioxidant; all subsequent measurements were then performed under 0 V.

Posteriorly, the effect of the following buffer solutions: phosphate, McIlvaine, and Britton–Robinson on the response of the $Bi_2S_2/BiVO_4/FTO$ PEC platform in the presence of 0.03 mol L^{-1} AA were evaluated (Figure S2). Figure S2 showed no significative difference in the intensity of photocurrent among the different electrolytes studied; however, considering the high stability of the AA photocurrent response and the simplicity of preparing the buffer, a phosphate buffer solution was chosen for all subsequent assays of the PEC sensor. Furthermore, the effect of the donor molecule concentration on the platform response was also investigated by monitoring the photocurrent of the platform for AA in the following concentrations: 0.01, 0.02, 0,03, 0.04, 0.05, and 0.06 mol L^{-1} (Figure S3). This study showed that the process of donor molecule oxidation achieves maximum efficiency when the AA concentration reaches 0.04–0.05 mol L^{-1}. Above this concentration range, there is a tendency to obtain a lower photocurrent because the PEC platform hinders the oxidation of the analyte. In this context, the donor molecule concentration was kept at 0.04 mol L^{-1} for all subsequent studies. Subsequently, the influence of the concentration of the immobilized anti-cTnI antibodies on the $Bi_2S_3/BiVO_4/FTO$ PEC platform was evaluated, as was the interaction time between the antibody and the antigen (cTnI) immobilized on the PEC platform. Both studies were performed in the presence of 0.04 mol L^{-1} AA.

Figure 3A shows the photoelectrochemical response of the $Bi_2S_3/BiVO_4/FTO$ PEC platform in a phosphate buffer containing 0.04 mol L^{-1} AA after the incubation of the platform with different concentrations of anti-troponin I (anti-cTnI) (1, 2, 5 and 7 µg mL^{-1}). Figure 3B shows the variation of the photocurrents ($\Delta I = I_0 - I$, where I_0 and I are the photocurrents obtained before and after incubation of the platform with anti-cTnI, respectively) obtained from Figure 3A. According to Figure 3B, it can be observed that the photocurrent presented a high increase from 1 to 2 µg mL^{-1}. At concentrations of 2, 5, and 7 µg mL^{-1}, the photocurrent presented a percentage of decrease in relation to the initial photocurrent value, I_0 (without antibody immobilization) of approximately 50, 40, and 35%, respectively. Based on these results, it was considered that any of the three concentrations (2, 5, or 7 µg mL^{-1}) could be used for the preparation of the anti-cTnI/$Bi_2S_3/BiVO_4/FTO$ photoelectrochemical immunosensor. In this context, an intermediate concentration of 5 µg mL^{-1} was chosen for further experiments.

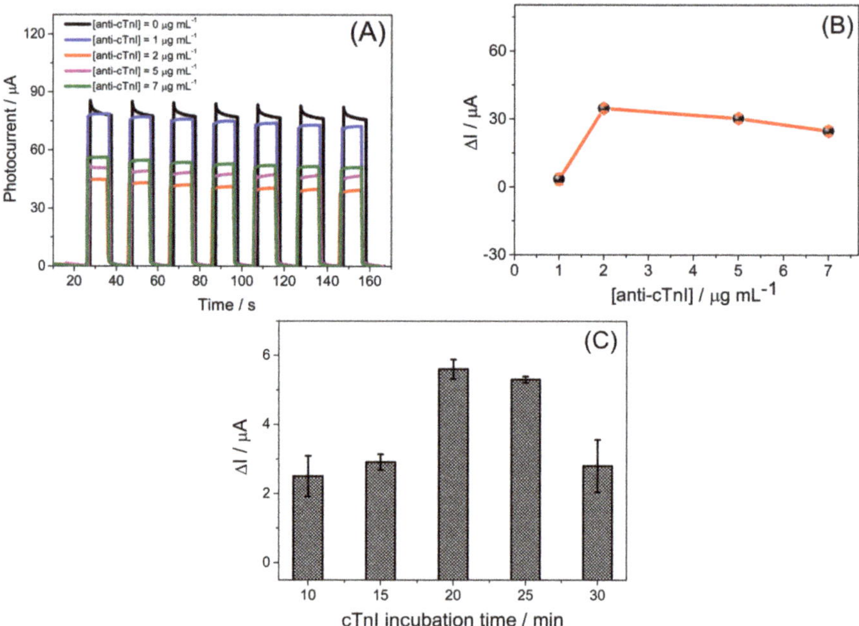

Figure 3. (**A**) Photoelectrochemical response of the Bi_2S_3/$BiVO_4$/FTO PEC platform for different anti-cTnI concentrations and (**B**) plot of the photocurrent variation vs. [anti-cTnI]. Data obtained from Figure 3(**A**). (**C**) Effect of incubation time of the anti-cTnI/Bi_2S_3/$BiVO_4$/FTO PEC platform with 1 ng mL^{-1} cTnI on the variation of the photocurrent. All measurements were carried in 0.1 mol L^{-1} phosphate buffer, pH 7.4, containing 0.04 mol L^{-1} AA. E_{appl} = 0 V vs. Ag/AgCl/KCl$_{sat}$.

In order to evaluate the effects of the interaction time between the antibodies (anti-cTnI) immobilized on the platform and the antigens (cTnI) from the incubation solution, the immunosensor platform was incubated with 1 ng mL^{-1} of cTnI antigens in a phosphate buffer solution containing 0.04 mol L^{-1} AA at several incubation times (10, 15, 20, 25, and 30 min). Figure 3C shows the variation of the photocurrent of the anti-cTnI/Bi_2S_3/$BiVO_4$/FTO PEC immunosensor before, I_0, and after, I, the interaction with the cTnI antigens ($\Delta I = I_0 - I$). According to the results shown in Figure 3C, the inhibition of the photocurrent of the anti-cTnI/Bi_2S_3/$BiVO_4$/FTO PEC immunosensor to the donor molecule increased significantly when the incubation time increased from 10 min to 20 min, with a low decrease observed for 25 min of interaction. As the standard deviation of the photocurrent was lower at 25 min, this interaction time between the antibodies and the Troponin I antigens was selected for all subsequent assays. The amperograms shown in Figure 3C are presented in Figure S4 in the supporting information. Additionally, it is important to emphasize that the results presented in this figure suggest that it is possible to monitor the cTnI antigen concentrations with the anti-cTnI/Bi_2S_3/$BiVO_4$/FTO PEC immunosensor from the decrease in the analytical signal. However, in order to propose a possible mechanism for the detection of troponin I, the effects of different concentrations of cTnI on the variation of photocurrent of the immunosensor were evaluated.

3.4. Analytical Performance of Bi_2S_3/$BiVO_4$/FTO PEC Immunosensor

Under the optimum immunoassay conditions, the PEC platform response to the AA donor molecule was evaluated after incubation with different cTnI concentrations. Figure 4 exhibits the photocurrent responses for different concentrations. The immunocomplex formation on the platform surface decreased the AA photocurrent, whose inhibition is expressed in ΔI values ($\Delta I = I_0 - I_n$) in which I_0 and I_n correspond to the AA photocurrent before and after interaction with the immunosensor with cTnI, respectively. As can be seen in the inset of Figure 4, an analytical curve was obtained for the concentration range from 1 pg mL^{-1} to 1000 ng mL^{-1} cTnI, linear equation for which was ΔI (μA) = 5.92 ($\pm$0.04) + 1.70 ($\pm$0.02) log[cTnI] (ng mL^{-1}), with the correlation coefficient of 0.998 (n = 12). An experimental limit of detection (LOD) of 1 pg mL^{-1} was obtained from a signal-to-noise ratio equal to three. The LOD obtained was significantly lower than the maximum limits allowed for a clinical diagnosis of myocardial infarction at approximately 0.1 ng mL^{-1} [29]. In combination with the linear range of response, this result was compared to further reported PEC immunosensors for cTnI (Table 1: references [3,15,24–39]). As can be seen, the anti-cTnI/Bi$_2$S$_3$/BiVO$_4$/FTO immunosensor presents some interesting features for the determination of this biomarker in comparison to previously reported PEC sensors.

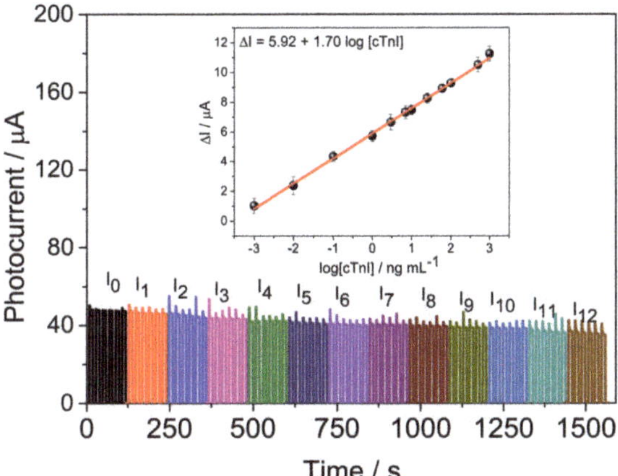

Figure 4. Photoelectrochemical response of the anti-cTnI/Bi$_2$S$_3$/BiVO$_4$/FTO immunosensor after incubation with different cTnI concentrations. [AA]= 0.04 mol L^{-1}. Inset: Analytical curve obtained from of the data of Figure 4. [cTnI] from of 1 pg mL^{-1} to 1000 ng mL^{-1}. E$_{appl}$ = 0 V vs. Ag/AgCl/KCl$_{sat}$.

Considering the anti-cTnI/Bi$_2$S$_3$/BiVO$_4$/FTO PEC immunosensor response to the decrease in the AA molecule with the increase in the cTnI antigen concentration, a schematic representation of the PEC determination of cTnI (Scheme 1) under the incidence of light was proposed. As shown in Scheme 1, the BiVO$_4$ and Bi$_2$S$_3$ harvest photons of energy higher than their band gap, promoting electrons from the valence to the conduction band and giving rising to e$^-$/h$^+$ couples. The electron photogenerated at the conduction band of Bi$_2$S$_3$ can be injected into the conduction band of the BiVO$_4$, while the hole photogenerated in the valence band of the Bi$_2$S$_3$ can be transferred to the AA molecule. The AA molecule acts as an ideal electron donor to capture the photogenerated holes in the valence band (VB) of Bi$_2$S$_3$, inhibiting the recombination of electron–hole pairs [30] and generating an anodic photocurrent. The cTnI biomarker can then interact with the immobilized anti-cTnI/Bi$_2$S$_3$/BiVO$_4$/FTO, decreasing the efficiency of the system to produce a photocurrent since the cTnI biomarker/anti-cTnI interaction reduces the efficiency of the photoactive material to transfer holes to donor molecules, an inhibition that is proportional to the amount of cTnI antigens immobilized on the immunosensor surface.

Table 1. Comparison of the analytical parameters of different PEC sensors for detection of cTnI.

Modified Electrode	Technique	Linear Range (ng mL^{-1})	LOD (pg mL^{-1})	Reference
Bi$_2$S$_3$/BiVO$_4$/FTO	PEC	0.0010–1000	1.0	This work
[a] NAC-CdAgTe QDs/AuNPs/GCE	PEC	0.0050–20	1.70	[3]
[b] Mn:CdS@Cu$_2$MoS$_4$/G/ITO	PEC	0.0050–1000	0.18	[15]
[c] CM-dextran/Au/TiO$_2$ NTA/Ti	PEC	0.00484–0.484	2.20	[31]
[d] SnO$_2$/NCQDs/BiOI/ITO	PEC	0.00100–100	0.30	[32]
[e] Au NPs/ZIS/Bi$_2$Se$_3$/ITO-PET	PEC	0.080–40	26.00	[33]
[f] Zn$_2$SnO$_4$/N,S-GQDs/CdS/ITO	PEC	0.0010–50	0.30	[34]
[g] Cu^{2+}@Zr-MOF@TiO$_2$ NRs	PEC	0.01–10	8.60	[35]
[h] Pd/I:BiOBr-OVs/SOD-Au@PANI	PEC	0.0001–100	0.042	[36]
[i] Ag$_2$S/ZnO/ITO	PEC	0.00001–1	0.003	[37]
[j] SiNWs@PDA	PEC	0.005–10	1.47	[38]
[k] Ag@Cu$_2$O core-shell SPs/TiO$_2$/CdS	PEC	0.00002–50	0.0067	[39]

[a] N-acetyl-L-cysteine-capped CdAgTe quantum dots and dodecahedral Au nanoparticles; [b] Manganese doped CdS sensitized graphene/Cu$_2$MoS$_4$ composite; [c] Carboxymethylated dextran-coated and gold-modified TiO$_2$ nanotube arrays; [d] Nitrogen-doped carbon quantum dots–bismuth oxyiodide–flower-like SnO$_2$; [e] ZnIn$_2$S$_4$/Bi$_2$Se$_3$ Nanocomposite; [f] N,S-GQDs and CdS co-sensitized hierarchical Zn$_2$SnO$_4$ cube; [g] Zr-MOF coated onto TiO$_2$ NRs on fluorine-doped tin oxide; [h] Pd nanoparticles loading on the I-doped bismuth oxybromide with oxygen vacancies sensibilized by superoxide dismutase loaded on gold@polyaniline; [i] Ag$_2$S/ZnO Nanocomposites. [j] Silicon nanowire arrays (SiNWs) at Polydiacetylene (PDA); [k] Ag at Cu$_2$O core-shell submicron-particles on CdS QDs sensitized TiO$_2$ nanosheets.

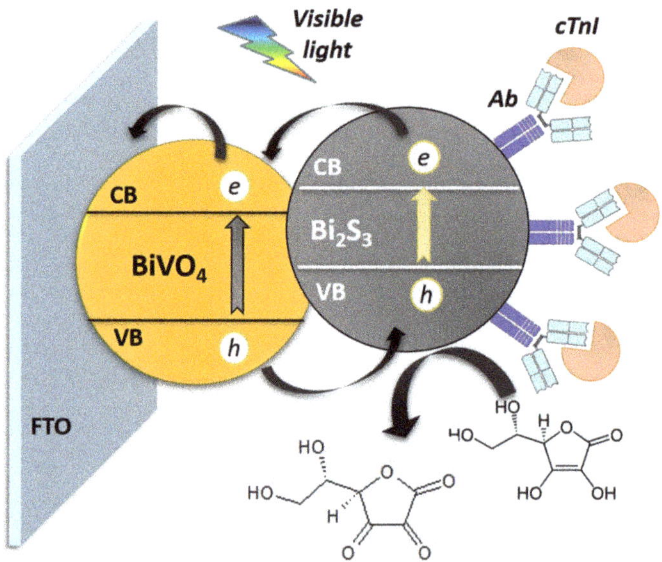

Scheme 1. Representation of the proposed mechanism for PEC determination of cTnI by label-free Bi$_2$S$_3$/BiVO$_4$/FTO immunosensor. Ab: anti-cTnI.

For practical application, repeatability, reproducibility, and selectivity are important features of the immunosensor [5,15]. Figure 5A presents the photocurrent response of the Bi$_2$S$_3$/BiVO$_4$/FTO platform to AA. As can be seen, no significant photocurrent changes were observed. The relative standard deviation (RSD) among signals was only 1.04%

(n = 15), indicating that the proposed sensing platform had good stability for the interaction with the AA donor molecule.

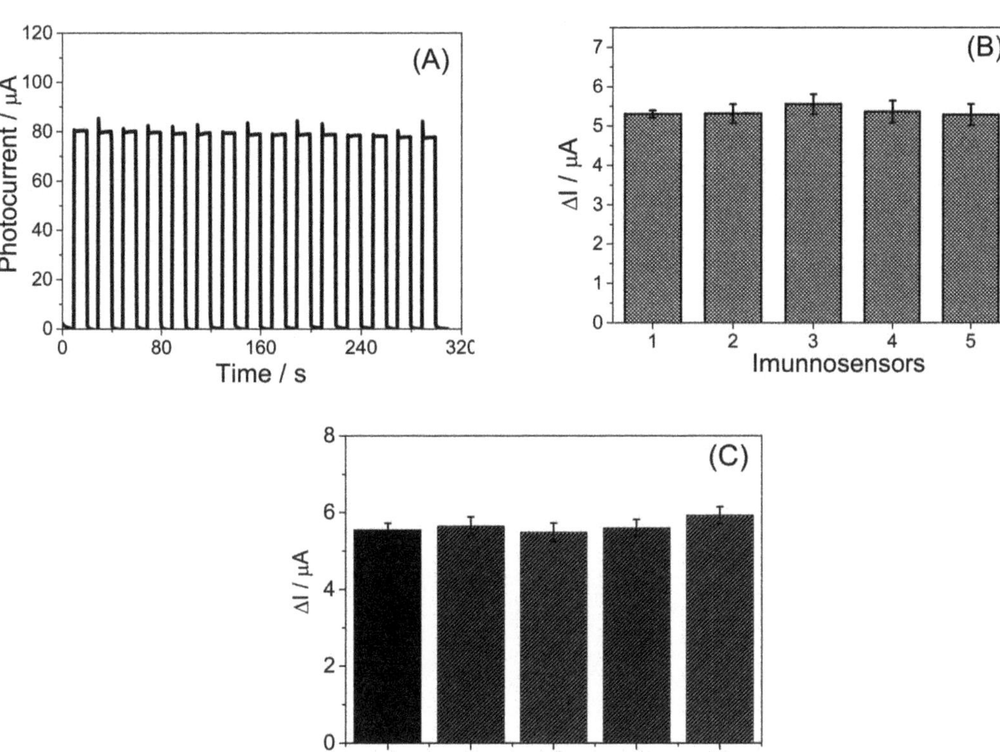

Figure 5. (**A**) Evaluation of repeatability of photocurrent of the Bi_2S_3/$BiVO_4$/FTO PEC platform. (**B**) Reproducibility of photocurrent of the Bi_2S_3/$BiVO_4$/FTO PEC platform incubated with 1 ng mL^{-1} cTnI antigen solution. (**C**) Evaluation of selectivity of the anti-cTnI/Bi_2S_3/$BiVO_4$/FTO PEC immunosensor. (**a**) cTnI, (**b**) cTnI + albumin, (**c**) cTnI + C-reactive protein, (**d**) cTnI + glucose, and (**e**) cTnI + myoglobin. [cTnI] = 1 ng mL^{-1}. [Foreign species] = 100 ng mL^{-1}. Measurements performed in 0.1 mol L^{-1} phosphate buffer, pH 7.4. [AA]= 0.04 mol L^{-1}, E_{appl} = 0 V vs. Ag/AgCl/KCl$_{sat}$.

The reproducibility of the proposed immunosensor also was evaluated (Figure 5B). This parameter was assessed using five different electrodes for the same concentration of cTnI and under optimal experimental conditions. The RSD value obtained for this study was 2.1%, indicating a satisfactory reproducibility of the immunosensor.

In order to appraise the selectivity of the PEC immunosensor for cTnI detection, some potential interfering substances were investigated. Therefore, a solution containing 1 ng mL^{-1} cTnI (a) and solutions containing 1 ng mL^{-1} cTnI and 100 ng mL^{-1} of potential interfering substances such as albumin (b), C-reactive protein (c), glucose (d), and myoglobin (e) were tested, respectively, under optimized experimental conditions. The results are shown in Figure 5C. As can be seen, there was no significant change in the photocurrent presented by the immunosensor after the addition of those substances, and the RSD of the measurement was within 5%, indicating that the designed immunosensor possesses a remarkable selectivity for detection of cTnI. The data from Figure 5 are presented in Figures S5 and S6 in the supporting information.

3.5. Detection of cTnI in the Artificial Blood Plasma Samples

To demonstrate the accuracy and the potential of application of the developed immunosensor in clinical samples, the anti-cTnI/Bi_2S_3/$BiVO_4$/FTO electrode was tested to determine cTnI at different concentrations in artificial blood plasma samples. The samples were spiked with cTnI at three concentration levels (0.05, 2.0, and 50 ng mL^{-1}), and quantification performed using an external calibration method. The recovery values (Table 2) were found between 98.0% and 98.5% with very low values of relative standard deviation, indicating good accuracy and precision. Thus, the results suggest that the immunosensing platform based on Bi_2S_3/$BiVO_4$/FTO can be used as a promising strategy for the detection of cTnI in clinical blood plasma samples.

Table 2. Recovery values obtained from detection of cTnI in artificial blood plasma samples using the proposed immunosensor.

Sample	Additioned/ng mL^{-1}	Found/ng mL^{-1}	Recovery/%	RSD/%
1	0.05	0.049 ($\pm$0.001)	98.0	2.04
2	2.0	1.96 ($\pm$0.01)	98.0	0.51
3	50	49.24 ($\pm$ 0.05)	98.5	0.10

4. Conclusions

In this paper, the feasibility of the sensitization of a $BiVO_4$ semiconductor material with a Bi_2S_3 electrodeposited film for the development of a PEC immunosensing platform to determine cTnI as a biomarker of myocardial infarction was reported. The proposed immunosensor presented high photoelectrochemical efficiency under visible LED light irradiation, a wide linear range of response to cTnI, high sensitivity, a low limit of detection, and good selectivity and stability. Moreover, the immunosensor demonstrated good accuracy and good precision with excellent application for the termination of the cTnI biomarker in artificial blood plasma samples.

Supplementary Materials: The following supporting information can be downloaded at: https://www.mdpi.com/article/10.3390/bios13030379/s1, Figure S1. (A) Photoelectrochemical response of the Bi_2S_3/$BiVO_4$/FTO platform obtained at different potentials. Amperometric measurements performed in 0.1 mol L^{-1} phosphate buffer (pH 7.4) containing 0.03 mol L^{-1} AA. (B) Plot of photocurrent vs. E_{appl}. Data obtained from the Figure S1A; Figure S2. (A) Photoelectrochemical response of the Bi_2S_3/$BiVO_4$/FTO platform obtained at different buffer solutions. (B) Plot of photocurrent vs. different buffer solutions. Amperometric measurements performed in 0.1 mol L^{-1} of buffer (pH 7.4) containing 0.03 mol L^{-1} AA. E_{appl} = 0 V vs. Ag/AgCl/KCl$_{sat}$; Figure S3. (A) Photoelectrochemical response of the Bi_2S_3/$BiVO_4$/FTO platform obtained at different AA concentrations (0.01–0.06 mol L^{-1}). (B) Amperometric measurements performed in 0.1 mol L^{-1} phosphate buffer (pH 7.4) containing 0.04 mol L^{-1} AA. E_{appl} = 0 V vs. Ag/AgCl/KCl$_{sat}$; Figure S4. Photoelectrochemical responses of the anti-cTnI/Bi_2S_3/$BiVO_4$/FTO PEC immunosensor before (black amperogram) and after incubation with cTnI antigens (red amperograms) at different incubation times. The measurements were performed in 0.1 mol L^{-1} phosphate buffer, pH 7.4, containing 0.04 mol L^{-1} AA. E_{appl} = 0 V vs. Ag/AgCl/KCl$_{sat}$. [anti-cTnI] = 5 µg mL^{-1}; [cTnI] = 1 ng mL^{-1}; Figure S5. Photoelectrochemical responses obtained with 5 (five) different anti-cTnI/Bi_2S_3/$BiVO_4$/FTO PEC immunosensors under optimized conditions before (black amperograms) and after (red amperograms) incubation with cTnI. [cTnI]=1 ng mL^{-1}, $t_{incubation}$= 25 min; Figure S6. Photoelectrochemical responses obtained with the anti-cTnI/Bi_2S_3/$BiVO_4$/FTO PEC immunosensor under optimized conditions before (black amperogram) and after (red amperogram) incubation with cTnI (1 ng mL^{-1}) in absence and presence of different species (albumin, C-reactive protein, glucose, and myoglobin). [Foreign specie] = 100 ng ng mL^{-1}; $t_{incubation}$= 25 min.

Author Contributions: Conceptualization: T.O.M., A.G.d.S.N., A.S.d.M., F.S.D., R.d.C.S.L. and O.F.-F.; methodology: T.O.M., F.S.D., R.d.C.S.L. and O.F.-F.; formal analysis: T.O.M.; A.G.d.S.N., F.S.D., R.d.C.S.L. and O.F.-F.; investigation: T.O.M., A.G.d.S.N., A.S.d.M., F.S.D., R.d.C.S.L. and O.F.-F.; writing—original draft preparation: T.O.M., R.d.C.S.L. and O.F.-F.; and writing—review and editing: T.O.M., A.G.d.S.N., A.S.d.M., F.S.D., R.d.C.S.L. and O.F.-F.; supervision: F.S.D., R.d.C.S.L. and O.F.-F. All authors have read and agreed to the published version of the manuscript.

Funding: The authors are grateful to the Instituto Nacional de Ciência e Tecnologia em Bioanalítica (465389/2014-7); FAPESP (Grant: 2020/01050-5); FAPEMA (Grants: INFRA-02021/21; INFRA-02050/21; INFRA-02203/2021; UNIVERSAL-06535/22; POS-GRAD-02432/21), CNPq (Grants: 308204/2018-2; 309828/2020-1; 305806/2020-3; 313324/2021-2), and FINEP.

Institutional Review Board Statement: Not applicable.

Informed Consent Statement: Not applicable.

Data Availability Statement: Not applicable.

Acknowledgments: We are also thankful to Multiuser Centre for Research in Materials and Biosystems (CeMatBio), of the Federal University of Maranhão (UFMA), for the support with the XRD and SEM measurement.

Conflicts of Interest: The authors declare no conflict of interest.

References

1. Fathil, M.F.M.; Md Arshad, M.K.; Gopinath, S.C.B.; Hashim, U.; Adzhri, R.; Ayub, R.M.; Ruslinda, A.R.; Nuzaihan, M.N.M.; Azman, A.H.; Zaki, M.; et al. Diagnostics on Acute Myocardial Infarction: Cardiac Troponin Biomarkers. *Biosens. Bioelectron.* **2015**, *70*, 209–220. [CrossRef]
2. World Health Organization. Cardiovascular Diseases. Available online: http://www.who.int/cardiovascular_diseases/en/ (accessed on 26 November 2022).
3. Tan, Y.; Wang, Y.; Li, M.; Ye, X.; Wu, T.; Li, C. Enhanced Photoelectrochemical Immunosensing of Cardiac Troponin I Based on Energy Transfer between N-Acetyl-L-Cysteine Capped CdAgTe Quantum Dots and Dodecahedral Au Nanoparticles. *Biosens. Bioelectron.* **2017**, *91*, 741–746. [CrossRef]
4. Campu, A.; Muresan, I.; Craciun, A.M.; Cainap, S.; Astilean, S.; Focsan, M. Cardiac Troponin Biosensor Designs: Current Developments and Remaining Challenges. *Int. J. Mol. Sci.* **2022**, *23*, 7728. [CrossRef] [PubMed]
5. Bahadır, E.B.; Sezgintürk, M.K. Applications of Electrochemical Immunosensors for Early Clinical Diagnostics. *Talanta* **2015**, *132*, 162–174. [CrossRef]
6. Song, S.Y.; Han, Y.D.; Kim, K.; Yang, S.S.; Yoon, H.C. A Fluoro-Microbead Guiding Chip for Simple and Quantifiable Immunoassay of Cardiac Troponin I (CTnI). *Biosens. Bioelectron.* **2011**, *26*, 3818–3824. [CrossRef] [PubMed]
7. Seo, S.-M.; Kim, S.-W.; Park, J.-N.; Cho, J.-H.; Kim, H.-S.; Paek, S.-H. A Fluorescent Immunosensor for High-Sensitivity Cardiac Troponin I Using a Spatially-Controlled Polymeric, Nano-Scale Tracer to Prevent Quenching. *Biosens. Bioelectron.* **2016**, *83*, 19–26. [CrossRef] [PubMed]
8. Miao, L.; Jiao, L.; Tang, Q.; Li, H.; Zhang, L.; Wei, Q. A Nanozyme-Linked Immunosorbent Assay for Dual-Modal Colorimetric and Ratiometric Fluorescent Detection of Cardiac Troponin I. *Sens. Actuators B Chem.* **2019**, *288*, 60–64. [CrossRef]
9. Chen, F.; Wu, Q.; Song, D.; Wang, X.; Ma, P.; Sun, Y. Fe$_3$O$_4$@PDA Immune Probe-Based Signal Amplification in Surface Plasmon Resonance (SPR) Biosensing of Human Cardiac Troponin I. *Colloids Surf. B Biointerfaces* **2019**, *177*, 105–111. [CrossRef] [PubMed]
10. Sinha, R.K. Wavelength Modulation Based Surface Plasmon Resonance Sensor for Detection of Cardiac Marker Proteins Troponin I and Troponin T. *Sens. Actuators A Phys.* **2021**, *332*, 113104. [CrossRef]
11. Schneck, N.A.; Phinney, K.W.; Lee, S.B.; Lowenthal, M.S. Quantification of Cardiac Troponin I in Human Plasma by Immunoaffinity Enrichment and Targeted Mass Spectrometry. *Anal. Bioanal. Chem.* **2018**, *410*, 2805–2813. [CrossRef]
12. Khlebtsov, B.N.; Bratashov, D.N.; Byzova, N.A.; Dzantiev, B.B.; Khlebtsov, N.G. SERS-Based Lateral Flow Immunoassay of Troponin I by Using Gap-Enhanced Raman Tags. *Nano Res.* **2019**, *12*, 413–420. [CrossRef]
13. Li, H.; Qiao, Y.; Li, J.; Fang, H.; Fan, D.; Wang, W. A Sensitive and Label-Free Photoelectrochemical Aptasensor Using Co-Doped ZnO Diluted Magnetic Semiconductor Nanoparticles. *Biosens. Bioelectron.* **2016**, *77*, 378–384. [CrossRef]
14. Zhang, N.; Ma, Z.-Y.; Ruan, Y.-F.; Zhao, W.-W.; Xu, J.-J.; Chen, H.-Y. Simultaneous Photoelectrochemical Immunoassay of Dual Cardiac Markers Using Specific Enzyme Tags: A Proof of Principle for Multiplexed Bioanalysis. *Anal. Chem.* **2016**, *88*, 1990–1994. [CrossRef]
15. Chi, H.; Han, Q.; Chi, T.; Xing, B.; Ma, N.; Wu, D.; Wei, Q. Manganese Doped CdS Sensitized Graphene/Cu$_2$MoS$_4$ Composite for the Photoelectrochemical Immunoassay of Cardiac Troponin I. *Biosens. Bioelectron.* **2019**, *132*, 1–7. [CrossRef] [PubMed]

16. Raja, A.; Rajasekaran, P.; Selvakumar, K.; Arunpandian, M.; Kaviyarasu, K.; Asath Bahadur, S.; Swaminathan, M. Visible Active Reduced Graphene Oxide-BiVO$_4$-ZnO Ternary Photocatalyst for Efficient Removal of Ciprofloxacin. *Sep. Purif. Technol.* **2020**, *233*, 115996. [CrossRef]
17. Okoth, O.K.; Yan, K.; Liu, Y.; Zhang, J. Graphene-Doped Bi$_2$S$_3$ Nanorods as Visible-Light Photoelectrochemical Aptasensing Platform for Sulfadimethoxine Detection. *Biosens. Bioelectron.* **2016**, *86*, 636–642. [CrossRef] [PubMed]
18. Liu, C.; Yang, Y.; Li, W.; Li, J.; Li, Y.; Chen, Q. In Situ Synthesis of Bi$_2$S$_3$ Sensitized WO$_3$ Nanoplate Arrays with Less Interfacial Defects and Enhanced Photoelectrochemical Performance. *Sci. Rep.* **2016**, *6*, 23451. [CrossRef]
19. Yin, H.; Sun, B.; Zhou, Y.; Wang, M.; Xu, Z.; Fu, Z.; Ai, S. A New Strategy for Methylated DNA Detection Based on Photoelectrochemical Immunosensor Using Bi$_2$S$_3$ Nanorods, Methyl Bonding Domain Protein and Anti-His Tag Antibody. *Biosens. Bioelectron.* **2014**, *51*, 103–108. [CrossRef]
20. Sun, B.; Qiao, F.; Chen, L.; Zhao, Z.; Yin, H.; Ai, S. Effective Signal-on Photoelectrochemical Immunoassay of Subgroup J Avian Leukosis Virus Based on Bi$_2$S$_3$ Nanorods as Photosensitizer and in Situ Generated Ascorbic Acid for Electron Donating. *Biosens. Bioelectron.* **2014**, *54*, 237–243. [CrossRef] [PubMed]
21. Monteiro, T.O.; dos Santos, C.C.; do Prado, T.M.; Damos, F.S.; de Luz, R.C.S.; Fatibello-Filho, O. Highly Sensitive Photoelectrochemical Immunosensor Based on Anatase/Rutile TiO$_2$ and Bi$_2$S$_3$ for the Zero-Biased Detection of PSA. *J. Solid State Electrochem.* **2020**, *24*, 1801–1809. [CrossRef]
22. Prado, T.M.; Carrico, A.; Cincotto, F.H.; Fatibello-Filho, O.; Moraes, F.C. Bismuth Vanadate/Graphene Quantum Dot: A New Nanocomposite for Photoelectrochemical Determination of Dopamine. *Sens. Actuators B Chem.* **2019**, *285*, 248–253. [CrossRef]
23. Saitou, M.; Yamaguchi, R.; Oshikawa, W. Novel Process for Electrodeposition of Bi$_2$S$_3$ Thin Films. *Mater. Chem. Phys.* **2002**, *73*, 306–309. [CrossRef]
24. Liu, L.; Qiu, C.L.; Chen, Q.; Zhang, S.M. Corrosion Behavior of Zr-Based Bulk Metallic Glasses in Different Artificial Body Fluids. *J. Alloy. Compd.* **2006**, *425*, 268–273. [CrossRef]
25. Han, Q.; Wang, R.; Xing, B.; Zhang, T.; Khan, M.S.; Wu, D.; Wei, Q. Label-Free Photoelectrochemical Immunoassay for CEA Detection Based on CdS Sensitized WO$_3$@BiOI Heterostructure Nanocomposite. *Biosens. Bioelectron.* **2018**, *99*, 493–499. [CrossRef] [PubMed]
26. Shaislamov, U.; Krishnamoorthy, K.; Kim, S.J.; Chun, W.; Lee, H.-J. Facile Fabrication and Photoelectrochemical Properties of a CuO Nanorod Photocathode with a ZnO Nanobranch Protective Layer. *RSC Adv.* **2016**, *6*, 103049–103056. [CrossRef]
27. Jiang, D.; Zhang, L.; Yue, Q.; Wang, T.; Huang, Q.; Du, P. Efficient Suppression of Surface Charge Recombination by CoP-Modified Nanoporous BiVO4 for Photoelectrochemical Water Splitting. *Int. J. Hydrog. Energy* **2021**, *46*, 15517–15525. [CrossRef]
28. Liu, D.; Qian, Y.; Xu, R.; Zhang, Y.; Ren, X.; Ma, H.; Wei, Q. A Dual-Signal Amplification Photoelectrochemical Immunosensor for Ultrasensitive Detection of CYFRA 21-1 Based on the Synergistic Effect of SnS$_2$/SnS/Bi$_2$S$_3$ and ZnCdS@NPC-ZnO. *Sens. Actuators B Chem.* **2021**, *346*, 130456. [CrossRef]
29. Chekin, F.; Vasilescu, A.; Jijie, R.; Singh, S.K.; Kurungot, S.; Iancu, M.; Badea, G.; Boukherroub, R.; Szunerits, S. Sensitive Electrochemical Detection of Cardiac Troponin I in Serum and Saliva by Nitrogen-Doped Porous Reduced Graphene Oxide Electrode. *Sens. Actuators B Chem.* **2018**, *262*, 180–187. [CrossRef]
30. Zhu, S.-R.; Qi, Q.; Zhao, W.-N.; Fang, Y.; Han, L. Enhanced Photocatalytic Activity in Hybrid Composite Combined BiOBr Nanosheets and Bi$_2$S$_3$ Nanoparticles. *J. Phys. Chem. Solids* **2018**, *121*, 163–171. [CrossRef]
31. Guo, W.; Wang, J.; Guo, W.; Kang, Q.; Zhou, F. Correction to: Interference-Free Photoelectrochemical Immunoassays Using Carboxymethylated Dextran-Coated and Gold-Modified TiO$_2$ Nanotube Arrays. *Anal. Bioanal. Chem.* **2021**, *413*, 5921–5922. [CrossRef]
32. Fan, D.; Liu, X.; Shao, X.; Zhang, Y.; Zhang, N.; Wang, X.; Wei, Q.; Ju, H. A Cardiac Troponin I Photoelectrochemical Immunosensor: Nitrogen-Doped Carbon Quantum Dots–Bismuth Oxyiodide–Flower-like SnO$_2$. *Microchim. Acta* **2020**, *187*, 332. [CrossRef]
33. Dong, W.; Mo, X.; Wang, Y.; Lei, Q.; Li, H. Photoelectrochemical Immunosensor Based on ZnIn$_2$S$_4$/Bi$_2$Se$_3$ Nanocomposite for the Determination of Cardiac Troponin I. *Anal. Lett.* **2020**, *53*, 1888–1901. [CrossRef]
34. Fan, D.; Bao, C.; Khan, M.S.; Wang, C.; Zhang, Y.; Liu, Q.; Zhang, X.; Wei, Q. A Novel Label-Free Photoelectrochemical Sensor Based on N,S-GQDs and CdS Co-Sensitized Hierarchical Zn$_2$SnO$_4$ Cube for Detection of Cardiac Troponin I. *Biosens. Bioelectron.* **2018**, *106*, 14–20. [CrossRef]
35. Gao, Y.; Li, M.; Zeng, Y.; Liu, X.; Tang, D. Tunable Competitive Absorption-Induced Signal-On Photoelectrochemical Immunoassay for Cardiac Troponin I Based on Z-Scheme Metal–Organic Framework Heterojunctions. *Anal. Chem.* **2022**, *94*, 13582–13589. [CrossRef]
36. Feng, J.; Li, N.; Du, Y.; Ren, X.; Wang, X.; Liu, X.; Ma, H.; Wei, Q. Ultrasensitive Double-Channel Microfluidic Biosensor-Based Cathodic Photo-Electrochemical Analysis via Signal Amplification of SOD-Au@PANI for Cardiac Troponin I Detection. *Anal. Chem.* **2021**, *93*, 14196–14203. [CrossRef]
37. Liao, X.-J.; Xiao, H.-J.; Cao, J.-T.; Ren, S.-W.; Liu, Y.-M. A Novel Split-Type Photoelectrochemical Immunosensor Based on Chemical Redox Cycling Amplification for Sensitive Detection of Cardiac Troponin I. *Talanta* **2021**, *233*, 122564. [CrossRef] [PubMed]

38. Li, H.-J.; Zhi, S.; Zhang, S.; Guo, X.; Huang, Y.; Xu, L.; Wang, X.; Wang, D.; Zhu, M.; He, B. A Novel Photoelectrochemical Sensor Based on SiNWs@PDA for Efficient Detection of Myocardial Infarction. *Biomater. Sci.* **2022**, *10*, 4627–4634. [CrossRef]
39. Chen, J.; Kong, L.; Sun, X.; Feng, J.; Chen, Z.; Fan, D.; Wei, Q. Ultrasensitive Photoelectrochemical Immunosensor of Cardiac Troponin I Detection Based on Dual Inhibition Effect of Ag@Cu$_2$O Core-Shell Submicron-Particles on CdS QDs Sensitized TiO$_2$ Nanosheets. *Biosens. Bioelectron.* **2018**, *117*, 340–346. [CrossRef] [PubMed]

biosensors

Communication

A Novel Minidumbbell DNA-Based Sensor for Silver Ion Detection

Jiacheng Zhang [1,2,†], Yuan Liu [3,†], Zhenzhen Yan [1,2], Yue Wang [4,*] and Pei Guo [2,*]

[1]   School of Biology and Biological Engineering, South China University of Technology, Guangzhou 510006, China
[2]   Zhejiang Cancer Hospital, Institute of Basic Medicine and Cancer (IBMC), Chinese Academy of Sciences, Hangzhou 310022, China
[3]   South China Advanced Institute for Soft Matter Science and Technology, School of Emergent Soft Matter, Guangdong Provincial Key Laboratory of Functional and Intelligent Hybrid Materials and Devices, South China University of Technology, Guangzhou 510640, China
[4]   Institute of Molecular Medicine, Renji Hospital, School of Medicine, Shanghai Jiao Tong University, Shanghai 200127, China
*   Correspondence: wangyue515@mails.ucas.ac.cn (Y.W.); guopei@ibmc.ac.cn (P.G.)
†   Authors who contribute equally to this work.

Abstract: Silver ion (Ag^+) is one of the most common heavy metal ions that cause environmental pollution and affect human health, and therefore, its detection is of great importance in the field of analytical chemistry. Here, we report an 8-nucleotide (nt) minidumbbell DNA-based sensor (*M-DNA*) for Ag^+ detection. The minidumbbell contained a unique reverse wobble C·C mispair in the minor groove, which served as the binding site for Ag^+. The *M-DNA* sensor could achieve a detection limit of 2.1 nM and sense Ag^+ in real environmental samples with high accuracy. More importantly, the *M-DNA* sensor exhibited advantages of fast kinetics and easy operation owing to the usage of an ultrashort oligonucleotide. The minidumbbell represents a new and minimal non-B DNA structural motif for Ag^+ sensing, allowing for the further development of on-site environmental Ag^+ detection devices.

Keywords: DNA sensor; silver ion detection; minidumbbell; non-B DNA; C·C mismatch

1. Introduction

Citation: Zhang, J.; Liu, Y.; Yan, Z.; Wang, Y.; Guo, P. A Novel Minidumbbell DNA-Based Sensor for Silver Ion Detection. *Biosensors* **2023**, *13*, 358. https://doi.org/10.3390/bios13030358

Received: 10 January 2023
Revised: 2 March 2023
Accepted: 3 March 2023
Published: 8 March 2023

Silver ion (Ag^+) has been widely used as an antiseptic in cosmetics, building materials, and medical products owing to its antibacterial properties [1–4]. However, overuse of Ag^+ inevitably leads to environmental pollution. Human exposure to Ag^+ pollution mainly comes from the release of airborne silver nanoparticles and natural water contaminated by industrial sources [5,6]. The tolerable concentration of Ag^+ in drinking water is ~927 nM as recommended by the World Health Organization [7]. Excessive Ag^+ ingestion can cause certain serious health consequences, such as respiratory system injury, organ failure, and even cancer [6,8–11]. Various methods have been developed for detecting low concentrations of Ag^+ in environmental samples and drinking water sources. At present, Ag^+ detection is mainly carried out by conventional analytical methods such as inductively coupled plasma mass spectrometry [12], optical emission spectrometry [13], atomic absorption spectrometry [14,15], and laser ablation microwave plasma torch optical emission spectrometry [16]. These conventional methods are sensitive and selective, but they rely on expensive instruments and intensive labor.

In recent years, nucleic acid molecules have gained prominence in the fields of sensing and material science because of their programmability and predictability by forming complementary base pairs [17]. DNA molecules have been used to design sensors for detecting metal ions such as Ag^+, UO_2^{2+}, Cu^{2+}, Ca^{2+}, Mg^{2+}, Hg^{2+}, and Pb^{2+} [18–26]. In general, there are mainly two DNA-based strategies for Ag^+ detection. The first strategy utilizes an Ag^+-dependent DNAzyme that can irreversibly cleave an RNA or DNA substrate in the presence of Ag$^+$ [22]. The second strategy is based on the well-established knowledge that Ag$^+$ binds to cytosine (C) at the N3 site to coordinate and stabilize a C·C mismatch [27,28]. Ag$^+$ can induce the formation of DNA i-motif or hairpin structures that contain C·C mismatch(es), thus giving reporting signals upon DNA conformational change [26,29–33]. Moreover, the duplex or hairpin-forming strands can also be assembled onto nanomaterials for signal amplification [34–38]. The second strategy can achieve a low detection limit, but the reported ones generally used relatively long oligonucleotides, which might make the Ag$^+$-induced DNA conformational change slow. For instance, a DNA sensor based on a 20-nucleotide (nt) hairpin required an incubation time of at least 10 min for Ag$^+$ detection. Therefore, a DNA sensor using a short oligonucleotide is expected to have advantages of fast response, easy operation, and probably anti-interference capability in a complex environment, which allow for the further development of on-site environmental detection devices [33,39].

Minidumbbell (MDB) is a type of non-B DNA structure formed by 8–10-nt sequences [40,41]. The MDB structure was initially found to form in CCTG tetranucleotide repeats, which are associated with the neurodegenerative disease of myotonic dystrophy type 2 [40,41]. The CCTG MDB is simply composed of two repeats, i.e., 5′-CCTG CCTG-3′, and each repeat folds into a type II tetraloop. The C1-G4 and C5-G8 adopt Watson-Crick loop-closing base pairs; C2 and C6 fold into the minor groove, whereas T3 and T7 stack on the C1-G4 and C5-G8, respectively (Figure 1) [40]. One of the most interesting features of this MDB is that the two minor groove residues formed a unique reverse wobble C2·C6 mispair containing one/two hydrogen bond(s) or Na$^+$-mediated electrostatic interactions at neural pH [42], or a C2$^+$·C6 mispair containing three hydrogen bonds with C2 being protonated at acidic pH (Figure 1) [43]. Upon lowering the pH from 7 to 5, the melting temperature (T_m) of the CCTG MDB was increased from ~20 °C to 46 °C [43]. Apart from pH, we wondered if Ag$^+$ could coordinate the C2·C6 mispair to stabilize the MDB and then induce a DNA conformational change for Ag$^+$ sensing. Here we report a novel and minimal DNA sensor, based on the CCTG MDB, for Ag$^+$ detection with high sensitivity and fast kinetics.

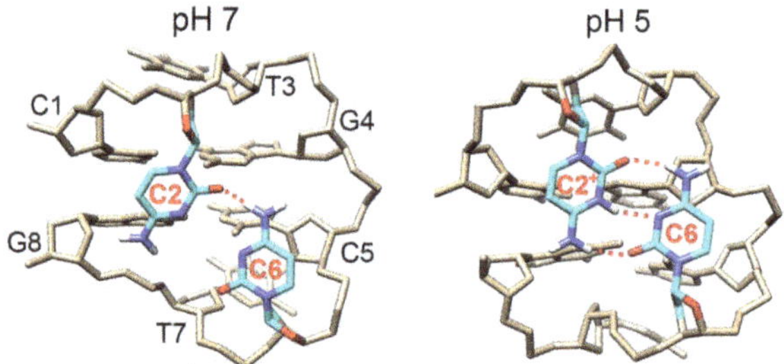

Figure 1. The averaged solution nuclear magnetic resonance (NMR) structure of the CCTG MDB at pH 7 (PDB ID: 5GWL) and pH 5 (PDB ID: 7D0Z). C2 and C6 formed predominantly a one-hydrogen-bond mispair at pH 7, whereas they formed a stable three-hydrogen-bond mispair at pH 5 with C2 being protonated.

2. Materials and Methods

2.1. DNA Sequence Design and Sample Preparation

Our designed *M-DNA* sensor was a duplex formed by the CCTG MDB strand (5′-CCTG CCTG-3′) and its complementary strand (5′-CAGG CAGG-3′), which were named $CCTG_2$ and $CAGG_2$, respectively. As a control, a self-complementary 8-bp duplex formed by 5′-GCAGCTGC-3′ was used. The high-performance liquid chromatography (HPLC)-purified DNA samples were purchased from Sangon Biotech (Shanghai, China), and they were further purified in our laboratory using diethylaminoethyl sephacel anion exchange column chromatography and Amicon Ultra-4 centrifugal filter devices. The ultra-violet (UV) absorbance at 260 nm was measured for DNA quantitation.

2.2. Preparation of SYBR Green I (SGI) and Metal Ion Stock Solutions

SGI (10,000×) was purchased from Beijing Solarbio Science and Technology Co., Ltd. (Beijing, China) and diluted using DMSO to a final concentration of 100× or 10× as the stock solution. It is noted that SGI 1× was equivalent to a concentration of 1.96 μM. The analytical-grade $AgNO_3$, KCl, LiCl, $CaCl_2$, $MgCl_2$, $MnCl_2$, $CoCl_2$, $CuSO_4$, $BaCl_2$, and $NiSO_4$ were purchased from Sinopharm Chemical Reagent Co., Ltd. (Beijing, China) and dissolved using DI water to a final concentration of 50 μM as the stock solutions.

2.3. NMR Experiments

To monitor the binding of Ag^+ to the CCTG MDB, NMR experiments were performed using a Bruker AVANCE NEO 400 MHz spectrometer. One-dimensional (1D) 1H NMR experiments were conducted at 25 °C using the excitation sculpting pulse sequence to suppress the water signal.

2.4. Circular Dichroism (CD) Experiments

CD experiments were performed using a Chirascan V100 CD spectrometer with a bandwidth of 1 nm at room temperature, unless otherwise specified. The CD samples (~100 μL) were placed in a cuvette of 0.5 mm path length, and the CD spectra were collected from 200 to 350 nm with a step size of 1 nm. For each sample, three sets of scans were acquired, and an average value was taken. CD spectra were background-corrected using the corresponding buffer solution.

2.5. Fluorescence Experiments

Fluorescence experiments, except for the kinetic study of Ag^+ sensing, were performed using a Shimadzu RF-6000 spectrometer at room temperature. The fluorescence samples (~2 mL) were placed in a 10 mm four-sided glazed quartz cuvette, and the fluorescence spectra were collected from 512 to 650 nm with a step size of 1 nm. Fluorescence intensity was recorded at 520 nm with an excitation wavelength of 492 nm. The excitation and emission band widths were 5 nm. For a kinetic study of Ag^+ sensing, fluorescence experiments were performed using an Edinburgh FLS1000 photoluminescence spectrometer at room temperature. The sample containing a DNA sensor in the absence of Ag^+ (~2.5 mL) was first placed in a 10 mm four-sided glazed quartz cuvette, and the fluorescence intensity at 520 nm was recorded from 0 to 180 s with a step time of 2 s. Ag^+ was then added to this sample, and the fluorescence intensity was immediately recorded from 0 to 180 s with a step time of 2 s. The excitation and emission band widths were 2 nm.

The detailed sample conditions for NMR, CD, and fluorescence experiments are stated in the figure legends.

3. Results

3.1. Ag^+ Induces a Conformational Change from Duplex to MDB

One-dimensional (1D) ^{1}H NMR experiments were first performed to investigate if Ag^+ could bind to C2·C6 mispair of the CCTG MDB. It showed that upon adding Ag^+ to the CCTG MDB, the H6 proton signals of C2 and C6 became broadened while those of T3 and T7 remained sharp and almost unchanged, suggesting that Ag^+ bound to the C2·C6 mispair (Figure 2). Besides, C1 H6, G4 H8, C5 H6, and G8 H8 peaks were also found to be broadened, as it has been reported that Ag^+ could also bind to C-G base pairs [44].

We then tested if Ag^+ could promote MDB formation to induce a DNA conformational change, which is the prerequisite of most DNA sensors. For this aim, we prepared a DNA duplex formed by the CCTG MDB strand (5′-CCTGCCTG-3′), namely $CCTG_2$, and its complementary strand (5′-CAGGCAGG-3′), namely $CAGG_2$, at pH 8/7/6 and collected CD spectra to monitor DNA conformational change upon Ag^+ titration at 25 °C. These two strands formed a duplex in the absence of Ag^+, as indicated by a positive CD band at 265 nm (Figure 3A–C, black lines) [45]. Upon adding Ag^+ to the duplex, a new major band at 290 nm was observed at pH 6, but not obvious at pH 7 and 8, when the DNA:Ag^+

ratio was 1:2 (Figure 3A–D, red lines). The CD band at 290 nm was characteristic of the CCTG MDB [46], suggesting that Ag^+ efficiently induced a conformational change from the duplex to the MDB at pH 6. Notably, the DNA:Ag^+ ratio of 1:2 showed the maximum population of Ag^+-induced MDB (Figure 3C). This may because Ag^+ is also non-selectively bound to C-G base pairs in the MDB (Figure 2), and thus more Ag^+ is required to promote MDB formation.

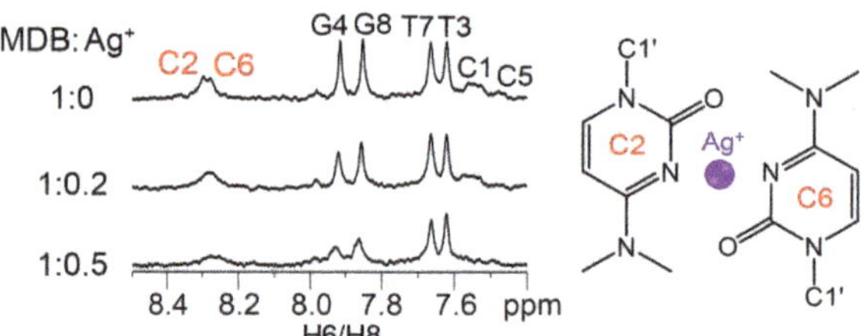

Figure 2. NMR spectra of 0.1 mM CCTG MDB in 1 mM sodium cacodylate (pH 6), 90% H_2O/10% D_2O, with various Ag^+ concentrations at 25 °C. Peak broadenings of C2 H6 and C6 H6 in the presence of Ag^+ suggest that Ag^+ is bound to C2·C6 mispair.

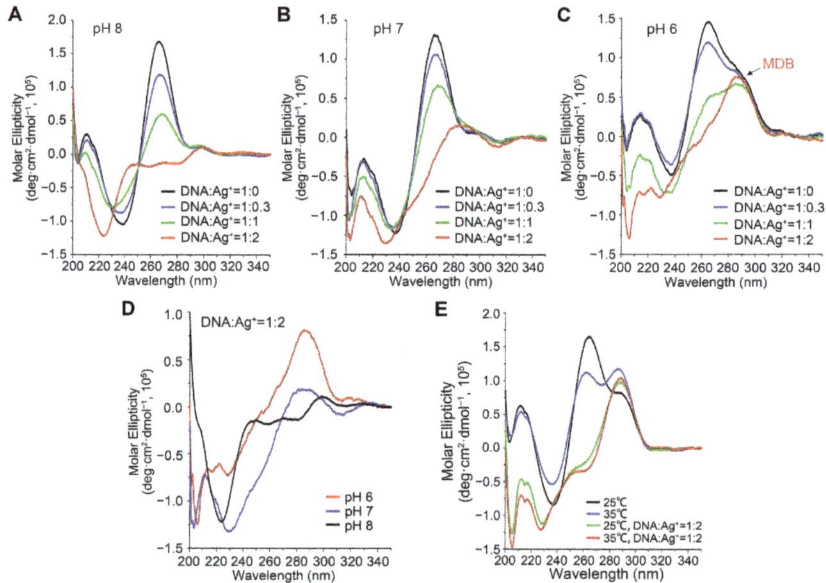

Figure 3. CD spectra of 15 µM $CCTG_2$ and $CAGG_2$ with 0, 5, 15, and 30 µM Ag^+ in 10 mM NaPi at pH 8 (**A**), pH 7 (**B**), and pH 6 (**C**). (**D**) CD spectra of 15 µM $CCTG_2$ and $CAGG_2$ in 30 µM Ag^+ at pH 8, 7, and 6 at 25 °C. (**E**) CD spectra of 15 µM $CCTG_2$ and $CAGG_2$ without Ag^+ and with 30 µM Ag^+ (pH 6) at 25 °C and 35 °C. Absorbance at 290 nm is characteristic of the free CCTG MDB.

We did not further lower the pH as previous work has demonstrated that the CCTG MDB completely dissociated from the duplex owing to its much higher thermodynamic stability than the duplex at pH 5 [43], therefore there would not be further conformational change upon adding Ag^+. We also performed the Ag^+ titration at 35 °C to examine if this system could function at an elevated temperature. However, the CD signal of MDB was observed without adding Ag^+ (Figure 3E), which could be attributed to the relatively higher thermodynamic stability of MDB than duplex at 35 °C and pH 6. Zhang et al. have also reported that a higher temperature leads to partial melting of the initial DNA duplex and thus a lower sensitivity [47].

3.2. Design and Optimization of the CCTG MDB-Based DNA (M-DNA) Sensor

Based on the Ag^+-induced formation of CCTG MDB at pH 6 (Figure 3C,D), we designed the *M-DNA* sensor, which was simply composed of the 8-bp duplex formed by $CCTG_2$ and $CAGG_2$. SYBR Green I (SGI) was used as a fluorescence reporter and it was expected to emit strong fluorescence when bound to the duplex in the absence of Ag^+ while giving weak fluorescence when the duplex was converted to MDB in the presence of Ag^+ (Figure 4A). To ensure SGI will not affect the DNA conformational change, CD spectra were collected without and with adding SGI, and the results showed that Ag^+-induced conformational change still effectively occurred (Figure S1).

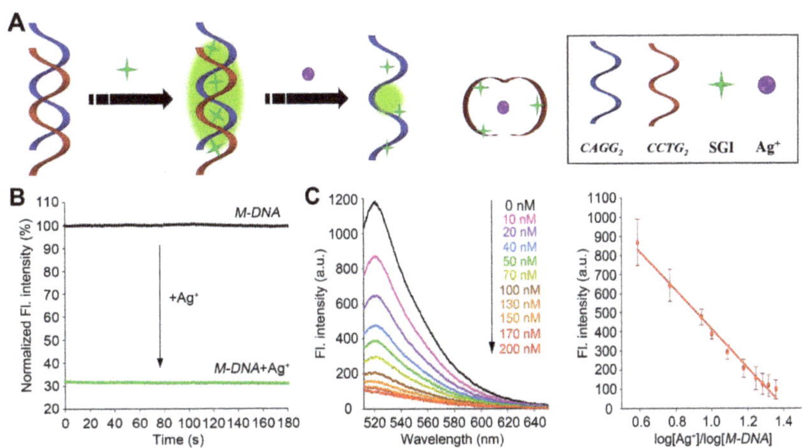

Figure 4. (**A**) Schematic of the *M-DNA* sensor for Ag^+ detection. (**B**) Normalized fluorescence intensity at 520 nm as a function of time for the *M-DNA* sensor in the absence of Ag^+ (black) and after adding 50 nM Ag^+ (green). (**C**) Fluorescence spectra of the *M-DNA* upon titrating Ag^+ ranging from 0 to 200 nM (**left**) and the fitting curve constructed using fluorescence intensity at 520 nm and $\log[Ag^+]/\log[M\text{-}DNA]$ ($R^2 = 0.99$) (**right**). Error bars were standard deviations obtained from three replicative experiments.

At pH 6, the *M-DNA* concentration and SGI:*M-DNA* ratio were further optimized. Two *M-DNA* concentrations (50 and 200 nM) and four SGI:*M-DNA* ratios (0.1:1, 0.5:1, 1:1, and 5:1) were tested to find the condition that would give the largest fluorescence change in response to Ag^+. The DNA concentration and SGI:*M-DNA* ratio were finally optimized to be 50 nM and 1:1, respectively (Figure S2). Therefore, the *M-DNA* used for Ag^+ sensing in the following experiments contained 50 nM $CCTG_2$, 50 nM $CAGG_2$, and 50 nM SGI in 10 mM NaPi at pH 6, unless otherwise specified.

To further verify whether the CCTG MDB played an important role in the *M-DNA* sensor for Ag^+ detection, we also performed Ag^+ titration on a controlled DNA (named *C-DNA*), which was an 8-bp self-complementary duplex. When the mixture of 50 nM *C-DNA* and 50 nM SGI in 10 mM NaPi at pH 6 was titrated with Ag^+, there was only a little change in fluorescence intensity (Figure S3), suggesting that the CCTG MDB played an irreplaceable role in Ag^+ sensing.

3.3. Kinetics, Sensitivity, and Selectivity of the M-DNA Sensor

One of the most interesting features of this *M-DNA* sensor is using an ultrashort 8-nt oligonucleotide, which is expected to undergo a much faster conformational change than longer i-motif and hairpin sequences [26,29,30,32]. Therefore, we also evaluated the kinetics of this *M-DNA* for Ag^+ sensing. The fluorescence intensity (520 nm) of the *M-DNA* sensor without Ag^+ was recorded from 0 to 180 s with a step time of 2 s. Ag^+ was then added to the same sample, and the fluorescence intensity was immediately recorded from 0 to

180 s with a step time of 2 s. Figure 4B shows that immediately after adding Ag^+, the fluorescence intensity drastically decreased and remained almost unchanged through the entire monitoring process for 180 s. Therefore, it is safe to conclude that the reaction was completed within the acquisition time for the first data point, i.e., 2 s. It was reported that the Ag^+-triggered conformational change from a single-stranded DNA to a 21-nt i-motif was completed in ~15 s [26,29,30,32], therefore it is reasonable that the conformational change to an 8-nt MDB was much faster.

The *M-DNA* was then used to sense Ag^+ at various concentrations ranging from 0 to 200 nM (Figure 4C). There was a good linear correlation between the fluorescence intensity and log[Ag^+]/log[M-DNA]. Following the rule of three times the standard deviation over the blank response [48], the Ag^+ detection limit was determined to be ~2.1 nM. As the tolerable level of Ag^+ in drinking water is ~927 nM [7], the detection limit of the *M-DNA* sensor should be sufficient for detecting Ag^+ in real samples containing Ag^+.

The anti-interference capability of the *M-DNA* sensor for Ag^+ detection in a complex environment was also evaluated. As the drinking water source may also contain other metal ions, we evaluated the fluorescence response of *M-DNA* to K^+, Li^+, Ca^{2+}, Mg^{2+}, Mn^{2+}, Co^{2+}, Cu^{2+}, Ba^{2+}, and Ni^{2+}, and the result showed only tiny fluorescence changes upon adding these ions (Figure 5A). Furthermore, an additional experiment was also performed to examine if the *M-DNA* could detect Ag^+ in the presence of these interfering metal ions. Upon adding 50 nM Ag^+ to the solutions containing the respective interfering metal ions, the fluorescence change became significant and achieved a similar level to that of only 50 nM Ag^+ (Figure 5B). Na^+ was not included as an interference ion in this study because the buffering system contained 10 mM NaPi. Approximately 10 to 200 mM Na^+ are also commonly used in buffering systems for many DNA-based sensors to neutralize the negatively charged phosphodiester backbones [26,29–31,34]. The concentrations of non-Ag^+ ions vary in different water samples, e.g., few mM Na^+ in most China river and lake basins [49] and hundreds mM Na^+ in sea water [50]. The *M-DNA* sensor should be applicable for detecting Ag^+ in common river and lake basins, and its performance may need to be further improved for sensing Ag^+ in water samples containing high concentrations of interfering ions (e.g., sea waters).

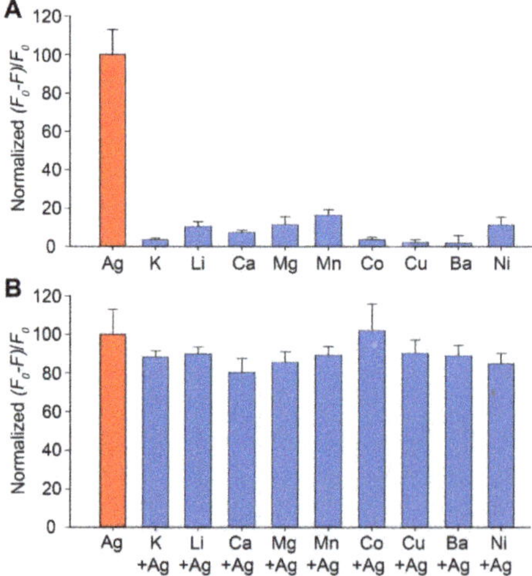

Figure 5. Fluorescence changes at 520 nm of the *M-DNA* in the presence of (**A**) 50 nM non-Ag^+ metal ions (blue) and (**B**) 50 nM non-Ag^+ metal ions plus adding 50 nM Ag^+ (blue). The fluorescence change in the presence of only 50 nM Ag^+ was shown as a reference (red). Error bars were standard deviations obtained from three replicative experiments. F_0: initial fluorescence intensity in the absence of Ag^+; F: fluorescence intensity after adding 50 nM $AgNO_3$ or other metal ions.

3.4. Ag⁺ Detection in Tap Water and Lake Water Samples Using the M-DNA Sensor

To examine the performance of the *M-DNA* sensor for Ag^+ detection in other water sources, we detected Ag^+ in tap water samples and two different lake water samples. The local tap water and lake water samples were collected and boiled for 5 min to remove chlorine, and lake water samples were further filtered with a 0.22 μm membrane following the reported procedures in the literature [26]. The *M-DNA* sensor was prepared using the treated tap and lake water samples instead of laboratory DI water, and no Ag^+ was detectable in these samples. We then added Ag^+ with known concentrations to the *M-DNA* sensor and recorded the fluorescence intensity. The Ag^+ concentration was calculated using the calibration curve shown in Figure 4C. The recovery ranged from 93.3% to 98.5% in tap water samples and 96.7% to 107.8% in lake water samples (Table 1), revealing a good accuracy of the *M-DNA* sensor for Ag^+ detection in environmental water sources.

Table 1. Ag^+ detection in tap and lake waters using the *M-DNA* sensor.

Water Source	C_{real} (nM)	C_{cal} (nM) [a]	Recovery (%)
Tap water	45	42 ± 4	93.3
	90	86 ± 4	95.6
	130	128 ± 5	98.5
	150	143 ± 6	95.3
Lake water 1	45	46 ± 1	102.2
	90	97 ± 8	107.8
	130	133 ± 6	102.3
	150	152 ± 6	101.3
Lake water 2	45	46 ± 4	102.2
	90	87 ± 7	96.7
	130	133 ± 25	102.3
	150	155 ± 17	103.3

[a] The standard deviations were obtained from three replicative experiments.

3.5. Discussions on DNA-Based Ag⁺ Sensors

As surveyed from the literature, DNA-based Ag^+ sensors can be generally classified into three types: (i) mismatch-containing DNA functionalized with nanomaterials [34–37], (ii) mismatch-containing DNA only [26,29,30,32], and (iii) DNAzyme [22] (Table 2). The ensemble of mismatch-containing DNA and nanomaterials is an effective strategy to improve the detection limit by taking advantage of amplified local DNA concentration and interaction surfaces. Recently, Pal et al. have reported an electrochemical Ag^+ sensor based on DNA hairpin-functionalized nanoflakes with a detection limit of 0.8 pM [38]. Comparing with the detection limits of other sensors using only mismatch-containing DNA (i-motifs and hairpins), detection limit of the *M-DNA* sensor was the lowest. In addition, the *M-DNA* sensor exhibited a response time of less than 2 s, which is kinetically much faster than those using i-motifs and hairpins (Table 2). However, the *M-DNA* sensor requires a controlled acidic pH to work, and this limitation may be further improved by chemical modification, such as cytosine methylation, to enhance the thermodynamic stability of the CCTG MDB. Overall, the *M-DNA* sensor uses an ultrashort oligonucleotide to achieve a high sensitivity and fast response for Ag^+ detection.

Table 2. Literature survey on DNA-based sensors for Ag^+ detection.

DNA Sensor	DNA Length (nt)	Kinetics	Detection Limit	Ref.
DNA/graphene oxide	32	[b]	5 nM	[34]
DNA/silver nanoclusters	12	<1 min [a]	10 nM	[35]
DNA/gold nanoparticle	27	[b]	3.5 nM	[36]
DNA/Fe3O4-gold nanoparticle	49	[b]	3.4 nM	[37]
DNA/nanoflakes	20	[c]	0.8 pM	[38]

Table 2. *Cont.*

DNA Sensor	DNA Length (nt)	Kinetics	Detection Limit	Ref.
DNAzyme	83	60 min [a]	24.9 nM	[22]
DNA hairpin	32	5 min [a]	59.9 nM	[26]
DNA hairpin	20	10 min [a]	32 nM	[29]
DNA hairpin	32	30 min [a]	4.3 nM	[30]
DNA i-motif	21	15 s [a]	17 nM	[32]
DNA minidumbbell	8	<2 s [a]	2.1 nM	This work

[a] The kinetic data was derived from time-dependent fluorescence spectra. [b] There was no kinetic data available. [c] The kinetic data was derived from time-dependent electrochemical change.

4. Conclusions

In sum, we have designed a smart DNA sensor for Ag^+ detection using a new form of non-B DNA, i.e., a minidumbbell, apart from the previously used hairpins and i-motifs. Owing to its small size, it shows fast response, high sensitivity, high selectivity, and good anti-interference capability for Ag^+ sensing. The performance of this *M-DNA* sensor may be further improved by chemical modification to further enhance the thermodynamic stability of the CCTG MDB. A successful demonstration of this *M-DNA* sensor provides new insights into Ag^+ detection, and paves the way for designing DNA-based tools to sense other metal ions and molecules.

Supplementary Materials: The following supporting information can be downloaded at: https://www.mdpi.com/article/10.3390/bios13030358/s1, Figure S1: CD changes of *M-DNA* with SGI in 10 mM NaPi at pH 6 before and after adding Ag^+; Figure S2: Fluorescence changes of various-concentration *M-DNA* in 10 mM NaPi at pH 6, with different SGI:*M-DNA* ratios before and after adding Ag^+; Figure S3: Normalized fluorescence intensity at 520 nm of the *M-DNA* and *C-DNA* upon titrating various concentrations of Ag^+.

Author Contributions: J.Z.: methodology, investigation, formal analysis, data curation, and writing—original draft; Y.L.: methodology, investigation, formal analysis, and writing—original draft; Z.Y.: investigation, formal analysis, and writing—original draft; Y.W.: conceptualization, methodology, formal analysis, and writing—review and editing; P.G.: conceptualization, methodology, formal analysis, writing—review and editing, supervision, and project administration. All authors have read and agreed to the published version of the manuscript.

Funding: This work was supported by the National Natural Science Foundation of China (22004038), the Natural Science Foundation of Guangdong Province, China (2021A1515010102), the Guangdong Basic and Applied Basic Research Foundation (2021A1515111174), the Science and Technology Project of Guangzhou (202201010471), the China Postdoctoral Science Foundation (2022M722112), and a start-up fund from the Institute of Basic Medicine and Cancer (IBMC) of the Chinese Academy of Sciences (2022QD13).

Institutional Review Board Statement: Not applicable.

Informed Consent Statement: Not applicable.

Data Availability Statement: The data presented in this study are available in supplementary material.

References

1. Sreekumari, K.; Nandakumar, K.; Takao, K.; Kikuchi, Y. Silver containing stainless steel as a new outlook to abate bacterial adhesion and microbiologically influenced corrosion. *ISIJ Int.* **2003**, *43*, 1799–1806. [CrossRef]
2. Kokura, S.; Handa, O.; Takagi, T.; Ishikawa, T.; Naito, Y.; Yoshikawa, T. Silver nanoparticles as a safe preservative for use in cosmetics. *Nanomedicine* **2010**, *6*, 570–574. [CrossRef] [PubMed]
3. Chernousova, S.; Epple, M. Silver as antibacterial agent: Ion, nanoparticle, and metal. *Angew. Chem. Int. Ed.* **2013**, *52*, 1636–1653. [CrossRef]
4. Gonzalez-Fernandez, S.; Lozano-Iturbe, V.; Garcia, B.; Andres, L.J.; Menendez, M.F.; Rodriguez, D.; Vazquez, F.; Martin, C.; Quiros, L.M. Antibacterial effect of silver nanorings. *BMC Microbiol.* **2020**, *20*, 172. [CrossRef] [PubMed]

5. Long, F.; Gao, C.; Shi, H.C.; He, M.; Zhu, A.N.; Klibanov, A.M.; Gu, A.Z. Reusable evanescent wave DNA biosensor for rapid, highly sensitive, and selective detection of mercury ions. *Biosens. Bioelectron.* **2011**, *26*, 4018–4023. [CrossRef] [PubMed]
6. Quadros, M.E.; Marr, L.C. Environmental and human health risks of aerosolized silver nanoparticles. *J. Air Waste Manag. Assoc.* **2010**, *60*, 770–781. [CrossRef] [PubMed]
7. Guidelines for Drinking-Water Quality. Available online: https://fctc.who.int/publications/i/item/9789241549950 (accessed on 11 February 2023).
8. Dorea, J.G.; Donangelo, C.M. Early (in uterus and infant) exposure to mercury and lead. *Clin. Nutr.* **2006**, *25*, 369–376. [CrossRef]
9. Mijnendonckx, K.; Leys, N.; Mahillon, J.; Silver, S.; Van Houdt, R. Antimicrobial silver: Uses, toxicity and potential for resistance. *Biometals* **2013**, *26*, 609–621. [CrossRef]
10. Chitpong, N.; Husson, S.M. Nanofiber ion-exchange membranes for the rapid uptake and recovery of heavy metals from water. *Membranes* **2016**, *6*, 59. [CrossRef]
11. Wang, Y.W.; Wang, M.; Wang, L.; Xu, H.; Tang, S.; Yang, H.H.; Zhang, L.; Song, H. A simple assay for ultrasensitive colorimetric detection of Ag^+ at picomolar levels using platinum nanoparticles. *Sensors* **2017**, *17*, 2521. [CrossRef]
12. Laborda, F.; Jiménez-Lamana, J.; Bolea, E.; Castillo, J.R. Selective identification, characterization and determination of dissolved silver(I) and silver nanoparticles based on single particle detection by inductively coupled plasma mass spectrometry. *J. Anal. At. Spectrom.* **2011**, *26*, 1362–1371. [CrossRef]
13. Hong, A.; Tang, Q.; Khan, A.U.; Miao, M.; Xu, Z.; Dang, F.; Liu, Q.; Wang, Y.; Lin, D.; Filser, J.; et al. Identification and speciation of nanoscale silver in complex solid matrices by sequential extraction coupled with inductively coupled plasma optical emission spectrometry. *Anal. Chem.* **2021**, *93*, 1962–1968. [CrossRef] [PubMed]
14. Resano, M.; Aramendia, M.; Garcia-Ruiz, E.; Crespo, C.; Belarra, M.A. Solid sampling-graphite furnace atomic absorption spectrometry for the direct determination of silver at trace and ultratrace levels. *Anal. Chim. Acta* **2006**, *571*, 142–149. [CrossRef] [PubMed]
15. Feichtmeier, N.; Leopold, K. Detection of silver nanoparticles in parsley by solid sampling high-resolution–continuum source atomic absorption spectrometry. *Anal. Bioanal. Chem.* **2014**, *406*, 3887–3894. [CrossRef] [PubMed]
16. Zeng, L.; Wu, M.; Chen, S.; Zheng, R.; Rao, Y.; He, X.; Duan, Y.; Wang, X. Direct and sensitive determination of Cu, Pb, Cr and Ag in soil by laser ablation microwave plasma torch optical emission spectrometry. *Talanta* **2022**, *246*, 123516–123524. [CrossRef]
17. Ouyang, P.; Fang, C.; Han, J.; Zhang, J.; Yang, Y.; Qing, Y.; Chen, Y.; Shang, W.; Du, J. A DNA electrochemical sensor via terminal protection of small-molecule-linked DNA for highly sensitive protein detection. *Biosensors* **2021**, *11*, 451. [CrossRef]
18. Lee, J.H.; Wang, Z.; Liu, J.; Lu, Y. Highly sensitive and selective colorimetric sensors for uranyl (UO_2^{2+}): Development and comparison of labeled and label-free DNAzyme-Gold nanoparticle systems. *J. Am. Chem. Soc.* **2008**, *130*, 14217–14226. [CrossRef]
19. Liu, M.; Zhao, H.; Chen, S.; Yu, H.; Zhang, Y.; Quan, X. Label-free fluorescent detection of Cu(II) ions based on DNA cleavage-dependent graphene-quenched DNAzymes. *Chem. Commun.* **2011**, *47*, 7749–7751. [CrossRef]
20. Zhang, X.B.; Kong, R.M.; Lu, Y. Metal ion sensors based on DNAzymes and related DNA molecules. *Annu. Rev. Anal. Chem.* **2011**, *4*, 105–128. [CrossRef]
21. Zhuang, J.; Fu, L.; Xu, M.; Zhou, Q.; Chen, G.; Tang, D. DNAzyme-based magneto-controlled electronic switch for picomolar detection of lead (II) coupling with DNA-based hybridization chain reaction. *Biosens. Bioelectron.* **2013**, *45*, 52–57. [CrossRef]
22. Saran, R.; Liu, J. A silver DNAzyme. *Anal. Chem.* **2016**, *88*, 4014–4020. [CrossRef]
23. Zhou, W.; Zhang, Y.; Ding, J.; Liu, J. In vitro selection in serum: RNA-cleaving DNAzymes for measuring Ca^{2+} and Mg^{2+}. *ACS Sens.* **2016**, *1*, 600–606. [CrossRef]
24. Memon, A.G.; Xing, Y.; Zhou, X.; Wang, R.; Liu, L.; Zeng, S.; He, M.; Ma, M. Ultrasensitive colorimetric aptasensor for Hg^{2+} detection using Exo-III assisted target recycling amplification and unmodified AuNPs as indicators. *J. Hazard. Mater.* **2020**, *384*, 120948–120953. [CrossRef] [PubMed]
25. Zhang, X.; Zhang, G.; Wei, G.; Su, Z. One-pot, in-situ synthesis of 8-armed poly(ethylene glycol)-coated Ag nanoclusters as a fluorescent sensor for selective detection of Cu^{2+}. *Biosensors* **2020**, *10*, 131. [CrossRef]
26. Zhou, X.; Memon, A.G.; Sun, W.; Fang, F.; Guo, J. Fluorescent probe for Ag^+ detection using SYBR GREEN I and C-C mismatch. *Biosensors* **2020**, *11*, 6. [CrossRef]
27. Torigoe, H.; Okamoto, I.; Dairaku, T.; Tanaka, Y.; Ono, A.; Kozasa, T. Thermodynamic and structural properties of the specific binding between Ag^+ ion and C:C mismatched base pair in duplex DNA to form C-Ag-C metal-mediated base pair. *Biochimie* **2012**, *94*, 2431–2440. [CrossRef] [PubMed]
28. Dairaku, T.; Furuita, K.; Sato, H.; Sebera, J.; Nakashima, K.; Kondo, J.; Yamanaka, D.; Kondo, Y.; Okamoto, I.; Ono, A.; et al. Structure determination of an Ag^I-mediated cytosine-cytosine base pair within DNA duplex in solution with $^1H/^{15}N/^{109}Ag$ NMR spectroscopy. *Chem. Eur. J.* **2016**, *22*, 13028–13031. [CrossRef] [PubMed]
29. Lin, Y.H.; Tseng, W.L. Highly sensitive and selective detection of silver ions and silver nanoparticles in aqueous solution using an oligonucleotide-based fluorogenic probe. *Chem. Commun.* **2009**, 6619–6621. [CrossRef] [PubMed]
30. Pu, W.; Zhao, H.; Huang, C.; Wu, L.; Xua, D. Fluorescent detection of silver(I) and cysteine using SYBR Green I and a silver(I)-specific oligonucleotide. *Microchim Acta* **2012**, *177*, 137–144. [CrossRef]
31. Shi, Y.; Sun, H.; Xiang, J.; Yu, L.; Yang, Q.; Li, Q.; Guan, A.; Tang, Y. I-motif-modulated fluorescence detection of silver(I) with an ultrahigh specificity. *Anal. Chim. Acta* **2015**, *857*, 79–84. [CrossRef]

32. Kang, B.H.; Gao, Z.F.; Li, N.; Shi, Y.; Li, N.B.; Luo, H.Q. Thiazole orange as a fluorescent probe: Label-free and selective detection of silver ions based on the structural change of i-motif DNA at neutral pH. *Talanta* **2016**, *156–157*, 141–146. [CrossRef] [PubMed]
33. Shen, F.; Mao, S.; Mathivanan, J.; Wu, Y.; Chandrasekaran, A.R.; Liu, H.; Gan, J.; Sheng, J. Short DNA oligonucleotide as a Ag^+ binding detector. *ACS Omega* **2020**, *5*, 28565–28570. [CrossRef] [PubMed]
34. Wen, Y.; Xing, F.; He, S.; Song, S.; Wang, L.; Long, Y.; Li, D.; Fan, C. A graphene-based fluorescent nanoprobe for silver(I) ions detection by using graphene oxide and a silver-specific oligonucleotide. *Chem. Commun.* **2010**, *46*, 2596–2598. [CrossRef] [PubMed]
35. Lee, J.; Park, J.; Hee Lee, H.; Park, H.; Kim, H.I.; Kim, W.J. Fluorescence switch for silver ion detection utilizing dimerization of DNA-Ag nanoclusters. *Biosens. Bioelectron.* **2015**, *68*, 642–647. [CrossRef] [PubMed]
36. Guo, Z.; Zheng, Y.; Xu, H.; Zheng, B.; Qiu, W.; Guo, Z. Lateral flow test for visual detection of silver (I) based on cytosine-Ag(I)-cytosine interaction in C-rich oligonucleotides. *Microchim. Acta.* **2017**, *184*, 4243–4250. [CrossRef] [PubMed]
37. Miao, P.; Tang, Y.; Wang, L. DNA modified Fe_3O_4@Au magnetic nanoparticles as selective probes for simultaneous detection of heavy metal ions. *ACS Appl. Mater. Interfaces* **2017**, *9*, 3940–3947. [CrossRef]
38. Pal, C.; Kumar, A.; Majumder, S. Fabrication of ssDNA functionalized MoS_2 nanoflakes based label-free electrochemical biosensor for explicit silver ion detection at sub-pico molar level. *Colloids Surf. A Physicochem. Eng. Asp.* **2022**, *655*, 130241. [CrossRef]
39. Tian, C.; Zhao, L.; Zhu, J.; Zhang, S. Simultaneous detection of trace Hg^{2+} and Ag^+ by SERS aptasensor based on a novel cascade amplification in environmental water. *Chem. Eng. J.* **2022**, *435*, 133879–133887. [CrossRef]
40. Guo, P.; Lam, S.L. Minidumbbell: A new form of native DNA structure. *J. Am. Chem. Soc.* **2016**, *138*, 12534–12540. [CrossRef]
41. Guo, P.; Lam, S.L. Minidumbbell structures formed by ATTCT pentanucleotide repeats in spinocerebellar ataxia type 10. *Nucleic Acids Res.* **2020**, *48*, 7557–7568. [CrossRef]
42. Wan, L.; Lam, S.L.; Lee, H.K.; Guo, P. Rational design of a reversible Mg^{2+}/EDTA-controlled molecular switch based on a DNA minidumbbell. *Chem. Commun.* **2020**, *56*, 10127–10130. [CrossRef]
43. Guo, P.; Lam, S.L. An extraordinarily stable DNA minidumbbell. *J. Phys. Chem. Lett.* **2017**, *8*, 3478–3481. [CrossRef] [PubMed]
44. Kondo, J.; Tada, Y.; Dairaku, T.; Hattori, Y.; Saneyoshi, H.; Ono, A.; Tanaka, Y. A metallo-DNA nanowire with uninterrupted one-dimensional silver array. *Nat. Chem.* **2017**, *9*, 956–960. [CrossRef] [PubMed]
45. Kypr, J.; Kejnovska, I.; Renciuk, D.; Vorlickova, M. Circular dichroism and conformational polymorphism of DNA. *Nucleic Acids Res.* **2009**, *37*, 1713–1725. [CrossRef]
46. Zhang, J.; Wang, Y.; Wan, L.; Liu, Y.; Yi, J.; Lam, S.L.; Guo, P. A pH and Mg^{2+}-responsive molecular switch based on a stable DNA minidumbbell bearing 5′ and 3′-overhangs. *ACS Omega* **2021**, *6*, 28263–28269. [CrossRef] [PubMed]
47. Zhang, L.; Mi, N.; Zhang, Y.; Wei, M.; Li, H.; Yao, S. Label-free DNA sensor for Pb^{2+} based on a duplex–quadruplex exchange. *Anal. Methods* **2013**, *5*, 6100–6105. [CrossRef]
48. Alies, B.; Ouelhazi, M.A.; Noireau, A.; Gaudin, K.; Barthelemy, P. Silver ions detection via nucleolipids self-assembly. *Anal. Chem.* **2019**, *91*, 1692–1695. [CrossRef]
49. Gao, B.; Gao, L.; Gao, J.; Xu, D.; Wang, Q.; Sun, K. Simultaneous evaluations of occurrence and probabilistic human health risk associated with trace elements in typical drinking water sources from major river basins in China. *Sci. Total Environ.* **2019**, *666*, 139–146. [CrossRef] [PubMed]
50. Dong, Y.; Rosenbaum, R.K.; Hauschild, M.Z. Assessment of metal toxicity in marine ecosystems: Comparative toxicity potentials for nine cationic metals in coastal seawater. *Environ. Sci. Technol.* **2016**, *50*, 269–278. [CrossRef]

 *biosensors*

Communication

Development of Fluorescent Turn-On Probes for CAG-RNA Repeats

Matthew Ho Yan Lau [1], Chun-Ho Wong [2], Ho Yin Edwin Chan [2,3,4] and Ho Yu Au-Yeung [1,4,5,*]

[1] Department of Chemistry, The University of Hong Kong, Pokfulam Road, Hong Kong, China
[2] School of Life Sciences, The Chinese University of Hong Kong, Hong Kong, China
[3] Gerald Choa Neuroscience Centre, The Chinese University of Hong Kong, Hong Kong, China
[4] Nexus of Rare Neurodegenerative Diseases, School of Life Sciences, The Chinese University of Hong Kong, Hong Kong, China
[5] State Key Laboratory of Synthetic Chemistry, The University of Hong Kong, Pokfulam Road, Hong Kong, China
* Correspondence: hoyuay@hku.hk

Abstract: Fluorescent sensing of nucleic acids is a highly sensitive and efficient bioanalytical method for their study in cellular processes, detection and diagnosis in related diseases. However, the design of small molecule fluorescent probes for the selective binding and detection of RNA of a specific sequence is very challenging because of their diverse, dynamic, and flexible structures. By modifying a bis(amidinium)-based small molecular binder that is known to selectively target RNA with CAG repeats using an environment-sensitive fluorophore, a turn-on fluorescent probe featuring aggregation-induced emission (AIE) is successfully developed in this proof-of-concept study. The probe (**DB-TPE**) exhibits a strong, 19-fold fluorescence enhancement upon binding to a short CAG RNA, and the binding and fluorescence response was found to be specific to the overall RNA secondary structure with A·A mismatches. These promising analytical performances suggest that the probe could be applied in pathological studies, disease progression monitoring, as well as diagnosis of related neurodegenerative diseases due to expanded CAG RNA repeats.

Keywords: fluorescent probe; AIE; RNA; CAG repeats

Citation: Lau, M.H.Y.; Wong, C.-H.; Chan, H.Y.E.; Au-Yeung, H.Y. Development of Fluorescent Turn-On Probes for CAG-RNA Repeats. *Biosensors* **2022**, *12*, 1080. https://doi.org/10.3390/bios12121080

Received: 8 November 2022
Accepted: 21 November 2022
Published: 25 November 2022

1. Introduction

Polyglutamine (PolyQ) diseases, including Huntington's disease, spinobulbar muscular atrophy and spinocerebellar ataxias, are a group of inheritable neurodegenerative disorders characterized by the expansion of a CAG trinucleotide repeat of the affected genes and the expression of polyQ proteins with a glutamine-enriched tract, in which the number of CAG repeats in patients or carriers can reach from 40 to over 100 as compared to that of 35–50 in healthy individuals [1–5]. The polyQ proteins are neurotoxic and can cause a wide range of cellular dysfunctions in neurons. Taking Huntington's disease as an example, the mutant *HTT* gene encodes the toxic and aggregation-prone mutant huntingtin (mHTT) protein with an expanded polyQ tract translated from the expanded CAG repeats, and the accumulation of the mHTT protein will lead to neuronal cell death due to altered neural circuitry, mitochondrial dysfunction, transcriptional dysregulation, disrupted protein homeostasis, impaired protein degradation, and aberrant activation of stress responses [6–10]. In addition to the toxicity due to the misfolding, accumulation, and aggregation of the polyQ proteins, recent studies are showing that the expanded CAG RNA repeats are also neurotoxic [11–14]. A recent work has demonstrated that cellular expression of short CAG (sCAG) RNAs can hybridize with and silence the CUG-containing mRNA in the *Nudix hydrolase 16* (*NUDT16*) gene and lead to DNA damages and cellular apoptosis [15]. Small molecule inhibitors that selectively bind to the CAG RNA repeats are therefore implicated in not only the development of therapeutic strategies for treating polyQ diseases, but also their rapid detection and diagnosis.

Designing small molecule binders and related fluorescent probes for specific RNA sequences is, however, very challenging because of the highly diverse, dynamic and flexible structures of RNA [16–18]. Recently, we have identified the bis(amidinium) compound, **DB213**, as a small molecule drug candidate for treating polyQ RNA toxicity. Detail binding and NMR studies showed that **DB213** binds with a K_d of 3.8 μM to the major groove of an sCAG hairpin (*rGG-(CAG)₆-CC*) and interacts specifically with two consecutive A·A mismatches via a combination of hydrogen bonds, electrostatic, and other interactions (Figure 1) [15]. The binding is selective, and a much weaker binding was observed for CUG RNA [19–21] and the rCAG/CUG heteroduplex formed from *rGG-(CAG)₄-(CUG)₂-CC* (K_d = 19.2 μM) [15]. Further cellular studies and biochemical analysis also showed that **DB213** did not affect the transcription level of the mRNA of a CAG repeat disease gene, and that it could suppress toxicity due to CAG RNA but not polyQ proteins, showing that **DB213** is specific to the CAG RNA over the corresponding DNA and proteins [15]. With these promising binding properties, we propose that **DB213** could be turned into a fluorescent probe for CAG RNA repeats by appending an environment sensitive fluorophore on the bis(amidinium) core, such that the probe would elicit a fluorescent response upon binding to the target CAG RNA as a result of a change in the local environment. In this work, we present our results on this proof-of-concept study including the design, synthesis and solution characterizations of three fluorophore-appended **DB213** derivatives as potential fluorescent probes for CAG RNA. In particular, **DB-TPE** which features a tetraphenylethylene fluorophore was found to exhibit a strong fluorescence enhancement upon binding to CAG RNA via aggregation-induced emission (AIE).

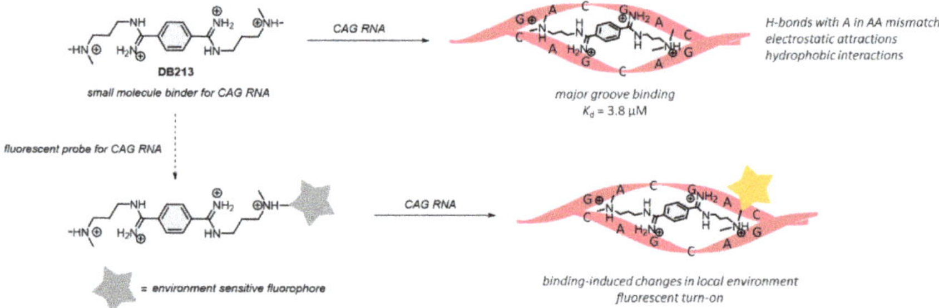

Figure 1. Chemical structure of **DB213** that binds selectively to CAG RNA, and the design of a fluorescent probe for CAG RNA.

2. Materials and Methods

All reagents were purchased from commercial suppliers (Aldrich, Dkmchem, Energy, J & K) and used without further purification unless otherwise specified. All solvents were of analytical grade (ACL Labscan, DUKSAN Pure Chemicals). Dried solvents were distilled over CaH_2 ($CHCl_3$, EtOH, MeOH). Thin layer chromatography was performed on silica gel 60 F254 (Merck) and column chromatography was carried out using silica gel 60F (Silicycle), neutral or basic aluminium oxide Brockmann I (Acros Organics). Detailed synthetic procedures and characterization of all other starting materials are described in the Supplementary File S1. ^{1}H and ^{13}C{1H} NMR spectra were obtained from a Brucker DPX 400 or Brucker DPX 500 spectrometer and signals were referenced to solvent residues. LC-MS analyses were performed on a UPLC-MS system with a Waters UPLC coupled to a 2489 UV/Vis detector and an ACQUITY QDa MS detector. High resolution ESI-MS data was obtained from a Waters Micromass Premier Q-ToF tandem mass spectrometer.

Synthesis of **DB-Res** was carried out as follows. Under an argon atmosphere, a mixture of **DB-N₃** (10 mg, 0.03 mmol), **Res-alkyne** (8 mg, 0.03 mmol), and [Cu(CH₃CN)₄](PF₆) (1 mg, 0.003 mmol) in DMSO (3 mL) was stirred at room temperature for overnight.

The resulting mixture was purified by a neutral alumina column with CH_2Cl_2/MeOH (v/v = 8:2) as the eluent. Fractions containing the product was combined and concentrated using a rotary evaporator, to which a saturated HCl solution in dry EtOH (1 mL) was added. The mixture was stirred at room temperature for 3 h and centrifuged (4000 rpm, 3 min). The solution was decanted and the remaining solid was washed with dry EtOH (10 mL × 1) and dry diethyl ether (10 mL × 3). The orange-red solid obtained was dried under vacuum. Yield = 15 mg, 73%. ^{1}H NMR (500 MHz, CD$_3$OD, 298 K): δ 8.27 (s, 1H), 7.99–7.95 (m, 2H), 7.95–7.91 (m, 2H), 7.78 (d, J = 8.9 Hz, 1H,), 7.54 (d, J = 9.8 Hz, 1H), 7.17 (d, J = 2.6 Hz, 1H), 7.13 (dd, J = 8.9, 2.7 Hz, 1H), 6.85 (dd, J = 9.8, 2.1 Hz, 1H), 6.28 (d, J = 2.1 Hz, 1H), 5.35 (s, 2H), 4.62 (t, J = 6.0 Hz, 2H), 3.57 (t, J = 6.7 Hz, 2H), 3.26 (t, J = 6.7 Hz, 2H), 2.96 (t, J = 6.0 Hz, 2H), 2.84–2.78 (m, 2H), 2.56 (s, 9H), 2.38–2.34 (m, 2H), 2.06 (quint, J = 7.2 Hz, 2H), and 1.85 (quint, J = 6.8 Hz, 2H). ^{13}C{^{1}H} NMR (125 MHz, CD$_3$OD, 298 K): δ 188.3, 164.9, 164.6, 164.2, 151.9, 147.1, 146.6, 143.9, 136.6, 135.0, 134.8, 134.7, 133.0, 130.1, 130.0, 130.0, 126.4, 115.7, 106.9, 102.3, 63.3, 57.2, 56.7, 55.1, 44.6, 44.5, 42.3, 42.3, 42.2, 26.2, and 25.5. HR-ESI-MS (+ve) calculated for $C_{34}H_{42}N_{10}O_3$ [M+H]$^+$ (m/z): 639.3514, found: 639.3504.

Synthesis of **DB-Dan** was carried out as follows. Under an argon atmosphere, a mixture of **DB-N$_3$** (43 mg, 0.11 mmol), **Dan-alkyne** (35 mg, 0.12 mmol), and [Cu(CH$_3$CN)$_4$](PF$_6$) (4 mg, 0.01 mmol) in MeCN/MeOH (2 mL/0.5 mL, v/v = 8:2) was stirred at room temperature for overnight. Solvents were removed using a rotary evaporator, and the residue was purified by a basic alumina column using a gradient mixture of CH_2Cl_2/MeOH (v/v = 9:1 to 8:2) as the eluent. Fractions containing the product were combined and concentrated using a rotary evaporator, to which a saturated HCl solution in dry ethanol (1 mL) was added. The mixture was heated at 60 °C for 3 h, cooled to room temperature, and centrifuged (4000 rpm, 3 min). The solution was decanted and the remaining solid was washed with dry ethanol (10 mL × 1) and dry diethyl ether (10 mL × 3). The resulting solid was dried under vacuum to obtain the product as a pale-yellow solid. Yield = 20 mg, 21%. ^{1}H NMR (400 MHz, CD$_3$OD, 298 K): δ 8.52 (d, J = 8.4 Hz, 1H), 8.32 (d, J = 8.7 Hz, 1H), 8.15 (d, J = 7.2 Hz, 1H), 7.71 (s, 4H), 7.60–7.48 (m, 3H), 7.25 (d, J = 7.5 Hz, 1H), 4.31 (t, J = 6.1 Hz, 2H), 4.11 (s, 2H), 3.33 (t, J = 7.1 Hz, 2H), 3.18 (t, J = 6.7 Hz, 2H), 2.87 (s, 6H), 2.74 (t, J = 6.2 Hz, 2H), 2.54–2.38 (m, 4H), 2.27 (s, 6H), 2.25 (s, 3H), 1.87 (quint, J = 7.2 Hz, 2H), and 1.75 (quint, J = 6.9 Hz, 2H). ^{13}C{^{1}H} NMR (125 MHz, CD$_3$OD, 298 K): δ 163.8, 157.3, 153.1, 146.0, 139.0, 138.6, 137.2, 131.1, 131.0, 130.9, 130.2, 129.1, 128.5, 128.4, 124.8, 124.3, 120.6, 116.3, 58.2, 57.6, 56.0, 45.8, 45.4, 42.9, 42.5, 42.2, 39.2, 27.6, and 27.3. HR-ESI-MS (+ve) calculated for $C_{34}H_{49}N_{11}O_2S$ [M+H]$^+$ (m/z): 676.3864, found: 676.3870.

Synthesis of **DB-TPE** was carried out as follows. Under an argon atmosphere, a mixture of **DB-alkyne** (18 mg, 0.05 mmol), **TPE-N$_3$** (24 mg, 0.06 mmol), and [Cu(CH$_3$CN)$_4$](PF$_6$) (2 mg, 0.005 mmol) in MeCN/MeOH/CH_2Cl_2 (2 mL/1 mL/1 mL, v/v = 2:1:1) was stirred at room temperature for overnight. The solvents were removed using a rotary evaporator, and the residue was washed with dry diethyl ether (10 mL × 2). A saturated HCl solution in dry ethanol (1 mL) was added to the residue, and the mixture was heated at 60 °C for 3 h, cooled to room temperature, and centrifuged (4000 rpm, 3 min). The solution was decanted and the remaining solid was washed with dry ethanol (10 mL × 1) and dry diethyl ether (10 mL × 3). The resulting solid was dried under vacuum to a obtain the product as a yellow solid. Yield = 42 mg, 93%. ^{1}H NMR (400 MHz, CD$_3$OD, 298 K): δ 8.39 (s, 1H), 8.03 (s, 4H), 7.16–7.01 (m, 12H), 7.01–6.88 (m, 7H), 5.58 (s, 2H), 4.59–4.37 (m, 2H), 3.72–3.55 (m, 2H), 3.53–3.39 (m, 4H), 2.94 (s, 6H), 2.90 (s, 3H), 2.39–2.28 (m, 2H), and 2.28–2.18 (m, 2H). ^{13}C{^{1}H} NMR (125 MHz, CD$_3$OD, 298 K): δ 165.1, 144.8, 144.7, 143.1, 141.5, 134.8, 132.9, 132.3, 132.2, 130.3, 128.9, 128.8, 128.7, 127.7, 127.7, 56.5, 43.9, 41.7, and 24.2. HR-ESI-MS (+ve) calculated for $C_{47}H_{53}N_9$ [M+H]$^+$ (m/z): 744.4497, found: 744.4494.

Fluorescence spectra were recorded on an Edinburgh Instruments FS5 Spectrofluorometer equipped with a 150 W CW ozone-free xenon arc lamp and a Photomultiplier R928P detection unit with spectral coverage of 200–870 nm. Fluorescence studies were performed in quartz cuvettes of 1 cm path length and 1.5 mL cell volume, and double distilled water was used in all measurement. Stock solutions of **DB-Res**, **DB-Dan**, and **DB-TPE** in water were diluted to 10 µM with H_2O and/or THF to the required THF/H_2O ratios. **DB-Res**, **DB-Dan**, and **DB-TPE** were excited at 475 nm, 330 nm, and 300 nm, respectively. Synthetic RNAs, *rGG-(CAG)$_6$-CC* (*5′-rGG CAG CAG CAG CAG CAG CAG CC-3′*), *rGG-(CAG)$_3$-CC* (*5′-rGG CAG CAG CAG CC-3′*), and *rGG-(CUG)$_3$-CC* (*5′-rGG CUG CUG CUG CC-3′*) were purchased from Integrated DNA Technologies, and stock solutions at 100 µM were prepared by dissolving the RNA in UltraPure™ DNase/RNase-free distilled water. For measurements in the presence of RNA, RNA sample were diluted with water, heated at 95 °C for 2 min, and slowly cooled to room temperature and added to a sample solution containing 10 µM of the fluorescent probe in 20 mM MOPS buffer (pH 7.0) in the presence of 300 mM NaCl.

Dissociation constant (K_d) of **DB-TPE** towards *rGG-(CAG)$_6$-CC* was determined by non-linear curve fitting of the fluorescence data using the following equation:

$$F_R = F_0 + \frac{F_{MAX} - F_0}{2[P]_0} \times \left[([P]_0 + [R] + K_d) - \sqrt{([P]_0 + [R] + K_d)^2 - 4[P]_0[R]} \right]$$

where F_R is the fluorescence intensity at 463 nm for a given concentration of RNA $[R]$, F_0 is the fluorescence intensity in the absence of RNA, F_{MAX} is the maximum fluorescence intensity at saturation, $[P]_0$ is the total concentration of **DB-TPE** (10 µM), and $[R]$ is the total concentration of RNA.

3. Results and Discussions

Three bis(amidinium) derivatives have been designed and synthesized as potential fluorescent probes for CAG RNA based on the structure of **DB213** (Figure 2). In particular, three different fluorophores, namely resorufin, dansyl, and tetraphenylethylene (TPE), were chosen as the fluorescent reporters that are sensitive to the local environment for obtaining a fluorescence response upon RNA binding. The probes were synthesized via a copper-catalyzed azide–alkyne cycloaddition between **DB-N$_3$** or **DB-alkyne** and the corresponding fluorophore-containing click partners. Of note, compared to oligonucleotide-based probes that can engage in sequence-specific hybridization with the target DNA or RNA [22,23], small molecule probes are more versatile with diverse structure, and their photophysical, chemical, and biological properties can be readily tuned via chemical modifications [24,25]. However, due to the highly diverse and dynamic structures of RNA, small molecular probes that can selectively recognize a specific RNA sequence are scarce [16–18]. In addition to RNA binding, an appropriate mechanism to induce a change in the fluorescent properties upon binding is also essential for a successful RNA probe. While the resorufin and dansyl fluorophores are known to be sensitive to the local environment and could elicit a change in their emission due to a different degree of the intramolecular charge transfer (ICT) [26–29], TPE is a representative AIEgen that the restricted intramolecular motions upon binding will block the non-radiative decay pathway and result in an emission turn-on [30].

All the three probes have been characterized by ^{1}H NMR, ^{13}C NMR, HR-ESI-MS, and UV–Vis. Fluorescent properties of the probes in response to environment changes were evaluated to assess their applicability in fluorescent RNA sensing. Figure 3 shows the emission spectra of 10 µM solutions of **DB-Res**, **DB-Dan**, and **DB-TPE** in water/THF mixtures of various ratios and ionic strengths (from 3 mM to 3 M NaCl). For **DB-Res**, weaker fluorescence was observed when there is an increase in the percentage of THF, and a 7.5-fold decrease in the emission intensity at 574 nm was found at 90% THF when compared to that at pure water. The weaker emission can be explained by the formation of non-emissive aggregates of the probe when the solvent becomes more non-polar. Consistent with this, weaker fluorescence was also observed when solvent ionic strength increased, and there was a 3.2-fold decrease in fluorescence in the presence of 3 M NaCl. On the other hand, an increase in fluorescence, along with a blue shift of the emission maximum, was observed for **DB-Dan** upon increasing the THF percentage. At 90% THF, a 6.8-fold emission enhancement and a blue shift of the emission maximum by 41 nm (from 562 nm to 521 nm) was resulted, and these solvatochromic and fluorogenic effects are consistent with an ICT mechanism [31,32]. No significant change in the emission maximum and intensity was found for **DB-Dan** at different ionic strengths. Consistent with the AIE effect of TPE, a 10 µM solution of the aqueous soluble, cationic **DB-TPE** in water was weakly emissive and can be ascribed to the rapid non-radiative decay due to intramolecular rotations of the phenyl units [33,34]. Upon increasing the THF content to 90%, a strong emission enhancement by 85-fold and a blue shift in the emission maximum by 42 nm were observed. The decrease in solubility of **DB-TPE** upon addition of THF likely resulted in the aggregation of the probe and restricted the intramolecular rotations that activated the TPE emission. Compared to the emission intensity in the presence of 3 mM NaCl, a 24-fold increase in fluorescence was observed in the presence of 3 M NaCl, showing that the salting-out effect at high ionic strength can also trigger the AIE.

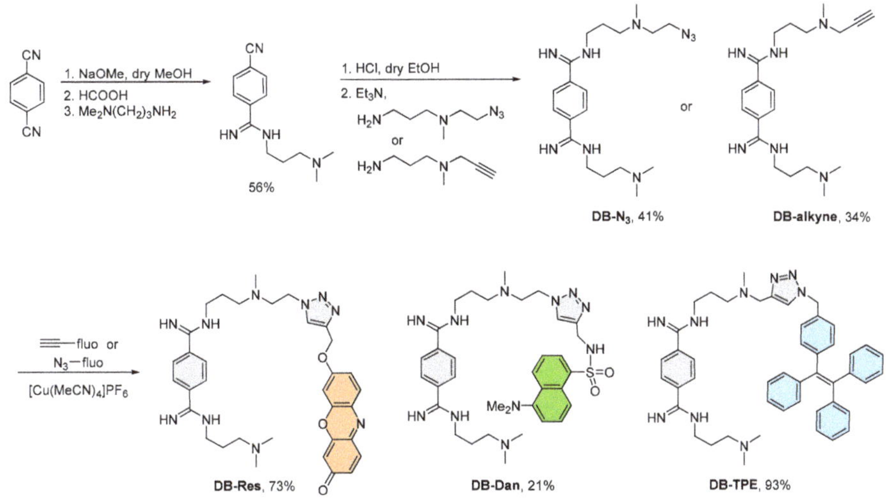

Figure 2. Synthesis of **DB-Res**, **DB-Dan**, and **DB-TPE**.

Fluorescence response of the probes was further tested using a synthetic RNA with the sequence *rGG-(CAG)$_6$-CC* RNA as a model of CAG repeats. Previous studies have shown that this particular sCAG RNA will form a hairpin structure in solution with which **DB213** can selectively bind with a strong affinity (K_d = 3.8 µM) [15,35]. Fluorescence response to the RNA was studied with 10 µM of the probe in the presence of 300 mM NaCl in 20 mM MOPS buffer at pH 7. As shown in Figure 3, only a minimal change in the emission was observed in the presence of 1 eq. of the (CAG)$_6$ RNA for **DB-Res** and **DB-Dan**, suggesting that the change in the local environment upon binding may not be significant enough to result in an obvious fluorescent response. On the contrary, a 19-fold emission enhancement of **DB-TPE** was observed under the same condition, suggesting that the AIE probe has a promising analytical potential for CAG RNA.

Response of **DB-TPE** towards the (CAG)$_6$ RNA was studied in more details. Figure 4 shows the concentration dependent response of the probe, and the relative emission intensity at 463 nm was found to increase from 3.5-fold to 18.6-fold when the amount of (CAG)$_6$ RNA increased from 0.05 eq. to 1 eq. Binding strength of **DB-TPE** to the (CAG)$_6$ RNA was obtained from the fluorescence titration data, and a K_d of 2 µM was found by fitting the binding isotherm to a 1:1 binding model. The K_d of **DB-TPE** is comparable to that of the binding of **DB213** to the same (CAG)$_6$ RNA, suggesting that both bis(amidinium) compounds share a similar binding mode that targets the major groove and overall secondary structure of the sCAG RNA with A·A mismatches in the hairpin structure. Consistent with this, a much weaker fluorescence, likely due to the non-specific electrostatic interactions between the cationic bis(amidinium) and anionic nucleic acid, was observed when the sCAG

RNA was denatured by urea (Figure 4b), or in the presence of 1 eq. of the canonically-paired duplex consisting of *rGG-(CAG)$_3$-CC* and *rGG-(CUG)$_3$-CC* with complementary A·U pairs (Figure 4c). Overall, these results clearly demonstrate the favorable analytical performance of **DB-TPE** as a turn-on fluorescent probe for the selective sensing of RNA with CAG repeats that contain A·A mismatches.

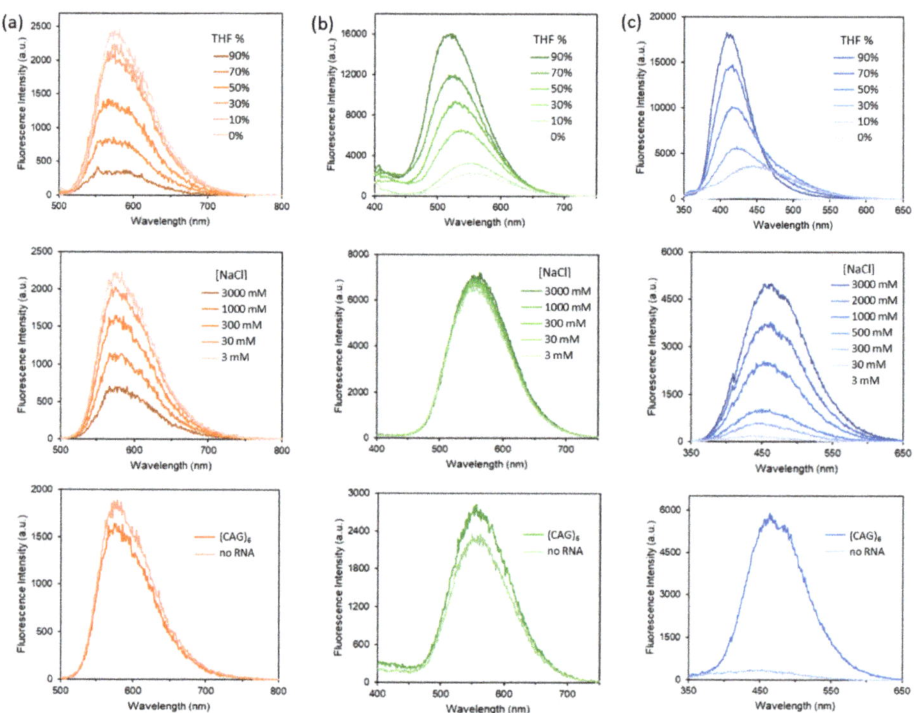

Figure 3. Fluorescence spectra of 10 µM (**a**) **DB-Res** (λ_{ex} = 475 nm), (**b**) **DB-Dan** (λ_{ex} = 330 nm), and (**c**) **DB-TPE** (λ_{ex} = 300 nm) in different water/THF ratios (upper panel), ionic strengths with various NaCl concentrations (middle panel), and the presence of 1 eq. of a (CAG)$_6$ RNA (lower panel).

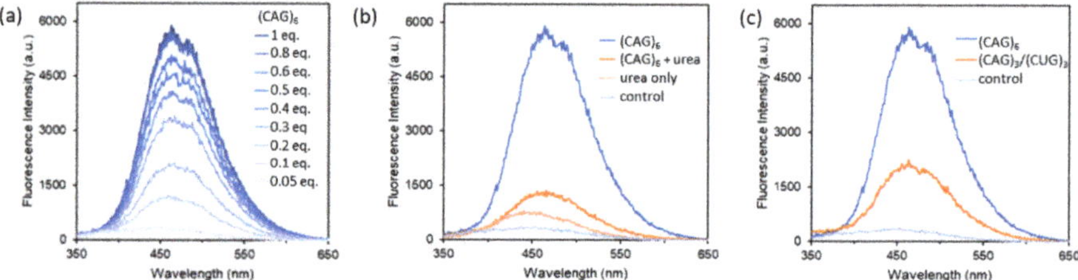

Figure 4. Fluorescence spectra of 10 µM **DB-TPE** (λ_{ex} = 300 nm) in the presence of (**a**) increasing amount of (CAG)$_6$ RNA from 0.05 to 1 eq., (**b**) 1 eq. of (CAG)$_6$ RNA denatured by 1 M urea, and (**c**) 1 eq. of complementary (CAG)$_3$/(CUG)$_3$ RNA duplex.

4. Conclusions

In summary, a fluorescent turn-on probe that selectively binds to sCAG RNA with a strong fluorescence enhancement is successfully developed. Despite the various challenges in specific RNA recognition due to their dynamic structures, by using tetraphenylethylene as an AIE-active fluorophore to decorate **DB213**, a known small molecule inhibitor that selectively binds to the major groove of CAG RNA with A·A mismatches, **DB-TPE** was obtained and found to display a strong (~19-fold) fluorescence enhancement upon binding to sCAG RNA with an apparent K_d of 2 μM. Comparative fluorescence studies showed that **DB-TPE** binds and reports the presence of sCAG RNA with a specific secondary structure. With this promising analytical performance, **DB-TPE** not only could be employed in the sensitive detection of CAG RNA for diagnostic applications of CAG repeats-related neurodegenerative diseases, but may also allow for the real-time monitoring of the CAG RNA for further understanding of the disease in cell models. Further works on applying the probe in biological samples and specimens are currently underway in our laboratory.

Supplementary Materials: The following supporting information can be downloaded at https://www.mdpi.com/article/10.3390/bios12121080/s1, File S1: synthetic procedures of precursors, NMR, MS and UV-Vis spectra. References [36,37] are citied in the Supplementary Materials.

Author Contributions: Conceptualization, M.H.Y.L., C.-H.W., H.Y.E.C. and H.Y.A.-Y.; methodology, M.H.Y.L. and H.Y.A.-Y.; investigation, M.H.Y.L. and H.Y.A.-Y.; resources, C.-H.W., H.Y.E.C. and H.Y.A.-Y.; writing—original draft preparation, M.H.Y.L. and H.Y.A.-Y.; writing—review and editing, M.H.Y.L., C.-H.W., H.Y.E.C. and H.Y.A.-Y. All authors have read and agreed to the published version of the manuscript.

Funding: This research was funded by the CAS-Croucher Funding Scheme for Joint Laboratories and a Collaborative Research Grant (C7075-21G). M.H.Y.L. acknowledges the support of the Hong Kong Postgraduate Fellowship.

Institutional Review Board Statement: Not applicable.

Informed Consent Statement: Not applicable.

Data Availability Statement: The data presented in this study are available in supplementary material.

Conflicts of Interest: The authors declare no conflict of interest.

References

1. Ross, C.A. Polyglutamine pathogenesis: Emergence of unifying mechanisms for Huntington's disease and related disorders. *Neuron* **2002**, *35*, 819–822. [CrossRef] [PubMed]
2. Orr, H.T.; Zoghbi, H.Y. Trinucleotide Repeat Disorders. *Annu. Rev. Neurosci.* **2007**, *30*, 575–621. [CrossRef] [PubMed]
3. Lieberman, A.P.; Shakkottai, V.G.; Albin, R.L. Polyglutamine Repeats in Neurodegenerative Diseases. *Annu. Rev. Pathol.* **2019**, *14*, 1–27. [CrossRef] [PubMed]
4. Shao, J.; Diamond, M.I. Polyglutamine Diseases: Emerging Concepts in Pathogenesis and Therapy. *Hum. Mol. Genet.* **2007**, *16*, R115–R123. [CrossRef] [PubMed]
5. Swinnen, B.; Robberecht, W.; Van Den Bosch, L. RNA Toxicity in Non-Coding Repeat Expansion Disorders. *EMBO J.* **2020**, *39*, e101112–e101114. [CrossRef]
6. Kouroku, Y.; Fujita, E.; Jimbo, A.; Kikuchi, T.; Yamagata, T.; Momoi, M.Y.; Kominami, E.; Kuida, K.; Sakamaki, K.; Yonehara, S.; et al. Polyglutamine aggregates stimulate ER stress signals and caspase-12 activation. *Hum. Mol. Genet.* **2002**, *11*, 1505–1515. [CrossRef] [PubMed]
7. Chen, Z.S.; Li, L.; Peng, S.; Chen, F.M.; Zhang, Q.; An, Y.; Lin, X.; Li, W.; Koon, A.C.; Chan, T.F.; et al. Planar cell polarity gene Fuz triggers apoptosis in neurodegenerative disease models. *EMBO Rep.* **2018**, *19*, e45409. [CrossRef] [PubMed]
8. Jiang, Y.; Chadwick, S.R.; Lajoie, P. Endoplasmic reticulum stress: The cause and solution to Huntington's disease? *Brain Res.* **2016**, *1648*, 650–657. [CrossRef]
9. Rusmini, P.; Crippa, V.; Cristofani, R.; Rinaldi, C.; Cicardi, M.E.; Galbiati, M.; Carra, S.; Malik, B.; Greensmith, L.; Poletti, A. The role of the protein quality control system in SBMA. *J. Mol. Neurosci.* **2016**, *58*, 348–364. [CrossRef]
10. Fardghassemi, Y.; Tauffenberger, A.; Gosselin, S.; Parker, J.A. Rescue of ATXN3 neuronal toxicity in Caenorhabditis elegans by chemical modification of endoplasmic reticulum stress. *Dis. Model. Mech.* **2017**, *10*, 1465–1480.
11. Bañez-Coronel, M.; Porta, S.; Kagerbauer, B.; Mateu-Huertas, E.; Pantano, L.; Ferrer, I.; Guzmán, M.; Estivill, X.; Martí, E. A Pathogenic Mechanism in Huntington's Disease Involves Small CAG-Repeated RNAs with Neurotoxic Activity. *PLoS Genet.* **2012**, *8*, e1002481–e1002495. [CrossRef] [PubMed]
12. Nalavade, R.; Griesche, N.; Ryan, D.P.; Hildebrand, S.; Krauss, S. Mechanisms of RNA-Induced Toxicity in CAG Repeat Disorders. *Cell Death Dis.* **2013**, *4*, e752–e762. [CrossRef] [PubMed]
13. Chan, H.Y.E. RNA-Mediated Pathogenic Mechanisms in Polyglutamine Diseases and Amyotrophic Lateral Sclerosis. *Front. Cell. Neurosci.* **2014**, *8*, 431–442. [CrossRef] [PubMed]
14. Martí, E. RNA Toxicity Induced by Expanded CAG Repeats in Huntington's Disease. *Brain Pathol.* **2016**, *26*, 779–786. [CrossRef]

15. Peng, S.; Guo, P.; Lin, X.; An, Y.; Sze, K.H.; Lau, M.H.Y.; Chen, Z.S.; Wang, Q.; Li, W.; Sun, J.K.-L.; et al. CAG RNAs induce DNA damage and apoptosis by silencing NUDT16 expression in polyglutamine degeneration. *Proc. Natl. Acad. Sci. USA* **2021**, *118*, e2022940118. [CrossRef]

16. Murata, A.; Sato, S.-I.; Kawazoe, Y.; Uesugi, M. Small-Molecule Fluorescent Probes for Specific RNA Targets. *Chem. Commun.* **2011**, *47*, 4712–4714. [CrossRef]

17. Cao, C.; Wei, P.; Li, R.; Zhong, Y.; Li, X.; Xue, F.; Shi, Y.; Yi, T. Ribosomal RNA-Selective Light-up Fluorescent Probe for Rapidly Imaging the Nucleolus in Live Cells. *ACS Sens.* **2019**, *4*, 1409–1416. [CrossRef]

18. Das, B.; Murata, A.; Nakatani, K. A Small-Molecule Fluorescence Probe ANP77 for Sensing RNA Internal Loop of C, U and A/CC Motifs and Their Binding Molecules. *Nucleic Acids Res.* **2021**, *49*, 8462–8470. [CrossRef]

19. Wong, C.-H.; Nguyen, L.; Peh, J.; Luu, L.M.; Sanchez, J.S.; Richardson, S.L.; Tuccinardi, T.; Tsoi, H.; Chan, W.Y.; Chan, E.H.Y.; et al. Targeting toxic RNAs that cause myotonic dystrophy type 1 (DM1) with a bisamidinium inhibitor. *J. Am. Chem. Soc.* **2014**, *136*, 6355–6361. [CrossRef]

20. Nguyen, L.; Luu, L.M.; Peng, S.; Serrano, J.F.; Chan, H.Y.E.; Zimmerman, S.C. Rationally Designed Small Molecules That Target Both the DNA and RNA Causing Myotonic Dystrophy Type 1. *J. Am. Chem. Soc.* **2015**, *137*, 14180–14189. [CrossRef]

21. Luu, L.M.; Nguyen, L.; Peng, S.; Lee, J.Y.; Lee, H.Y.; Wong, C.-H.; Hergenrother, P.J.; Chan, H.Y.E.; Zimmerman, S.C. A Potent Inhibitor of Protein Sequestration by Expanded Triplet (CUG) Repeats that Shows Phenotypic Improvements in a Drosophila Model of Myotonic Dystrophy. *ChemMedChem* **2016**, *11*, 1428–1435. [CrossRef] [PubMed]

22. Bao, G.; Rhee, W.J.; Tsourkas, A. Fluorescent Probes for Live-Cell RNA Detection. *Annu. Rev. Biomed. Eng.* **2009**, *11*, 25–47. [CrossRef]

23. Juskowiak, B. Nucleic Acid-Based Fluorescent Probes and Their Analytical Potential. *Anal. Bioanal. Chem.* **2011**, *399*, 3157–3176. [CrossRef] [PubMed]

24. Wang, L.; Frei, M.S.; Salim, A.; Johnsson, K. Small-Molecule Fluorescent Probes for Live-Cell Super-Resolution Microscopy. *J. Am. Chem. Soc.* **2019**, *141*, 2770–2781. [CrossRef] [PubMed]

25. Terai, T.; Nagano, T. Small-Molecule Fluorophores and Fluorescent Probes for Bioimaging. *Pflugers Arch.* **2013**, *465*, 347–359. [CrossRef] [PubMed]

26. Loving, G.S.; Sainlos, M.; Imperiali, B. Monitoring Protein Interactions and Dynamics with Solvatochromic Fluorophores. *Trends Biotechnol.* **2010**, *28*, 73–83. [CrossRef]

27. Klymchenko, A.S. Solvatochromic and Fluorogenic Dyes as Environment-Sensitive Probes: Design and Biological Applications. *Acc. Chem. Res.* **2017**, *50*, 366–375. [CrossRef]

28. Liese, D.; Haberhauer, G. Rotations in Excited ICT States—Fluorescence and Its Microenvironmental Sensitivity. *Isr. J. Chem.* **2018**, *58*, 813–826. [CrossRef]

29. Lakowicz, J.R. Solvent and Environmental Effects. In *Principles of Fluorescence Spectroscopy*; Lakowicz, J.R., Ed.; Springer: Boston, MA, USA, 2006; pp. 205–235.

30. Hong, Y.; Lam, J.W.Y.; Tang, B.Z. Aggregation-Induced Emission. *Chem. Soc. Rev.* **2011**, *40*, 5361–5388. [CrossRef]

31. Mocanu, S.; Ionita, G.; Ionescu, S.; Tecuceanu, V.; Enache, M.; Leonties, A.R.; Stavarache, C.; Matei, I. New Environment-Sensitive Bis-Dansyl Molecular Probes Bearing Alkyl Diamine Linkers: Emissive Features and Interaction with Cyclodextrins. *Chem. Phys. Lett.* **2018**, *713*, 226–234. [CrossRef]

32. Demeter, A.; Kovalenko, S.A. Photoinduced Intramolecular Charge Transfer and Relaxation Dynamics of 4-Dimethylaminopyridine in Water, Alcohols, and Aprotic Solvents. *J. Photochem. Photobiol. A Chem.* **2021**, *412*, 113246–113254. [CrossRef]

33. Leung, N.L.C.; Xie, N.; Yuan, W.; Liu, Y.; Wu, Q.; Peng, Q.; Miao, Q.; Lam, J.W.Y.; Tang, B.Z. Restriction of Intramolecular Motions: The General Mechanism behind Aggregation-Induced Emission. *Chemistry* **2014**, *20*, 15349–15353. [CrossRef] [PubMed]

34. Zhang, G.-F.; Chen, Z.-Q.; Aldred, M.P.; Hu, Z.; Chen, T.; Huang, Z.; Meng, X.; Zhu, M.-Q. Direct Validation of the Restriction of Intramolecular Rotation Hypothesis via the Synthesis of Novel Ortho-Methyl Substituted Tetraphenylethenes and Their Application in Cell Imaging. *Chem. Commun.* **2014**, *50*, 12058–12060. [CrossRef] [PubMed]

35. Marcheschi, R.J.; Tonelli, M.; Kumar, A.; Butcher, S.E. Structure of the HIV-1 frameshift site RNA bound to a small molecule inhibitor of viral replication. *ACS Chem. Biol.* **2011**, *6*, 857–864. [CrossRef] [PubMed]

36. Shi, H.; Kwok, R. T. K.; Liu, J.; Xing, B.; Tang, B. Z.; Liu, B. Real-Time Monitoring of Cell Apoptosis and Drug Screening Using Fluorescent Light-up Probe with Aggregation-Induced Emission Characteristics. *J. Am. Chem. Soc.* **2012**, *134*, 17972–17981. [CrossRef] [PubMed]

37. Chen, W.; Mohy Ei Dine, T.; Vincent, S. P. Synthesis of Functionalized Copillar[4+1]Arenes and Rotaxane as Heteromultivalent Scaffolds. *Chem. Commun.* **2021**, *57*, 492–495. [CrossRef] [PubMed]

Article

A Label-Free Gold Nanoparticles-Based Optical Aptasensor for the Detection of Retinol Binding Protein 4

Koena L. Moabelo [1,2], Teresa M. Lerga [3], Miriam Jauset-Rubio [3], Nicole R. S. Sibuyi [2], Ciara K. O'Sullivan [3], Mervin Meyer [2] and Abram M. Madiehe [1,2,*]

[1] Nanobiotechnology Research Group, Department of Biotechnology, University of Western Cape, Bellville 7535, South Africa
[2] Department of Science and Innovation (DSI)/Mintek Nanotechnology Innovation Centre, Biolabels Research Node, Department of Biotechnology, University of Western Cape, Bellville 7535, South Africa
[3] Interfibio Research Group, Departament d'Enginyeria Quimica, Universitat Rovira i Virgili, Avinguda Països Catalans 26, 43007 Tarragona, Spain
* Correspondence: amadiehe@uwc.ac.za; Tel.: +27-21-959-2468

Abstract: Retinol-binding protein 4 (RBP4) has been implicated in insulin resistance in rodents and humans with obesity and T2DM, making it a potential biomarker for the early diagnosis of T2DM. However, diagnostic tools for low-level detection of RBP4 are still lagging behind. Therefore, there is an urgent need for the development of T2DM diagnostics that are rapid, cost-effective and that can be used at the point-of-care (POC). Recently, nano-enabled biosensors integrating highly selective optical detection techniques and specificity of aptamers have been widely developed for the rapid detection of various targets. This study reports on the development of a rapid gold nanoparticles (AuNPs)-based aptasensor for the detection of RBP4. The retinol-binding protein aptamer (RBP-A) is adsorbed on the surface of the AuNPs through van der Waals and hydrophobic interactions, stabilizing the AuNPs against sodium chloride (NaCl)-induced aggregation. Upon the addition of RBP4, the RBP-A binds to RBP4 and detaches from the surface of the AuNPs, leaving the AuNPs unprotected. Addition of NaCl causes aggregation of AuNPs, leading to a visible colour change of the AuNPs solution from ruby red to purple/blue. The test result was available within 5 min and the assay had a limit of detection of 90.76 ± 2.81 nM. This study demonstrates the successful development of a simple yet effective, specific, and colorimetric rapid assay for RBP4 detection.

Keywords: retinol-binding protein 4; type 2 diabetes mellitus; aptasensor; colorimetric; gold nanoparticles; diagnosis; biosensing

Citation: Moabelo, K.L.; Lerga, T.M.; Jauset-Rubio, M.; Sibuyi, N.R.S.; O'Sullivan, C.K.; Meyer, M.; Madiehe, A.M. A Label-Free Gold Nanoparticles-Based Optical Aptasensor for the Detection of Retinol Binding Protein 4. *Biosensors* **2022**, *12*, 1061. https://doi.org/10.3390/bios12121061

Received: 3 October 2022
Accepted: 16 November 2022
Published: 22 November 2022

Publisher's Note: MDPI stays neutral with regard to jurisdictional claims in published maps and institutional affiliations.

1. Introduction

Type 2 diabetes mellitus (T2DM) is a chronic metabolic disease with debilitating effects on human health. It is a growing epidemic, that affects over 536.6 million adults worldwide [1]. If undiagnosed or untreated, T2DM can lead to microvascular (retinopathy, neuropathy and nephropathy) and macrovascular (stroke and acute coronary syndrome) complications [2]. Many of these complications can go un-noticed for years and are often subclinical at the onset, making the early diagnosis of T2DM difficult [3]. Therefore, an urgent need to develop novel and early diagnostic strategies to address this epidemic is warranted.

Retinol-binding protein 4 (RBP4), an adipokine responsible for obesity-induced insulin resistance, has been identified as one of the reliable biomarkers for the early diagnosis of T2DM [4]. RBP4 binds to retinol and transports it in the blood stream to the liver. Elevated levels of RBP4 have been found in insulin-resistant mice and humans with obesity and T2DM, causing dysfunctions in the production of glucose transporter 4 (GLUT4) and consequently leading to a failure of glucose uptake from the blood. This suggests that RBP4 could serve as a potential serological biomarker for the early diagnosis of T2DM [5–9].

The detection of RBP4 is usually performed through different biotechnological assays, such as enzyme-linked immunosorbent assay (ELISA) and Western blot [10–12], which rely on the interaction between an antibody and an analyte. However, the use of antibodies in diagnostic assays have several inherent limitations such as high production costs, instability, and poor specificity, and their use requires longer incubation periods and different substrates for detection [13]. Recently, the use of aptamers in diagnostic assays have been shown to alleviate these limitations. Aptamers, which are single-stranded deoxyribonucleic acid (DNA) or ribonucleic acid (RNA) molecules that bind with high affinity and specificity to their target molecules, exhibit significant advantages relative to antibodies in terms of their high specificity, selectivity, low molecular weight, and ease of production [13–15]. Furthermore, aptamers present a more favourable and desirable molecular recognition element (MRE) and have been employed in the development of various sensors for diagnostic purposes [16–20].

Lee et al. reported on the development of the first RBP-aptamer (RBP-A) and its application in a surface plasmon resonance (SPR) biosensor for the detection of RBP4 in serum. The SPR biosensor specifically detected RBP4 and had a limit of detection of 75 nM (1.58 µg/mL), which is sufficiently sensitive to probe for RBP4 in the serum of people who are at risk of developing T2DM [21]. For instance, it has been reported that the detection range of RBP4 in normal individuals is 23.0–24.3 µg/mL (1–1.06 µM) and ranges from 24.9–50.3 µg/mL (1.08–2.187 µM) in individuals who are diagnosed with T2DM [22]. Using the same aptamer, an enzyme-linked antibody-aptamer sandwich (ELAAS) method was developed for more convenient and sensitive detection of RBP4. Compared to the SPR-based biosensor, this method was between 20 and 68 times more sensitive [23]; however, the assay involved multiple incubation and washing steps, making it impossible for it to be applied in point-of-care testing (PoCT). Torabi et al. developed an ultrasensitive chemiluminescent aptasensor for the detection of RBP4 with an LOD of 951 fg/mL [24]. Similarly to the ELAAS, this assay involved multiple washing and incubation steps, rendering it unsuitable for use in PoCT [22].

Colorimetric assays have been extensively used for the rapid detection of various diseases due to their simplicity, quick response, and high sensitivity [25–31]. Nanomaterials, especially gold nanoparticles (AuNPs), exhibit strong localized surface plasmon resonance (LSPR) properties which have been widely leveraged for the fabrication of colorimetric sensors [32]. AuNPs give off a colourful signal that can be visualized with the naked eye and without using advanced instruments [33]. Generally, a solution of colloidal AuNPs have a ruby-red colour due to their LSPR phenomenon that is highly dependent on interparticle distance [32,34,35]. Upon addition of salt, the colour of the AuNP solution changes from ruby red to purple/blue due to the shift of the LSPR to a higher wavelength. In contrast, when the AuNPs are bound to molecules, such as MRE for example, they will retain their ruby-red colour in the presence of salt. If the analyte for the specific MRE is present, the MRE binds to the analyte instead of the AuNPs and the colour of the AuNPs will change from ruby red to purple/blue. Based on this principle, several AuNP-based optical aptasensors have been developed [36–40]. In this study, using this principle, a simple yet effective, specific, and rapid RBP-A-aptasensor for the detection of RBP4 was developed. The assay is user-friendly as it does not require any specialised instruments and it is capable of producing a test result within 5 min, highlighting the potential applicability of this assay in PoCT and in resource-limited areas.

2. Materials and Methods

2.1. Materials

Retinol-binding protein aptamer (RBP-A) 5′-ATA CCA GCT TAT TCA ATT ACA GTA GTG AGG GGT CC GTC GTG GGG TAG TTG GGT CGT GGA GAT AGT AAG TGC AAT CT-3′ [21], was synthesized by Biomers.net GmbH (Ulm, Germany). RBP4, alpha-2-macroglobulin (A2MG), leptin and bovine serum albumin (BSA) were purchased from Sino Biologicals (Beijing, China), ProsPec (Ness-Ziona, Israel), R&D Systems (Minneapolis,

MN, USA) and Merck (Pty) Ltd. (Rahway, NJ, USA), respectively. Chloroauric acid ($HAuCl_4 \cdot 3H_2O$), sodium chloride (NaCl), and sodium citrate were purchased from Sigma-Aldrich (St Louis, MO, USA). Purified Milli-Q water was used in all experiments (Millipore, MA, USA). All the chemicals were of analytical grade.

2.2. Synthesis and Characterization of AuNPs

All glassware used for the preparation of AuNPs was first immersed thoroughly in aqua regia (1:3 (v/v) HNO_3: HCl), then thoroughly washed with distilled water (dH_2O) and dried overnight at 70 °C in an oven. AuNPs were prepared via the reduction of $HAuCl_4 \cdot 3H_2O$ using sodium citrate as described by Stiolica et al. [41,42]. The spectroscopic characterization of the synthesized AuNPs was carried out using Ultraviolet–visible (UV–vis) spectrophotometry (POLARstar Omega plate reader, BMG Labtech, Offenburg, Germany). The concentration of the AuNPs was evaluated using the UV–vis spectra as described by Haiss et al. [43]. The size distribution and zeta potential (ζ- potential) characterization of the AuNPs were determined using a Malvern NanoZS90 Zetasizer (Malvern Instruments Ltd., Malvern, UK) at a scattering angle of 90° at 25 °C. For the characterization of the size and morphology of the AuNPs, one drop of the sample solution was loaded onto a carbon-coated copper grid. The grids were dried for a few minutes under a Xenon lamp. High-resolution transmission electron microscope (HR-TEM) images were captured using an FEI Tecnai G^2 20 field-emission gun (FEG) HRTEM (Hillsboro, OR, USA) operated in bright field mode at an accelerating voltage of 200 kV.

2.3. Development of the RBP-A-Aptasensor

2.3.1. Optimization of the Aptamer and NaCl Concentration

AuNPs are highly reactive and aggregate easily in the presence of salts; therefore, it is important to monitor the salt-induced AuNPs aggregation percentage for the development of a colorimetric aptasensor [44–46]. To optimize the performance of the developed assay, various conditions such as aptamer and NaCl concentrations were investigated. For this experiment, a total of 600 µL of different concentrations (0, 6.25, 12.5, 25, 50, 100 and 150 nM) of RBP-A was added to 1.08 mL of AuNPs in separate 2 mL Eppendorf tubes, mixed well, and incubated overnight at 25 °C. Then, 50 µL of the functionalized RBP-A-AuNPs (aptasensor) was added to a 96-well microtiter plate, containing 50 µL of dH_2O. This was followed by the addition of different concentrations of NaCl (0, 20, 40, 60, 80 and 100 mM). The samples were incubated at 25 °C for 5 min and observed for colour change. The UV–vis spectra were measured using a plate reader, and the absorbance ratios (A650/A520) were calculated to determine the aggregation percentage of the AuNPs [47]. The experiment was performed in triplicate and the average absorbance ratios were calculated and used to plot bar graphs.

2.3.2. RBP4 Detection Based on the Label-Free AuNPs-Based RBP-A-Aptasensor

For the detection of RBP4, 600 µL of 50 nM RBP-A and 1.08 mL of AuNPs were added to a 2 mL tube, mixed well, and incubated overnight at 25 °C. Subsequently, 50 µL of the aptasensor was added into the 96-well microtiter plate containing 25 µL of dH_2O. Then, 25 µL of different concentrations of RBP4 (0, 7.8, 15.6, 31.2, 62.5, 125, 250 nM) were added into the aptasensor, mixed thoroughly, and incubated for another 5 min at 25 °C, and observed for colour change This was followed by the addition of 6 µL of 1 M NaCl solution (60 mM final concentration) and incubation at 25 °C for 5 min. The UV–vis spectra were measured using a plate reader and the absorbance ratios (A650/A520) were calculated to determine the aggregation percentage of the AuNPs. The experiment was performed in triplicate and the average absorbance ratios were calculated and used to plot the linear curve.

2.3.3. Specificity of the Label-Free AuNPs-Based RBP-A-Aptasensor

To evaluate the specificity of the RBP-A-aptasensor, 50 μL of the aptasensor was added into the 96-well microtiter plate, containing 25 μL of dH$_2$O. Then, 25 μL of 250 nM RBP4, A2MG, leptin and BSA were added into the aptasensor, mixed thoroughly, and incubated for another 5 min at 25 °C and observed for colour change. This was followed by the addition of 6 μL of 1 M NaCl solution (60 mM final concentration) and incubation at 25 °C for 5 min. A water sample was used as a blank. The UV–vis spectra were measured using a plate reader and the absorbance ratios (A650/A520) were calculated to determine the aggregation percentage of the AuNPs. The experiment was performed in triplicate and the average absorbance ratios were calculated and used to plot bar graphs.

3. Results and Discussion

AuNPs-based aptasensors have been widely applied in the detection of various targets including chemical compounds, proteins, pesticides, toxins, and viruses [36,38–40,48–50]. Their simple preparation, rapid detection, and the fact that they do not require special instrumentation have made them the preferred choice for PoCT [51]. The sensors rely on the salt-induced aggregation of bare AuNPs, leading to a change of colour from ruby red to purple/blue. In the absence of the target or biomarker molecule, the detecting molecule or MRE stabilizes the AuNPs against salt-induced aggregation and no colour change is observed [52]. Similarly, this study investigated the feasibility of the RBP-A-aptasensor for detection of RBP4. Although at least two other studies demonstrated the development of aptasensors for RBP4, neither of these assays can be used at the PoCT as both of these assays require specialised instruments and must therefore be carried out in a laboratory setting. In contrast, the assay described herein does not require any specialised instruments and can be carried out in the field.

3.1. Characterization of the AuNPs

The synthesized AuNPs had an SPR peak at 519 nm (Figure 1a), indicating that the AuNPs were approximately 14 nm in size. The narrow peak suggested that the majority of the AuNPs were uniform and monodispersed [53]. The AuNPs had an absorbance value of 2.422 OD, and a concentration of 9.83 nM. DLS results revealed that the AuNPs had an average hydrodynamic diameter of 14.69 nm and a PDI of 0.175, the latter confirmed that the AuNPs were mostly monodispersed. The AuNPs showed a ζ- potential of −27.7 mV, indicating that the AuNPs were highly stable. TEM images revealed that the AuNPs were mostly spherical in shape and were relatively monodispersed (Figure 1b). This is due to electrostatic stabilization with a negatively charged layer of citrate ions [44]. The core diameters of AuNPs were found to be 13.98 ± 1.19 nm (Figure 1c), further corroborating the results obtained from the UV–vis spectrophotometry. The SPR results obtained in this study were similar to the results obtained by Stiolica et al., whereby 14 nm AuNPs also had an SPR of 519 nm. In contrast, the reported AuNPs had a ζ- potential of –51 mV [41], which was lower than the a ζ- potential obtained in this study.

3.2. Principles of AuNPs-Based RBP-A-Aptasensor

The principles of the AuNPs-based RBP-A-aptasensor for the detection of RBP4 is illustrated in Figure 2. Colloidal AuNPs produce a ruby-red colour in solution that is highly dependent on the interparticle distance [32,34,35]. In the presence of NaCl, the AuNPs undergo aggregation, leading to a visible colour change from ruby red to purple/blue. The presence of salt electrolytes decreased the interparticle distance of the AuNPs, thus reducing the ζ- potential and enabling electromagnetic interaction among the negatively charged AuNPs consequently leading to visible colour change [46]. In contrast, binding of RBP-A on the surface of AuNPs stabilized the AuNPs and prevented the NaCl-induced aggregation of the AuNPs [54,55]. The solution retained the ruby-red colour even at high NaCl concentrations (up to 60 mM). However, in the presence of RBP4, owing to the high affinity between RBP4 and RBP-A, RBP-A detached from the AuNPs and left the AuNPs unprotected. The bare AuNPs aggregated in the presence of NaCl, which led to a visible colour change from ruby red to purple/blue.

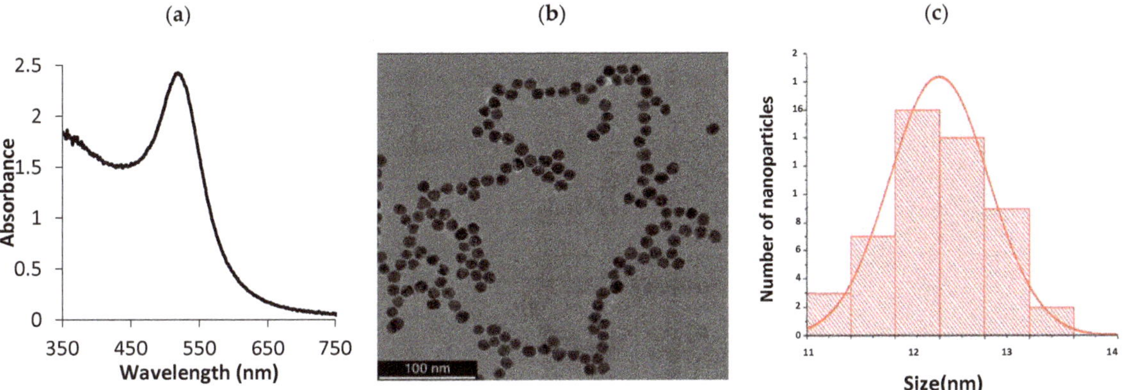

Figure 1. Characterization of AuNPs. (**a**) UV–vis spectrum of AuNPs; (**b**) a representative HR-TEM photomicrograph and (**c**) size distribution.

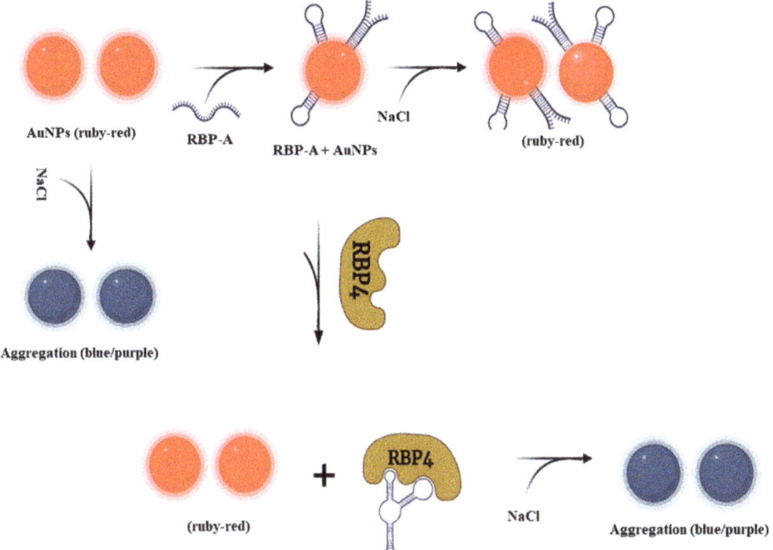

Figure 2. Schematic illustration of the colorimetric aptasensor for the detection of RBP4.

The compatibility of the AuNPs-based RBP-A-aptasensor for the detection of RBP4 was validated using four techniques: UV–vis spectroscopy, TEM, ζ- potential, and FTIR. As shown in Figure 3a, the AuNPs had an SPR peak at 519 nm which shifted to 560 nm upon the addition of NaCl, indicating aggregation. In the presence of RBP-A, the AuNPs were strongly protected against NaCl-induced aggregation and the SPR peak remained inert at 519 nm. Upon the addition of NaCl, the SPR peak shifted to 600 nm as a result of aggregation. This could be because, in the presence of RBP4, RBP-A detached from the AuNPs' surface and bound to RBP4, leaving the AuNPs' surface unprotected against salt-induced aggregation. HR-TEM was used to confirm these observations (Figure 3b). Bare AuNPs were monodispersed in the solution; however, in the presence of NaCl, the AuNPs aggregated and clumped together. The aptasensor without RBP4 showed no aggregation in the presence of NaCl, but in the presence of RBP4 and NaCl, the AuNPs aggregated.

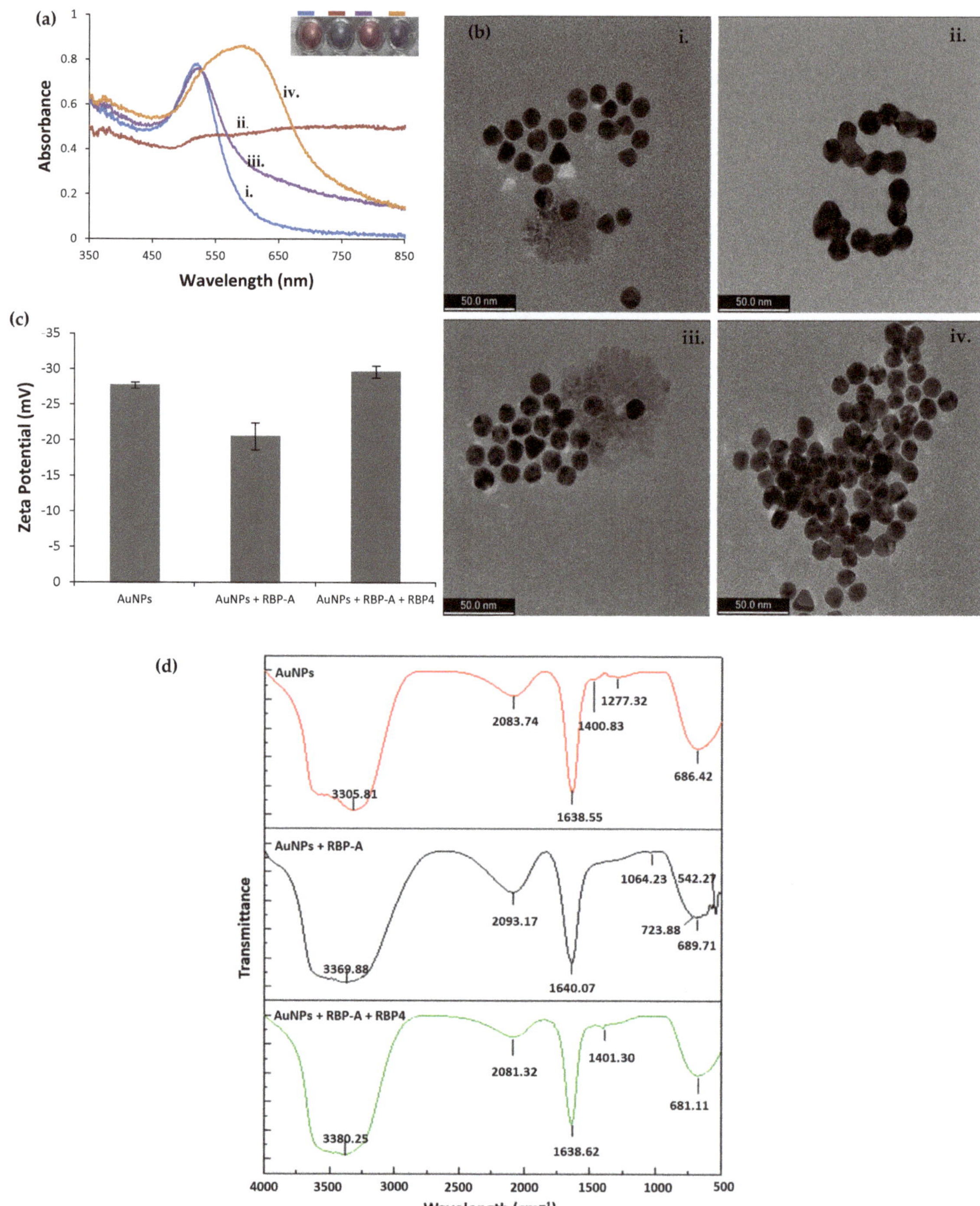

Figure 3. (**a**) UV–vis spectra, (**b**) HR-TEM of (i) AuNPs, (ii) AuNPs + NaCl, (iii) AuNPs + RBP-A + NaCl and (iv) AuNPs + RBP-A + RBP4 + NaCl, (**c**) ζ- potential and (**d**) FTIR analysis of the interaction between the RBP-A-aptasensor and RBP4. RBP4 was used at a concentration of 250 nM.

The ζ- potential was also measured to monitor the changes on the AuNPs' surface during the development of the aptasensor and when the aptasensor was incubated with RBP4 (Figure 3c). AuNPs exhibited a ζ- potential of -27.7 mV, which is in agreement with previously published data [56]. In contrast, the ζ- potential of -20.5 mV was obtained when RBP-A was incubated with the AuNPs solution, indicating that negative charges are partially neutralized by the aptamer covering the surface of the particles [47]. Finally, when RBP4 was added, the ζ- potential was restored to -29.6 mV, indicating that the RBP-A bound to RBP4 and left the AuNPs naked, thus returning them to their original ζ-potential. Using the same technique, Lerga et al. developed a AuNPs-based aptasensor for the detection of histamine. In their study, AuNPs exhibited a ζ- potential of -47.6 mV which decreased to -28.6 mV upon incubation of the aptamer with the AuNPs. The ζ-potential was restored to approximately -50 mV after the addition of histamine [47].

Finally, FTIR (PerkinElmer Spectrum One FTIR spectrophotometer) analysis was carried out to monitor elemental changes to the AuNPs during the development of the aptasensor and when the aptasensor was incubated with RBP4 (Figure 3d). The FTIR spectrum of AuNPs features characteristic peaks at 3305.81, 2082.74, 1638.55, 1400.83, 1277.32 and 686.42 cm^{-1}. The peak at 3305.81 cm^{-1} is attributed to the stretching vibration of OH and NH2. The peaks at 1400.82 cm^{-1} and 1638.55 cm^{-1} are attributed to the carboxylate symmetric and asymmetric stretching bonds of the carboxylate group ($COO-$) in citrate ions [57], respectively. The peak at 2083.74 cm^{-1} is attributed to the S-H group on the AuNPs [58], further validating successful synthesis of the AuNPs. Upon adsorption of the RBP-A, two weak peaks (1064.23 and 723.88 cm^{-1}) were observed in RBP-A-AuNPs conjugate spectra which were absent in the spectrum for the AuNPs. The peak at 1064.23 cm^{-1} is attributed to the symmetric C=O stretching band of the phosphate backbone and the peak at 723.88 cm^{-1} is attributed to the C–H out of plane base vibration of the RBP-A [59], indicating successful adsorption of RBP-A on the surface of the AuNPs. Finally, when RBP4 is added, the peak attributed to the carboxylate asymmetric stretching band appeared at 1460.05 cm^{-1}. This peak was also visible in the AuNPs, further supporting the assertion that RBP-A was bound to RBP4 and detached from the AuNPs.

3.3. Determination of the Optimum Concentration of NaCl and Aptamer

The flocculation assay was used to determine the stability of the AuNPs-based RBP-A-aptasensor in the presence of different concentrations of NaCl. From Figure 4, it can be deduced that AuNPs and the RBP-A-AuNPs conjugates at all RBP-A concentrations showed no aggregation in the absence of NaCl. The AuNPs and the RBP-A-AuNPs conjugates at all RBP-A concentrations showed minimal aggregation below 20 mM NaCl; whereas, some aggregation was observed at 40 mM NaCl. The minimal changes of the absorbance at low NaCl concentrations were due to the low ionic strength of the electrolyte. At these NaCl concentrations, the aggregation of AuNPs is slow and the mean size of the formed aggregate is close to the size of the original AuNPs, rendering the changes insignificant. However, the AuNPs and the RBP-A-AuNPs conjugate solutions at lower RBP-A concentrations (6.25–12.5 nM) changed colour from ruby red to purple/blue with increasing concentrations of NaCl (60–100 mM), with the AuNPs reaching saturation levels between 60 and 100 mM NaCl. This observation suggested that the Na^+ and Cl^- ions destroyed the ionic environment and led to the aggregation of AuNPs in the absence of aptamers [60]. In contrast, higher concentrations of RBP-A (25–150 nM) showed excellent protection efficiency against NaCl-induced aggregation of AuNPs, which was indicated by the retention of a ruby-red colour and a lower aggregation percentage. Overall, the results indicated that the addition of RBP-A caused a dose-dependent protection against NaCl-induced aggregation for all concentrations of NaCl. To ensure the assay is more cost effective, the optimal concentrations of RBP-A and NaCl for the development of the aptasensor were selected as 50 nM and 60 mM, respectively.

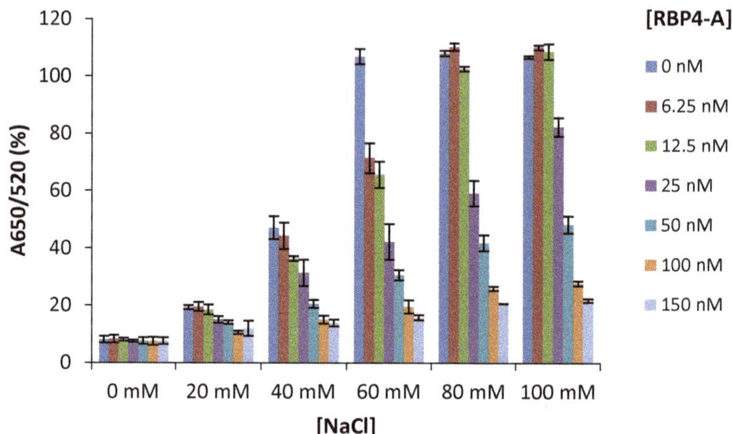

Figure 4. Stability of the AuNPs and RBP-A-aptasensor in the presence of increasing concentrations of NaCl. Final aptamer concentrations: 0, 6.25, 12.5, 25, 50, 100 and 150 nM; final NaCl concentrations: 0, 20, 40, 60, 80 and 100 mM.

3.4. Sensitivity of the Aptasensor for RBP4 Detection

The sensitivity of the AuNPs-based RBP-A-aptasensor for the detection of RBP4, under the optimized experimental conditions (50 nM RBP-A and 60 mM NaCl), was assessed by incubation with different concentrations of RBP4 and by calculating the absorbance ratios (A650/A520) to evaluate the aggregation of the AuNPs. Figure 5a and the insert indicate that the addition of NaCl caused a dose-dependent aggregation of the AuNPs, which is indicated by the gradual colour change of the AuNPs from ruby red to purple/blue and the shift in SPR from 519 to 600 nm. The results demonstrate that RBP-A bound to RBP4; thus, detaching from the AuNPs and leaving the AuNPs unprotected. In the presence of 60 mM NaCl, the AuNPs aggregated in a dose-dependent manner, with a more visible colour change between 62.5 and 250 nM. The assay was sensitive with a limit of detection (LOD) of 90.76 ± 2.81 nM. Moreover, the assay was rapid, and the test results were visually detectable within 5 min. The LOD was calculated as described by the International Union of Pure and Applied Chemistry (IUPAC) [44] as follows:

$$LOD = 3 \times SD/S$$

where SD is the standard deviation of the response (y-intercept of the regression line) and S is the slope of the calibration curve.

3.5. Validation of the Aptasensor

Over the years, there has been a greater focus on the development of colorimetric aptasensors due to their simplicity, which is based on the visual detection of a test result. However, they frequently have sensitivity and reproducibility issues; thus, the reproducibility of the RBP-A-aptasensor was evaluated. The assay was performed in triplicate (at a concentration of 250 nM), on three separate days. As shown in Table 1, the recovery rate of RBP4 was in the range from 90.36 ± 3.60% to 101.49 ± 1.9% and the intra-assay coefficients of variation (CV) calculated ranged from 1,88% to 3,99%. The results for the RBP-A-aptasensor were comparable with the results of a commercially available Human ELISA kit (ab108897) which has a recovery rate of 97% and an intra-assay CV of 4.8%. In contrast, our RPB-A-aptasensor had an intra-assay CV of 3.12%, which is better than that of the ELISA kit (8.5%), further confirming the exceptional reproducibility of the RBP-A-aptasensor.

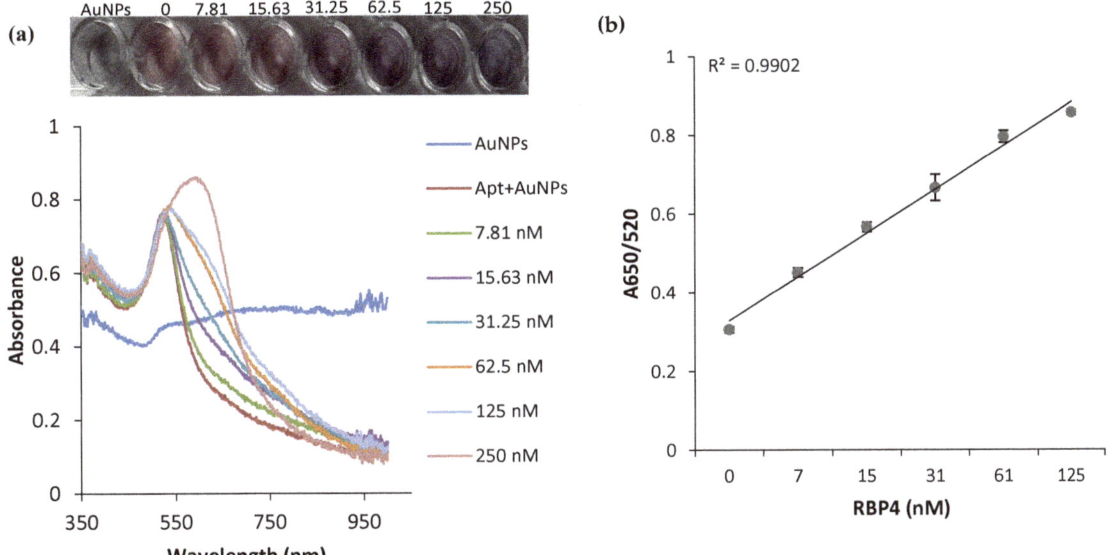

Figure 5. Analysis of the sensitivity of the AuNP-based colorimetric aptasensor. (**a**) Absorption spectra of RBP-A-AuNPs-aptasensor in the presence of increasing concentrations of RBP4. Insert: visual colour changes of the aptasensor incubated with increasing concentrations of RBP4 (final RBP4 concentrations: 0, 7.81, 15.63, 31.25, 62.5, 125 and 250 nM). (**b**) Linear calibration curve for different absorbance ratios (A620/A520) corresponding to different concentrations of RBP4 (0–125 nM).

Table 1. Mean Recoveries and Coefficients of Variation for RBP4.

Day	Mean Recovery $\pm$ SD (%)	CV (%)	Avarage CV (%)
1	90.36 $\pm$ 3.60	3.99	
2	94.95 $\pm$ 3.13	3.48	3.12
3	101.49 $\pm$ 1.91	1.88	

SD: Standard deviation.

3.6. Specificity of the Aptasensor for RBP4

The specificity of the developed RBP-A-aptasensor for the detection of RBP4 was evaluated against two diabetes-related biomarkers (A2MG and leptin) and one serum-abundant protein (BSA). The proteins (at a concentration of 250 nM) were individually incubated with the RBP-A-aptasensor, and the absorbance ratios of A620/A520 were calculated (Figure 6). A water solution was used as the blank. The RBP-A-aptasensor showed a colour change from ruby red to purple/blue in the presence of RBP4 and no colour change was observed in the presence of A2MG, leptin and BSA. This indicated that RBP4 had a very high response to the RBP-A-aptasensor; whereas, A2MG, leptin and BSA displayed no interaction with the RBP-A-aptasensor. Previous studies have also evaluated the specificity of RBP-A towards RBP4 using various proteins such as RBP4, adiponectin, vaspin, nampt, BSA, human serum albumin, human IgG, fibrinogen, insulin and anti-RBP4 antibody and no interference was detected [21,23], further corroborating our results that RBP-A is highly specific to RBP4.

3.7. Comparison of the Aptasensor with Other Aptamer-Based Approaches for RBP4 Detection

Since the identification of RBP-A in 2008 by Lee et al. [36], there have been only three studies published detailing aptamer-based assays for the detection of RBP4. A comparison of the results of these assays is provided in Table 2. The LODs for ELAAS, SPR and chemiluminescence were 3.39 nM, 75 nM, and 0.04 pM, respectively [36,53,54]. In contrast, the aptasensor developed in this study had a limit of detection of 90.76 nM. This indicated that the developed aptasensor was less sensitive when compared to the aptamer-based SPR, ELAAS and chemiluminescence sensors. Nevertheless, the results in this study were still within the normal detection range. More importantly, the results for this assay were obtained within 5 min, whereas the results for the other assays were obtained after approximately 2 h, indicating that the developed aptasensor was more rapid. Additionally, confirmation of the test results is based on visual detection of colour changes without the need for any advanced instruments, a laboratory or a power source. This further indicates that the aptasensor is suitable for PoCT.

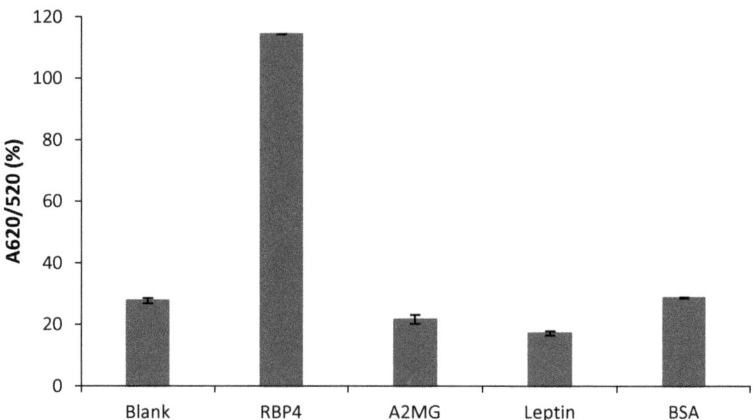

Figure 6. Specificity of the RBA-A-aptasensor for RBP4 detection. The specificity experiment was performed at a concentration of 250 nM for all proteins. The blank sample contained water.

Table 2. Comparison of the RBP-A-aptasensor with other methods used for RBP4 detection.

MRE	Method	Nanoparticles	LOD	Detection Time	Reference
Aptamer	SPR	None	75 nM	2 h and 20 min	[21]
Aptamer and antibodies	ELAAS	None	3.39 nM	2 h	[23]
Aptamer and antibodies	Chemiluminescence	AuNPs	0.04 pM	2 h	[24]
Aptamer	Colorimetric	AuNPs	90.76 nM	5 min	Present study

The incubation time of the aptasensor with RBP4 was also evaluated. This was carried out by incubating RBP4 (at a concentration of 250 nM) with the aptasensor for different time periods (0–30 min), followed by the UV–vis measurements and calculation of the absorbance ratios (A620/A520) to determine the degree of aggregation of the AuNPs (Figure 7). The results indicated that the aptasensor had lower aggregation in the presence of NaCl, as indicated by the lower absorbance ratio. When the aptasensor was incubated with RBP4 for 5 to 30 min, followed by the addition of NaCl, the aggregation increased and reached saturation levels between 5 and 30 min, indicating that 5 min incubation of the aptasensor and RBP4 was sufficient for binding. The interaction of RBP-A with RBP4 was further analyzed using the MicroScale Thermophoresis (MST) technique (Monolith NT.115, Nanotemper). As shown in Table 3, MST revealed that RBP-A bound with a dissociation constant (Kd) of 3.523 nM within 1.5 s and continued until it reached a Kd of 13.438 nM within 20 s. This is attributed to the high affinity of the aptamer for the target thus allowing rapid recognition. One of the major advantages of PoCT is that it provides much faster access to test results, allowing for more rapid clinical decision-making and more appropriate treatments and interventions. This means that standard RBP4 detection assays such as ELISAs, which require longer incubation periods, may be replaced by the aptasensor at the PoC, allowing the process from sample collection to data analysis to happen within 10 min.

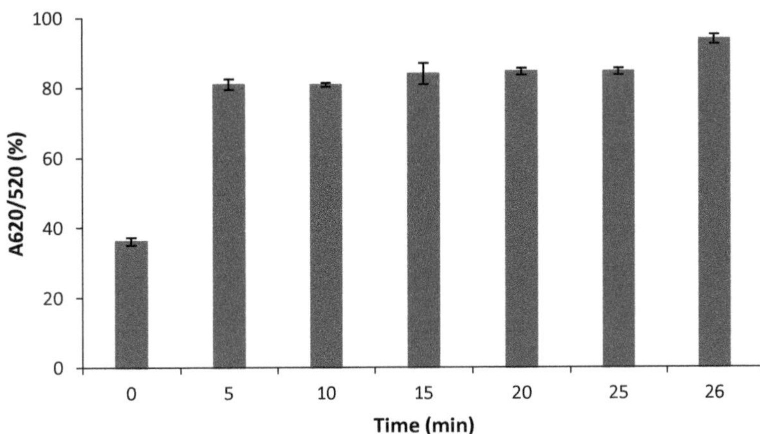

Figure 7. Aggregation of the RBP-A-aptasensor in the presence of 60 mM NaCl at various incubation periods.

Table 3. MST analysis of the interaction of RBP-A with RBP4.

Time (s)	1.5	2.5	5	10	15	20
Kd (nM)	3.523	4.6723	6.6014	8.313	11.501	13.438

4. Conclusions

In this study, a label-free colorimetric aptasensor for the detection of RBP4 using ssDNA aptamer and unmodified AuNPs was successfully developed. The assay is based on colour change from ruby red to blue/purple due to the binding of RBP-A to RBP4, which leaves AuNPs exposed to the phenomenon of salt-induced AuNP aggregation. The assay was sensitive, with an LOD of 90.76 ± 2.81 nM. Further development using clinical samples on non-diabetic and diabetic patients is required to assess the utility of the aptasensor for the early detection of T2DM.

Author Contributions: K.L.M., T.M.L. and M.J.-R.—data collection, data analysis, and validation; K.L.M. and N.R.S.S.—writing of manuscript; C.K.O., M.M. and A.M.M.—conceptualisation and student supervision; M.M. and A.M.M.—funding acquisition, resources and project administration. All authors have read and agreed to the published version of the manuscript.

Funding: The authors acknowledge the NRF Equipment Related Travel Grant (Ref. ERT180801351351) for funding K.L.M.'s travelling expenses to Spain (Universität Rovira I Virgili, Tarragona).

Institutional Review Board Statement: Not applicable.

Informed Consent Statement: Not applicable.

Data Availability Statement: The data generated in this study has been represented as Tables and Figures in the manuscript and available from the corresponding author upon request.

Acknowledgments: The authors acknowledge DSI/Mintek NIC and the NRF Masters Scholarship for the financial support. The authors also acknowledge the Electron Microscope Unit at the University of Cape Town for assistance with HR-TEM analysis, and the Drug Delivery Research Proto-Unit (Department of Pharmaceutical Sciences, University of KwaZulu-Natal) and Riziki Martin (Biolabels Research Node) for assistance with MST analysis.

Conflicts of Interest: The authors declare no conflict of interest.

References

1. Sun, H.; Saeedi, P.; Karuranga, S.; Pinkepank, M.; Ogurtsova, K.; Duncan, B.B.; Stein, C.; Basit, A.; Chan, J.C.; Mbanya, J.C. IDF diabetes Atlas: Global, regional and country-level diabetes prevalence estimates for 2021 and projections for 2045. *Diabetes Res. Clin. Pract.* **2022**, *183*, 109119. [CrossRef]
2. Chawla, A.; Chawla, R.; Jaggi, S. Microvasular and macrovascular complications in diabetes mellitus: Distinct or continuum? *Indian J. Endocrinol. Metab.* **2016**, *20*, 546. [CrossRef] [PubMed]
3. Al-Taie, A.; Elseidy, A.S.; Victoria, A.O.; Hafeez, A.; Ahmad, S. Diabetic microvascular complications and proposed interventions and approaches of management for patient care. *Biomed. Biotechnol. Res. J.* **2021**, *5*, 380. [CrossRef]
4. Li, Z.; Lu, X.; Liu, J.; Chen, L. Serum retinol-binding protein 4 levels in patients with diabetic retinopathy. *J. Int. Med. Res.* **2010**, *38*, 95–99. [CrossRef] [PubMed]
5. Wolf, G. Serum retinol-binding protein: A link between obesity, insulin resistance, and type 2 diabetes. *Nutr. Rev.* **2007**, *65*, 251–256. [CrossRef]
6. Mody, N.; Graham, T.E.; Tsuji, Y.; Yang, Q.; Kahn, B.B. Decreased clearance of serum retinol-binding protein and elevated levels of transthyretin in insulin-resistant ob/ob mice. *Am. J. Physiol. Endocrinol. Metab.* **2008**, *294*, E785–E793. [CrossRef] [PubMed]
7. Shaker, O.; El-Shehaby, A.; Zakaria, A.; Mostafa, N.; Talaat, S.; Katsiki, N.; Mikhailidis, D.P. Plasma visfatin and retinol binding protein-4 levels in patients with type 2 diabetes mellitus and their relationship to adiposity and fatty liver. *Clin. Biochem.* **2011**, *44*, 1457–1463. [CrossRef]
8. Park, S.E.; Park, C.-Y.; Sweeney, G. Biomarkers of insulin sensitivity and insulin resistance: Past, present and future. *Crit. Rev. Clin. Lab. Sci.* **2015**, *52*, 180–190. [CrossRef] [PubMed]
9. Paul, A.; Chiriacò, M.S.; Primiceri, E.; Srivastava, D.N.; Maruccio, G. Picomolar detection of retinol binding protein 4 for early management of type II diabetes. *Biosens. Bioelectron.* **2019**, *128*, 122–128. [CrossRef] [PubMed]
10. Lee, N.S.; Kim, H.S.; Park, S.E.; Blüher, M.; Park, C.-Y.; Youn, B.-S. Development of a mouse IgA monoclonal antibody-based enzyme-linked immunosorbent sandwich assay for the analyses of RBP4. *Sci. Rep.* **2018**, *8*, 2578. [CrossRef]
11. Von Eynatten, M.; Lepper, P.; Liu, D.; Lang, K.; Baumann, M.; Nawroth, P.; Bierhaus, A.; Dugi, K.; Heemann, U.; Allolio, B. Retinol-binding protein 4 is associated with components of the metabolic syndrome, but not with insulin resistance, in men with type 2 diabetes or coronary artery disease. *Diabetologia* **2007**, *50*, 1930–1937. [CrossRef] [PubMed]
12. Parkash, O.; Shueb, R.H. Diagnosis of dengue infection using conventional and biosensor based techniques. *Viruses* **2015**, *7*, 5410–5427. [CrossRef] [PubMed]
13. Thiviyanathan, V.; Gorenstein, D.G. Aptamers and the next generation of diagnostic reagents. *Proteom. -Clin. Appl.* **2012**, *6*, 563–573. [CrossRef] [PubMed]
14. Chen, A.; Yang, S. Replacing antibodies with aptamers in lateral flow immunoassay. *Biosens. Bioelectron.* **2015**, *71*, 230–242. [CrossRef]
15. Ştefan, G.; Hosu, O.; De Wael, K.; Lobo-Castañón, M.J.; Cristea, C. Aptamers in biomedicine: Selection strategies and recent advances. *Electrochim. Acta* **2021**, *376*, 137994. [CrossRef]
16. Jauset-Rubio, M.; Svobodová, M.; Mairal, T.; McNeil, C.; Keegan, N.; Saeed, A.; Abbas, M.N.; El-Shahawi, M.S.; Bashammakh, A.S.; Alyoubi, A.O. Ultrasensitive, rapid and inexpensive detection of DNA using paper based lateral flow assay. *Sci. Rep.* **2016**, *6*, 37732. [CrossRef] [PubMed]
17. Raston, N.H.A.; Nguyen, V.-T.; Gu, M.B. A new lateral flow strip assay (LFSA) using a pair of aptamers for the detection of Vaspin. *Biosens. Bioelectron.* **2017**, *93*, 21–25. [CrossRef]
18. Kottappara, R.; Paravannoor, A.; Vijayan, B.K. Nanosensors for virus detection. In *Nanosensors for Smart Agriculture*; Elsevier: Amsterdam, The Netherlands, 2022; pp. 531–546.
19. Das, S.; Jain, S.; Ilyas, M.; Anand, A.; Kumar, S.; Sharma, N.; Singh, K.; Mahlawat, R.; Sharma, T.K.; Atmakuri, K. Development of DNA Aptamers to Visualize Release of Mycobacterial Membrane-Derived Extracellular Vesicles in Infected Macrophages. *Pharmaceuticals* **2022**, *15*, 45. [CrossRef]
20. Saad, M.; Castiello, F.R.; Faucher, S.P.; Tabrizian, M. Introducing an SPRi-based titration assay using aptamers for the detection of Legionella pneumophila. *Sens. Actuators B Chem.* **2022**, *351*, 130933. [CrossRef]
21. Lee, S.J.; Youn, B.-S.; Park, J.W.; Niazi, J.H.; Kim, Y.S.; Gu, M.B. ssDNA aptamer-based surface plasmon resonance biosensor for the detection of retinol binding protein 4 for the early diagnosis of type 2 diabetes. *Anal. Chem.* **2008**, *80*, 2867–2873. [CrossRef]
22. Graham, T.E.; Yang, Q.; Blüher, M.; Hammarstedt, A.; Ciaraldi, T.P.; Henry, R.R.; Wason, C.J.; Oberbach, A.; Jansson, P.-A.; Smith, U. Retinol-binding protein 4 and insulin resistance in lean, obese, and diabetic subjects. *New Engl. J. Med.* **2006**, *354*, 2552–2563. [CrossRef] [PubMed]
23. Lee, S.J.; Park, J.-W.; Kim, I.-A.; Youn, B.-S.; Gu, M.B. Sensitive detection of adipokines for early diagnosis of type 2 diabetes using enzyme-linked antibody-aptamer sandwich (ELAAS) assays. *Sens. Actuators B Chem.* **2012**, *168*, 243–248. [CrossRef]
24. Torabi, R.; Ghourchian, H. Ultrasensitive nano-aptasensor for monitoring retinol binding protein 4 as a biomarker for diabetes prognosis at early stages. *Sci. Rep.* **2020**, *10*, 594. [CrossRef]
25. Moitra, P.; Alafeef, M.; Dighe, K.; Frieman, M.B.; Pan, D. Selective naked-eye detection of SARS-CoV-2 mediated by N gene targeted antisense oligonucleotide capped plasmonic nanoparticles. *ACS Nano* **2020**, *14*, 7617–7627. [CrossRef] [PubMed]
26. Aithal, S.; Mishriki, S.; Gupta, R.; Sahu, R.P.; Botos, G.; Tanvir, S.; Hanson, R.W.; Puri, I.K. SARS-CoV-2 detection with aptamer-functionalized gold nanoparticles. *Talanta* **2022**, *236*, 122841. [CrossRef] [PubMed]

27. Su, F.; Wang, L.; Sun, Y.; Liu, C.; Duan, X.; Li, Z. Highly sensitive detection of CpG methylation in genomic DNA by AuNP-based colorimetric assay with ligase chain reaction. *Chem. Commun.* **2015**, *51*, 3371–3374. [CrossRef] [PubMed]
28. Ye, H.; Yang, K.; Tao, J.; Liu, Y.; Zhang, Q.; Habibi, S.; Nie, Z.; Xia, X. An enzyme-free signal amplification technique for ultrasensitive colorimetric assay of disease biomarkers. *ACS Nano* **2017**, *11*, 2052–2059. [CrossRef]
29. Moabelo, K.L.; Martin, D.R.; Fadaka, A.O.; Sibuyi, N.R.; Meyer, M.; Madiehe, A.M. Nanotechnology-Based Strategies for Effective and Rapid Detection of SARS-CoV-2. *Materials* **2021**, *14*, 7851. [CrossRef] [PubMed]
30. Nguyen, D.K.; Jang, C.-H. Ultrasensitive colorimetric detection of amoxicillin based on Tris-HCl-induced aggregation of gold nanoparticles. *Anal. Biochem.* **2022**, *645*, 114634. [CrossRef] [PubMed]
31. Sargazi, S.; Mukhtar, M.; Rahdar, A.; Bilal, M.; Barani, M.; Díez-Pascual, A.M.; Behzadmehr, R.; Pandey, S. Opportunities and challenges of using high-sensitivity nanobiosensors to detect long noncoding RNAs: A preliminary review. *Int. J. Biol. Macromol.* **2022**, *205*, 304–315. [CrossRef] [PubMed]
32. Sibuyi, N.R.S.; Moabelo, K.L.; Fadaka, A.O.; Meyer, S.; Onani, M.O.; Madiehe, A.M.; Meyer, M. Multifunctional Gold Nanoparticles for Improved Diagnostic and Therapeutic Applications: A Review. *Nanoscale Res. Lett.* **2021**, *16*, 174. [CrossRef] [PubMed]
33. Yetisen, A.K.; Akram, M.S.; Lowe, C.R. based microfluidic point-of-care diagnostic devices. *Lab A Chip* **2013**, *13*, 2210–2251. [CrossRef] [PubMed]
34. Chang, C.-C.; Chen, C.-P.; Wu, T.-H.; Yang, C.-H.; Lin, C.-W.; Chen, C.-Y. Gold nanoparticle-based colorimetric strategies for chemical and biological sensing applications. *Nanomaterials* **2019**, *9*, 861. [CrossRef] [PubMed]
35. Zhao, W.; Brook, M.A.; Li, Y. Design of gold nanoparticle-based colorimetric biosensing assays. *ChemBioChem* **2008**, *9*, 2363–2371. [CrossRef] [PubMed]
36. Qi, X.; Zhao, Y.; Su, H.; Wang, L.; Li, L.; Ma, R.; Yan, X.; Sun, J.; Wang, S.; Mao, X. A label-free colorimetric aptasensor based on split aptamers-chitosan oligosaccharide-AuNPs nanocomposites for sensitive and selective detection of kanamycin. *Talanta* **2022**, *238*, 123032. [CrossRef] [PubMed]
37. Xu, J.; Li, Y.; Bie, J.; Jiang, W.; Guo, J.; Luo, Y.; Shen, F.; Sun, C. Colorimetric method for determination of bisphenol A based on aptamer-mediated aggregation of positively charged gold nanoparticles. *Microchim. Acta* **2015**, *182*, 2131–2138. [CrossRef]
38. Bai, W.; Zhu, C.; Liu, J.; Yan, M.; Yang, S.; Chen, A. Gold nanoparticle–based colorimetric aptasensor for rapid detection of six organophosphorous pesticides. *Environ. Toxicol. Chem.* **2015**, *34*, 2244–2249. [CrossRef] [PubMed]
39. Lee, E.-H.; Lee, S.K.; Kim, M.J.; Lee, S.-W. Simple and rapid detection of bisphenol A using a gold nanoparticle-based colorimetric aptasensor. *Food Chem.* **2019**, *287*, 205–213. [CrossRef]
40. Luan, Y.; Chen, Z.; Xie, G.; Chen, J.; Lu, A.; Li, C.; Fu, H.; Ma, Z.; Wang, J. Rapid visual detection of aflatoxin B1 by label-free aptasensor using unmodified gold nanoparticles. *J. Nanosci. Nanotechnol.* **2015**, *15*, 1357–1361. [CrossRef]
41. Stiolica, A.T.; Popescu, M.; Bubulica, M.V.; Oancea, C.N.; Nicolicescu, C.; Manda, C.V.; Neamtu, J.; Croitoru, O. Optimization of gold nanoparticles synthesis using design of experiments technique. *Rev. Chim.* **2017**, *68*, 1518–1523. [CrossRef]
42. Borse, V.; Konwar, A.N. Synthesis and characterization of gold nanoparticles as a sensing tool for the lateral flow immunoassay development. *Sens. Int.* **2020**, *1*, 100051. [CrossRef]
43. Haiss, W.; Thanh, N.T.; Aveyard, J.; Fernig, D.G. Determination of size and concentration of gold nanoparticles from UV−Vis spectra. *Anal. Chem.* **2007**, *79*, 4215–4221. [CrossRef]
44. Ma, Q.; Wang, Y.; Jia, J.; Xiang, Y. Colorimetric aptasensors for determination of tobramycin in milk and chicken eggs based on DNA and gold nanoparticles. *Food Chem.* **2018**, *249*, 98–103. [CrossRef] [PubMed]
45. Wei, X.; Wang, Y.; Zhao, Y.; Chen, Z. Colorimetric sensor array for protein discrimination based on different DNA chain length-dependent gold nanoparticles aggregation. *Biosens. Bioelectron.* **2017**, *97*, 332–337. [CrossRef] [PubMed]
46. Borghei, Y.-S.; Hosseini, M.; Dadmehr, M.; Hosseinkhani, S.; Ganjali, M.R.; Sheikhnejad, R. Visual detection of cancer cells by colorimetric aptasensor based on aggregation of gold nanoparticles induced by DNA hybridization. *Anal. Chim. Acta* **2016**, *904*, 92–97. [CrossRef] [PubMed]
47. Lerga, T.M.; Skouridou, V.; Bermudo, M.C.; Bashammakh, A.S.; El-Shahawi, M.S.; Alyoubi, A.O.; O'Sullivan, C.K. Gold nanoparticle aptamer assay for the determination of histamine in foodstuffs. *Microchim. Acta* **2020**, *187*, 452. [CrossRef] [PubMed]
48. Xiao, S.; Lu, J.; Sun, L.; An, S. A simple and sensitive AuNPs-based colorimetric aptasensor for specific detection of azlocillin. *Spectrochim. Acta Part A: Mol. Biomol. Spectrosc.* **2022**, *271*, 120924. [CrossRef]
49. Yin, X.; Wang, S.; Liu, X.; He, C.; Tang, Y.; Li, Q.; Liu, J.; Su, H.; Tan, T.; Dong, Y. Aptamer-based colorimetric biosensing of ochratoxin A in fortified white grape wine sample using unmodified gold nanoparticles. *Anal. Sci.* **2017**, *33*, 659–664. [CrossRef]
50. Pavase, T.R.; Lin, H.; Soomro, M.A.; Zheng, H.; Li, X.; Wang, K.; Li, Z. Visual detection of tropomyosin, a major shrimp allergenic protein using gold nanoparticles (AuNPs)-assisted colorimetric aptasensor. *Mar. Life Sci. Technol.* **2021**, *3*, 382–394. [CrossRef]
51. Lan, L.; Yao, Y.; Ping, J.; Ying, Y. Recent progress in nanomaterial-based optical aptamer assay for the detection of food chemical contaminants. *ACS Appl. Mater. Interfaces* **2017**, *9*, 23287–23301. [CrossRef]
52. Zhao, V.X.T.; Wong, T.I.; Zheng, X.T.; Tan, Y.N.; Zhou, X. Colorimetric biosensors for point-of-care virus detections. *Mater. Sci. Energy Technol.* **2020**, *3*, 237–249. [CrossRef] [PubMed]
53. Balasubramanian, S.K.; Yang, L.; Yung, L.-Y.L.; Ong, C.-N.; Ong, W.-Y.; Liya, E.Y. Characterization, purification, and stability of gold nanoparticles. *Biomaterials* **2010**, *31*, 9023–9030. [CrossRef] [PubMed]
54. Zhang, X.; Servos, M.R.; Liu, J. Surface science of DNA adsorption onto citrate-capped gold nanoparticles. *Langmuir* **2012**, *28*, 3896–3902. [CrossRef] [PubMed]

55. Liu, J. Adsorption of DNA onto gold nanoparticles and graphene oxide: Surface science and applications. *Phys. Chem. Chem. Phys.* **2012**, *14*, 10485–10496. [CrossRef] [PubMed]

56. Wusu, A.D.; Sibuyi, N.R.S.; Moabelo, K.L.; Goboza, M.; Madiehe, A.; Meyer, M. Citrate-capped gold nanoparticles with a diameter of 14 nm alter the expression of genes associated with stress response, cytoprotection and lipid metabolism in CaCo-2 cells. *Nanotechnology* **2021**, *33*, 105101. [CrossRef] [PubMed]

57. Shirani, M.; Kalantari, H.; Khodayar, M.J.; Kouchak, M.; Rahbar, N. A novel strategy for detection of small molecules based on aptamer/gold nanoparticles/graphitic carbon nitride nanosheets as fluorescent biosensor. *Talanta* **2020**, *219*, 121235. [CrossRef] [PubMed]

58. Qadami, F.; Molaeirad, A.; Alijanianzadeh, M.; Azizi, A.; Kamali, N. Localized surface plasmon resonance (LSPR)-based nanobiosensor for methamphetamin measurement. *Plasmonics* **2018**, *13*, 2091–2098. [CrossRef]

59. Banyay, M.; Sarkar, M.; Gräslund, A. A library of IR bands of nucleic acids in solution. *Biophys. Chem.* **2003**, *104*, 477–488. [CrossRef]

60. Wu, S.; Liu, L.; Duan, N.; Li, Q.; Zhou, Y.; Wang, Z. Aptamer-based lateral flow test strip for rapid detection of zearalenone in corn samples. *J. Agric. Food Chem.* **2018**, *66*, 1949–1954. [CrossRef] [PubMed]

 *biosensors*

Article

A Microfluidic Platform Revealing Interactions between Leukocytes and Cancer Cells on Topographic Micropatterns

Xin Cui [1,*,†], Lelin Liu [2,3,†], Jiyu Li [2,†], Yi Liu [2], Ya Liu [4], Dinglong Hu [4], Ruolin Zhang [5,6], Siping Huang [2], Zhongning Jiang [2], Yuchao Wang [2], Yun Qu [2], Stella W. Pang [5,7] and Raymond H. W. Lam [2,7,8,9,*]

[1] Key Laboratory of Biomaterials of Guangdong Higher Education Institutes, Department of Biomedical Engineering, Jinan University, Guangzhou 519070, China
[2] Department of Biomedical Engineering, City University of Hong Kong, Hong Kong 999077, China
[3] Research Center of Biological Computation, Zhejiang Laboratory, Hangzhou 311100, China
[4] BGI-Shenzhen, Shenzhen 518083, China
[5] Department of Electrical Engineering, City University of Hong Kong, Hong Kong 999077, China
[6] Department of Biomedical Engineering, The Hong Kong Polytechnic University, Hong Kong 999077, China
[7] Centre for Biosystems, Neuroscience, and Nanotechnology, City University of Hong Kong, Hong Kong 999077, China
[8] Centre for Robotics and Automation, City University of Hong Kong, Hong Kong 999077, China
[9] Shenzhen Research Institute, City University of Hong Kong, Shenzhen 518057, China
* Correspondence: cx2019@jnu.edu.cn (X.C.); rhwlam@cityu.edu.hk (R.H.W.L.); Tel.: +86-8522-0469 (X.C.); +86-852-3442-8577 (R.H.W.L.); Fax: +852-3442-0172 (R.H.W.L.)
† These authors contributed equally to this work.

Citation: Cui, X.; Liu, L.; Li, J.; Liu, Y.; Liu, Y.; Hu, D.; Zhang, R.; Huang, S.; Jiang, Z.; Wang, Y.; et al. A Microfluidic Platform Revealing Interactions between Leukocytes and Cancer Cells on Topographic Micropatterns. *Biosensors* 2022, 12, 963. https://doi.org/10.3390/bios12110963

Received: 14 October 2022
Accepted: 31 October 2022
Published: 2 November 2022

Publisher's Note: MDPI stays neutral with regard to jurisdictional claims in published maps and institutional affiliations.

Abstract: Immunoassay for detailed analysis of immune−cancer intercellular interactions can achieve more promising diagnosis and treatment strategies for cancers including nasopharyngeal cancer (NPC). In this study, we report a microfluidic live−cell immunoassay integrated with a microtopographic environment to meet the rising demand for monitoring intercellular interactions in different tumor microenvironments. The developed assay allows: (1) coculture of immune cells and cancer cells on tunable (flat or micrograting) substrates, (2) simultaneous detection of different cytokines in a wide working range of 5–5000 pg/mL, and (3) investigation of migration behaviors of mono- and co-cultured cells on flat/grating platforms for revealing the topography-induced intercellular and cytokine responses. Cytokine monitoring was achieved on-chip by implementing a sensitive and selective microbead-based sandwich assay with an antibody on microbeads, target cytokines, and the matching fluorescent-conjugated detection antibody in an array of active peristaltic mixer-assisted cytokine detection microchambers. Moreover, this immunoassay requires a low sample volume down to 0.5 µL and short assay time (30 min) for on-chip cytokine quantifications. We validated the biocompatibility of the co-culture strategy between immune cells and NPC cells and compared the different immunological states of undifferentiated THP-1 monocytic cells or PMA-differentiated THP-1 macrophages co-culturing with NP460 and NPC43 on topographical and planar substrates, respectively. Hence, the integrated microfluidic platform provides an efficient, broad-range and precise on-chip cytokine detection approach, eliminates the manual sampling procedures and allows on-chip continuous cytokine monitoring without perturbing intercellular microenvironments on different topographical ECM substrates, which has the potential of providing clinical significance in early immune diagnosis, personalized immunotherapy, and precision medicine.

Keywords: live-cell immunoassay; topographic micropatterns; on-chip cytokine detection; nasopharyngeal cancer; integrated microfluidics

1. Introduction

Nasopharyngeal carcinoma (NPC) is a major health problem for the Southeastern Asian and North African populations [1–3], which can be divided into Epstein-Barr virus

(EBV)-positive and negative subgroups. Although current treatments, including radiotherapy, chemotherapy or chemo-radiotherapy, can improve the survival rates, treatment resistance and tumor recurrence remains to date challenging, largely because of the unique and complex intercellular interactions between cancer cells and immune cells, involving various soluble factors released by the tumor microenvironment. The presence of EBVs can alter the biomolecular secretion of NPC cells as well as the immune responses to them. The non-keratinizing Epstein-Barr virus (EBV)-positive nasopharyngeal carcinoma represents a unique tumor microenvironment, characterized by dense infiltrating immune cells comprising macrophages [1]. For instance, EBV-infected NPC cells secreted a higher level of IL-1α (1561 pg/mL), IL-1β (16.6 pg/mL) and IL-8 (422.9 pg/mL) as compared to EBV-negative cells [4]. Macrophages secrete cytokines such as TNF (tumor necrosis factor), IL-1, IL-6, IL-8, and IL-12, while the stimulated ones produce more TNF-α, IL-12/IL-23p40, and IL-10 [5,6]. In addition, EBV-infected human monocytes induce the expression of MCP-1 (monocyte chemotactic protein-1) via Toll-like receptor (TLR) 2 [7]; and therefore such MCP-1 further inhibits the IL-12 production by inflammatory macrophages [8]. Clinical studies also highlighted that high serum levels of MCP-1, TNF-α, IL-6 and IL-8 are the most prominent cytokines associated with bone invasion, distant metastasis, and particularly poor outcome in NPC patients [1,7,9]. Hence, precisely investigating and quantifying intercellular interactions and cytokine secretions from immune and cancer cells would help in understanding the NPC developments and identifying potential targets for optimized NPC diagnose and treatments.

The enzyme-linked immunosorbent assay (ELISA) is widely used as a "gold standard" method for cytokine quantifications, which rely on repeated time-consuming sample incubation and washing procedures and cannot be applied for in-situ, real-time and multiplex cytokine profiling. Over the past few decades, microfluidic-based immunoassays have been developed for rapid analysis of cytokine secretion in complex fluidic bio-samples due to the significant advantages of microfluidics in fluid flow controlling and ultra-low reagent consumption [10–13]. For example, a strategy reported by Han et al. is capable of multidimensionally analyzing single cells cytokine secretion frequencies by quantitative micro-engraving [14]. Baraket et al. designed an integrated electrochemical biosensor platform which can perform highly sensitive multi-detection for IL-10 and IL-1β [15]. Choi et al. designed a microfluidic magnetic-beads-based device for protein analysis and bio-molecule detection [16]. Min et al. also developed a microfluidic immunoassay platform for biomolecular quantitative detection, which was based on acridine esterification chemiluminescence [17]. To illustrate the dynamic cytokine profiles of various immune cell subtypes, Junkin et al. developed an automated high-throughput microfluidic chip to rebuild the dynamics of single immune cell [18], which was only supposed to detect the cytokine secretion of single types of immune cells. To achieve the multiple cytokine profiling, our pervious study developed an automated microfluidic microbeads-based device for dynamic immunoassay, which profiled multiple cytokines secretion with a low detection limit and short testing time [19]. However, cells growing in regular culture wells cannot reflect the three-dimensional extracellular biochemical and morphologic environments and more physiological-relevant results [20,21]. Our previous studies showed that nasopharyngeal carcinoma cell migration dynamics and spreading directionality can be regulated by microenvironmental morphology [22], suggesting that the grating-like topography of pterygoid muscles can play a role in nasopharyngeal cancer spreading [23]. Andersson et al. demonstrated the effects of substrate morphology on epithelial cell morphologic behaviors and cytokine secretions [24]. Furthermore, the direct observation of cell behaviors such as migration, together with cytokine measurement, may facilitate the future development of new prognostic tools to reveal cancer-immune cell interactions.

Herein, we report a microfluidic immunoassay device by integrating the cell culture region with microtopographic substrates and our previously reported cytokine dynamics profiling scheme [25,26] by implementing a microbeads-based immunofluorescence assay for achieving more sensitive and parallel detection of multiple cytokines. Co-cultured immune cells (undifferentiated THP-1 monocytic cells or PMA-differentiated THP-1 macrophages) and immortal cells NP460/NPC43 were cultured in the device, in which cell culture medium was extracted from the cell culture at different time points, transferred and analyzed in the cytokine detection microchambers for quantifying secretions of TNF and IL-12p70 throughout the co-culture period. As IL-12p70 secretion of monocytes can reflect their EBV infection, these results can quantitatively reflect the role of EBV in immune responses upon nasopharyngeal carcinoma.

2. Materials and Methods

2.1. Device Fabrication

The reported microfluidic device (Figure 1a) consists of three layers of microstructures made of polydimethylsiloxane (PDMS; Dow Corning, Midland, MI, USA): a control layer (height: 20 μm), a flow layer (gas channel height: 20 μm) and a micro−grating layer, with the design layout as illustrated in Figure 1b. The device was fabricated by multilayer soft lithography [25] and the required replica molds were fabricated by photolithography, as described in Figure S1 in the Supplementary Material.

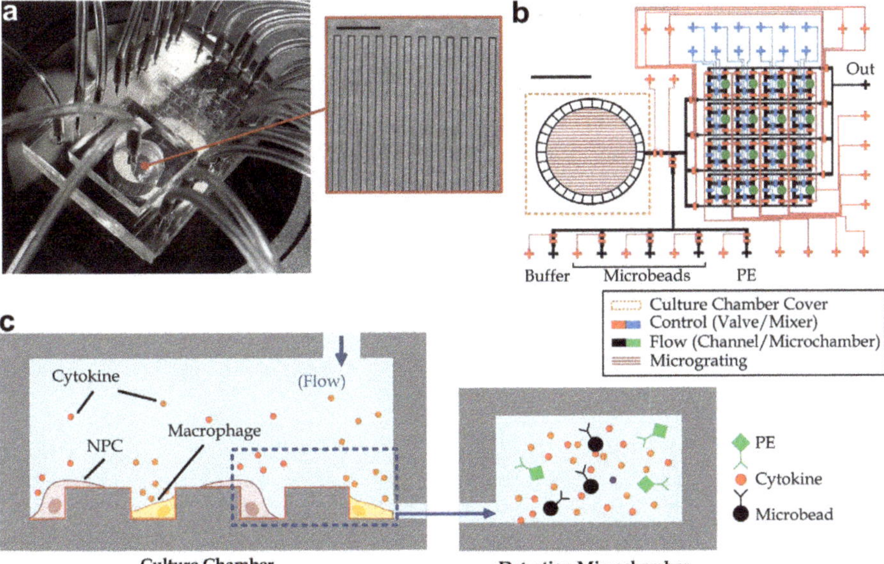

Figure 1. (**a**) Representative microscopic image of a microfluidic immunoassay device integrated with an ECM protein-coated topologic environment. Inset: microscopic image of the parallel grating array in cell culture chamber (grating width: 18 μm, the depth: 18 μm). Scale bar: 100 μm. (**b**) The integrated microfluidic immunoassay device microchannels in the control layer consist of two functional units: valves (*cyan*) for micromixers and valves (*red*) for flow control. The flow channels are composed of an array of microchambers (*green*) for microbead trapping, connecting channels (*black*) between the cell culture region and the downstream cytokine detection arrays. Specially, the connecting microchannels surrounding cell culture chamber contained multiple micro-gaps (height 4 μm; width 50 μm) to allow liquid flow while simultaneously confining the cells in the culture chamber. Scale bar: 4 mm. (**c**) Schematic of the integrated microfluidic immunoassay device operation. Immune cells (macrophages) and nasopharyngeal cells (NPC) were co-cultured in the grating array-embedded cell culture chamber, and the media with secreted cytokines from the cell culture chamber were extracted and incubated with pre-loaded antibody-conjugated microbeads for capturing cytokines in one of the cytokine detection-microchamber. Fluorescent-conjugated detection antibodies were used to label the captured cytokine for microscopic imaging-based multiple cytokine quantification.

The layout of the microfluidic device was designed with Adobe Illustrator CS6 software. Plastic photomasks for each layer were printed by Newway Photomask, Inc. for fabricating the molds. The mold of parallel micro-gratings (width: 18 µm; depth: 18 µm) was fabricated by deep reactive ion etching with an AZ50XT positive photoresist (AZ Electronic Materials, Somerville, NJ, USA) as the sacrificial layer. On the other hand, molds of the control layer and the flow layer were also fabricated by photolithography of SU−8 photoresist (SU−8 2010, Microchem, Westborough, MA, USA) on planar silicon wafers. Microchannel patterns of the flow layer mold were fabricated with AZ50XT with a height of 20 µm, reflowed at 120 °C for 1 min, while the microchambers and the filtering microstructure around the cell culture region (height: 30 µm) of the mold were aligned with the microchannel patterns and fabricated with an SU-8 2010 photoresist (Microchem, Westborough, MA, USA). All the molds were then silanized by trichloro (1H, 1H, 2H, 2H-perfluoro-octyl) silane (Sigma-Aldrich, St. Louis, MO, USA) for 12 h to facilitate the release of the molded PDMS layers in the later procedures.

Afterwards, PDMS pre-polymer was prepared by mixing the base and the curing agent with a weight ratio of 10:1. The microfluidic device with multiple PDMS layers were fabricated by replica molding of PDMS from the molds using the multiplayer soft lithography as described in Figure S1, with the PDMS applied on the molds with different thickness: 5 mm by pouring on the control layer mold, 35 µm by spin-coating over the flow layer mold, and 1 mm by pouring on the micro-grating mold. All the following bonding processes between PDMS layers were then achieved by air plasma (energy 10 kJ; Harrick plasma cleaner PDC-002, Ithaca, NY, USA). According to the layout of the microfluidic device, as shown in Figure 1a, the PDMS control layer was aligned and bonded onto the flow layer, with holes punched at the gas/liquid inlets and outlets (diameter: 1 mm; WHAWB100073, Sigma-Aldrich, St. Louis, MO, USA), followed by punching a hole (diameter: 6 mm; WHAWB100082, Sigma-Aldrich, St. Louis, MO, USA) at the cell culture region. The PDMS substrate was then bonded on the micro-grating layer, with the micro-grating structures (Figure 1a and Figure S1 in the Supplementary Material) facing the culture chamber. The culturing region was further covered by bonding with another PDMS layer (thickness: 5 mm), with two punched holes as the sample inlet and outlet. Finally, the entire multilayer PDMS substrate was bonded onto a glass slide (Cytoglass, Nanjing, China) for physical support.

2.2. Cell Preparation

A human monocytic cell line (THP-1, ATCC TIB-202, Manassas, VA, USA) was cultured in a complete RPMI-1640 culture medium supplemented with 10% fetal bovine serum and 1% penicillin. An immortal human nasopharyngeal epithelial cell line (NP 460) and a nasopharyngeal carcinoma cell line (NPC 43) were developed and donated by the research team of S. W. Tsao, from cell extracts of nasopharyngeal cancer patients [27]. NP 460 cells were cultured in a mixture of 50% complete Eplife medium (Thermo Fisher Scientific, Waltham, MA, USA), 50% complete defined keratinocyte−SFM (Thermo Fisher Scientific, Waltham, MA, USA) with 100 units/mL penicillin and 100 µg/mL streptomycin. NPC 43 cells were maintained in RPMI-1640 (Sigma-Aldrich, St. Louis, MO, USA) added with 10% fetal bovine serum, 4 µM Y27632 dihydrochloride (Alexis), 100 unit/mL penicillin and 100 µg/mL streptomycin. All the cell types were cultured in a 37 °C incubator (HERA cell 150, Thermo Fisher Scientific, Waltham, MA, USA) with a humidified and 5% CO_2 environment. Additionally, the macrophages were derived from THP−1 cells before experiments. THP−1 cells with a density of 1×10^6 cells/mL was treated with 50 ng/mL of phorbol 12-myristate 13-acetate [28] (PMA; Sigma-Aldrich, St. Louis, MO, USA) for 20 h. The treated cells were then trypsinized (Sigma-Aldrich, St. Louis, MO, USA) for 3 min at room temperature and replaced with fresh media before transferring to the device.

2.3. Automated Microscope and Cytokine Quantification

All the microfluidic manipulation for cell culture, monitoring and imaging was operated by an automated microscope platform developed in our laboratory [29]. It mainly includes an inverted fluorescence microscope (IX71, Olympus, Tokyo, Japan) integrated with a microscope camera (Zyla 4.2, Andor Technology Ltd., Belfast, UK) with computer-controlled compressed air supply manifolds and a confining shield mounted on a computer-controlled movable stage of the microscope, offering stable temperature (37 °C), humidity and gas (5% CO_2) conditions for cell culture (Figure S2 in the Supplementary Material). Cytokine concentrations were quantified using a commercial human inflammatory cytokine kit (Catalog No. 551811, BD Biosciences) in all samples.

2.4. Statistics

All experiments were conducted with at least four independent experiments. The p-values were calculated using Student's t-test in Excel (Microsoft), with $p < 0.05$ considered as statistically significant.

3. Results and Discussion

3.1. Device Design and Operation

We have developed an integrated microfluidic immunoassay device, which can implement in-situ and multiplex monitoring of time-lapsed cytokine secretions in an extracellular matrix (ECM) protein-coated topologic environment. This device consists of three main components: a cell culture chamber with parallel microgratings, cytokine detection arrays (four rows and four columns) and an array of active peristaltic mixers (Figure 1a,b). The parallel gratings (each with 18 μm in width and 18 μm in depth) fabricated by polydimethylsiloxane (PDMS) replica molding can mimic the microtopography of pterygoid muscles [23]. The optically transparent cell culture chamber allows the direct observation of cell behaviors and live cell staining under a microscope. Furthermore, it was surrounded by a barrier containing multiple micro-gaps (height: 4 μm; width: 50 μm) to confine cells inside the cell culture chamber, while allowing cell culture media and the secreted cytokines to flow to one of the cytokine detection chambers to achieve the microbeads-based immunoassay for multiple cytokines at different time points. The extracted media were incubated with antibody-conjugated microbeads for capturing cytokines, followed by mixing with phycoerythrin (PE)-conjugated detection antibodies to form fluorescent sandwich complexes for microscopic imaging and cytokine quantification (Figure 1c). In each cytokine detection chamber (volume: 160 nl), a bypass channel (volume: 30 nl) was integrated with a peristaltic mixer, which consisted of three peristaltic microvalves for active pumping and mixing to accelerate the cytokine detection based on the Taylor dispersion effect [30]. The whole assay can be conducted in less than 30 min.

Notably, we applied 20 μg/mL bovine collagen (Sigma Aldrich, St. Louis, MO, USA) [31] into the culture chamber for 2 h at room temperature before injecting the selected cells, because collagen is a key ECM protein of pterygoid muscles [32]. The nasopharyngeal cells (NP460 or NPC43) and macrophage cells were pre-mixed with a defined population ratio and seeded into the cell culture chamber through the culture inlet and incubated in a microscope incubator with stabilized temperature at 37 °C and 5% CO_2 to facilitate immune cells secreting cytokines in response to target cells. Simultaneously, the detection microbeads were loaded into the bypass channel aside the cytokine detection chamber by opening the corresponding control valves for defining the access from the 'Microbeads' inlet (Figure 1b) to the target detection chamber. We then waited for 10 min for the microbeads sitting on the bottom of the microchamber. The device operation for extracting and quantifying cytokine is shown in Figure 2a and Supplementary video S1. In each measurement, 0.5 μL of the media with secreted cytokines in the culture chamber was extracted from the culture chamber and transported into a detection microchamber with preloaded microbeads for cytokine detection. Meanwhile, fresh media flowed into the culture chamber to supplement the media being extracted. The bypass channel corresponding to the defined cytokine detection chamber was flushed with a PE-conjugated antibody solution, and then the peristaltic mixer was activated for 2 min to mix the PE solution and microbeads (Supplementary video S2). We removed the unconjugated antibody solution by flowing phosphate-buffered saline (PBS) along the detection chamber under gentle driving compressed air with pressure of 0.2 psi for 3 min. The microbeads expressed fluorescence at 647 nm from the bead bodies and the bound cytokine molecules induced fluorescence at 488 nm over the bead surfaces. After imaging the microbeads using an inverted fluorescent microscope, the measured fluorescence intensities at 488 nm were converted to the cytokine concentrations according to the calibration curves (Figure 2b).

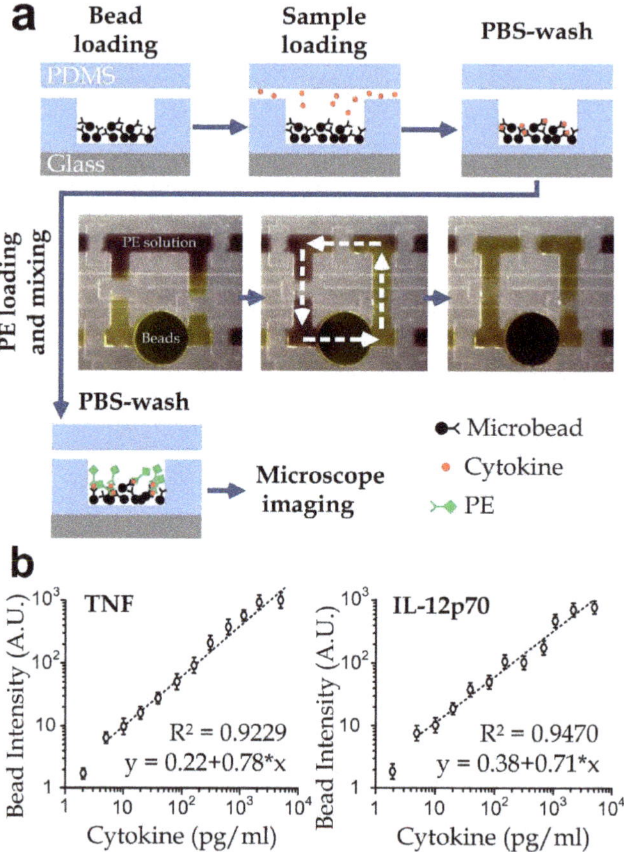

Figure 2. (**a**) Cytokine detection procedures of the integrated microfluidic immunoassay device. Cytokines produced by NPC and immune cells on-chip were transferred from the cell culture region to the cytokine detection arrays, and specifically captured by the pre-loaded antibody conjugated microbeads, followed by washing with PBS buffer. To quantify the cytokine concentrations on-chip, fluorescent phycoerythrin (PE)-conjugated detection antibodies were mixed with the cytokine-binding microbeads using an integrated micromixer (middle panel, representative microscope images), causing the fluorescence change on the microbeads as a readout of binding events. Finally, the fluorescence change was quantified by microscope imaging, and the cytokine concentration was calculated according to the calibration curves. (**b**) The calibration curve of two different cytometric microbeads for TNF and IL12p70 in the microfluidic device by challenging microbeads with different concentrations of cytokines. The data points for 2.4 pg/mL cytokine concentration are not included in the fitting line. Each data point was obtained from the average value of $n > 100$ from 3 repeated measurements. All error bars represent the standard errors.

To obtain the calibration curves for the selected cytokines, different concentrations (2.4 pg/mL, 4.9 pg/mL, 9.8 pg/mL, 19.5 pg/mL, 39.1 pg/mL, 78.1 pg/mL, 156.3 pg/mL, 312.5 pg/mL, 625 pg/mL, 1250 pg/mL, 2500 pg/mL, 5000 pg/mL) of each cytokine were prepared by standard PBS dilution of the stocking cytokine solution (5 ng/mL). Linear regression with $R^2 > 0.9$ was observed between fluorescence intensity (488 nm) and molecular concentrations for each cytokine type. The limit of detection was calculated as ~5 pg/mL for both cytokines based on the concentration, giving a signal equal to the blank signal (Y_0) plus three standard deviations of the blank (3σ), or 4 pg/mL for TNF and 3 pg/mL for IL12p70 based on $3\sigma/S$, where S is the slope of the calibration curve.

3.2. Viability and Cytokine Secretion of Single Cell-Type Cultures

To verify cell viability in the culture chamber, we applied live/dead-cell staining (L-3224, Thermo Fisher Scientific, Waltham, MA, USA) to the immune cells (THP-1 cells and differentiated macrophages) and nasopharyngeal cells (NP460 and NPC43) growing in the culture chamber at a cell density of 1×10^5 cells/mL. For instance, a stained NPC43 pseudopod attaching on a micro-grating can be captured as shown in Figure 3a, which suggests the good biocompatibility of the on-chip microtopographic environment; and three-dimensional fluorescence micrographs of stained NPC43 cells are shown in Figure 3b, where green cells are live cells and red cells are dead cells. Our results (Figure 3c) indicate that viabilities of NP460, NPC43, THP-1 and macrophages are maintained at >95% for the cells growing in the culture chamber for 24 h.

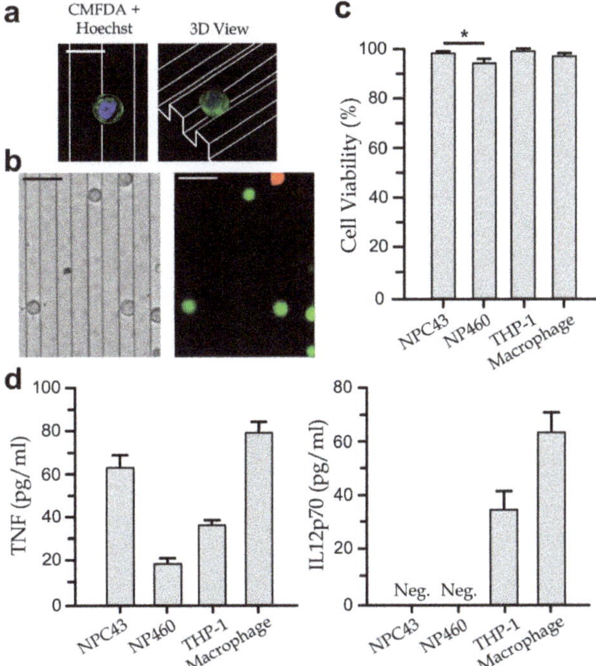

Figure 3. (**a**) Top view (*left* panel) and three-dimensional view (*right* panel) of the reconstructed confocal microscopy images of a NPC43 cell culture on a micro-grating array. Scale bar: 20 μm. (**b**) Brightfield (*left*) and live/dead-stained (*right*) images of NPC43 cultured on micro-gratings for 8 h. Scale bar: 30 μm. (**c**) Viability of NPC43, NP460, THP-1 and THP-1-derived macrophages in the culturing chamber for 8 h. Asterisk represents a *p*-value of <0.05 calculated by Student's *t*-test. (**d**) Cytokine secretions of mono cultured NPC43/NP460 cells and immune cells on micro-gratings after 8 h of culture. Error bars represent the standard errors.

Additionally, we have also measured the cytokine secretion levels (TNF and IL-12p70) of the single cell-type cultures as summarized in Figure 3d. As described above, the simulated macrophages produce more TNF-α and IL-12 [5,6]; and EBV-infection of the monocytes can cause suppressed IL-12 secretion [8] via the induced MCP-1 expression [7]. Therefore, we selected and quantified TNF and IL-12p70 (an active heterodimer of IL-12) for reflecting the simulated activity and the EBV-infection of the macrophages, respectively. Each of the cell types was seeded in the culture chamber at a density of 1×10^5 cells/mL and incubated for 8 h before the cytokine quantification. In brief, our results show that the unstimulated macrophages can secrete a measurable level of IL-12-p70 (65 pg/mL); whereas relatively larger portions of TNF can be contributed from macrophages (80 pg/mL) and NPC43 cells (62 pg/mL). Such measurements of the single cell-type cultures can determine the baselines of the cytokine levels for the immune-nasopharyngeal cell cocultures.

3.3. Cytokine Secretion Dynamics of Immune Cells Co-Cultured with Nasopharyngeal Cells

In the measurements of the cytokine dynamics, immune cells (1×10^5 cells/mL) and nasopharyngeal cells (1×10^5 cells/mL) were cocultured in the culture chamber, with either undifferentiated THP-1 cells or differentiated macrophages applied as the immune cells, and either EBV-positive nasopharyngeal carcinoma (NPC43) cells or EBV-negative nasopharyngeal (NP460) cells applied as the nasopharyngeal cells. Macrophage cells were differentiated from the THP-1 cells by PMA with a concentration of 50 ng/mL for 24 h before the measurements. The cytokine measurements of TNF and IL-12-p70 were performed prior to the culture and at 3, 4, 5, 6, 7 and 8 h of culture. The first three hours of culture could offer a stable environment for cell adaptation.

In the measurements as shown in Figure 4a, we have considered both NPC43 and NP460 in three conditions for different purposes: (1) cocultured with undifferentiated THP-1 on a flat surface as a control case, (2) cocultured with THP-1-differentiated macrophages on a flat surface for estimating the NPC43/NP460-stimulated cytokine secretion, and (3) cocultured with THP-1-derived macrophages on microgratings for revealing the topography-induced cytokine responses.

For the TNF measurements, our results show that the differentiated macrophages secrete a much higher level of the pro-inflammatory cytokine TNF than the undifferentiated THP-1 monocytic cells for both cocultured cases (NPC43 and NP460). The TNF secretions of the cell cocultures on microgratings have an increment of ~20% on average, compared to the cells on flat surfaces, implicating an underlying mechanism related to the microtopographic factors (Figure 4a). Nevertheless, the cytokine secretions of both macrophages and monocytic cells cocultured with NPC43 are less than those cocultured with NP460, suggesting that NPC43 may suppress the TNF secretion of immune cells. One possible explanation is that EBV can suppress TNF-α synthesis from lipopolysaccharide-treated monocytes at both protein and transcriptional levels as reported previously [33].

For the IL-12p70 measurements, it is noteworthy that a distinct secretion profile of IL-12p70 was observed compared to that of TNF (Figure 4b). Though similar trends as the TNF secretion cases have been shown that the IL-12p70 levels are, in general, higher (1) in the macrophages cocultures than in the monocyte cocultures and (2) on the microgratings than on the flat surface for the macrophage cocultures; the coculture with NPC43 does not necessarily induce higher IL-12p70 levels than the coculture with NP460. Furthermore, IL-12p70 secretions of the macrophage-NPC43 cocultures have increasing cytokine levels in the initial stage (0–6 h), reaching the maximum levels at ~6 h and gradually decreasing afterward. Notably, the coefficient of variation (CV), the ratio of the standard deviation to the mean, was investigated for repeatability, which was less than 15% for all data.

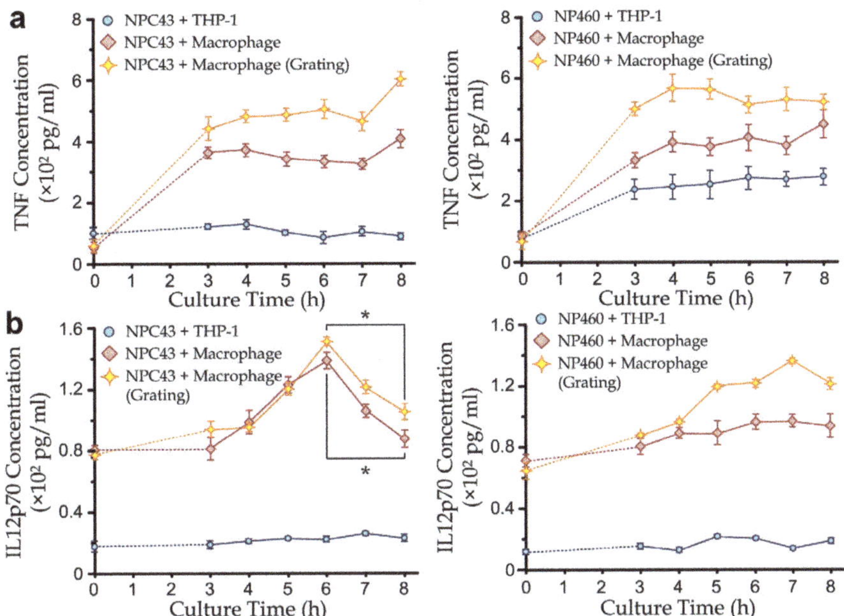

Figure 4. (**a**) The concentration of TNF secreted from cocultured undifferentiated THP-1 and THP-1-differentiated macrophages with NPC43 cells or NP460 cells, with and without a parallel grating array. (**b**) The concentration of IL-12p70 secreted from THP-1 and macrophages co-cultured with NPC43 cells or NP460 cells, with and without parallel grating array (*n* = 4 for each point). Error bars are the standard errors in all plots. Asterisk represents a *p*-value of <0.05.

These results reveal some insightful observations. Interestingly, the macrophage-NPC43 cocultures exhibit suppressed IL-12p70 expressions after 6 h of coculture, implicating an underlying related mechanism between macrophages and nasopharyngeal cancer cells. This agrees with previous clinical studies that nasopharyngeal carcinoma patients have a reduced level of IL-12p70 in their serum [9]. This indicates the suppression of IL-12 secretion, which can be caused by the induced MCP-1 expression of the EBV-infected macrophages [7]. EBV-infection of microphages can also promote polarization to the M2 macrophages [34], leaving a smaller portion of IL-12p70 secreting M1 macrophages [35]. On the other hand, our results show that the microgratings as a microtopographic factor can induce the cytokine secretions of TNF and IL-12p70 on top of the molecular interactions between the immune and nasopharyngeal cells. One possible explanation is that cells adhering to the parallel gratings have different cell behaviors, e.g., morphology and migration [23], which can then affect the immune-cancer cell interaction to some extents. Therefore, it is worthwhile to further apply the reported microfluidic immunoassay to investigate the cell-microenvironment dependency through the simultaneous monitoring of cell behaviors during the coculture periods.

3.4. Cell Migration

We further investigated the migration behaviors of NP460 and NPC43 single cell co-cultured with or without THP-1 derived macrophages on grating platforms. We seeded NP460/MPC43 cells at a density of 5×10^2 cells/cm^2 and each of the co-cultured cells at a density of 2.5×10^2 cells/cm^2, followed by culturing the cells for 8 h and monitoring their migration under a microscope. Migration trajectories of NP460 and NPC43 cells growing on the planar/microgratings substrates (along 90°/270°) are shown in Figure 5a. Clearly, the cells on planar surfaces display a random migration trajectory, whereas the cells on microgratings migrate with a direction along the microgratings.

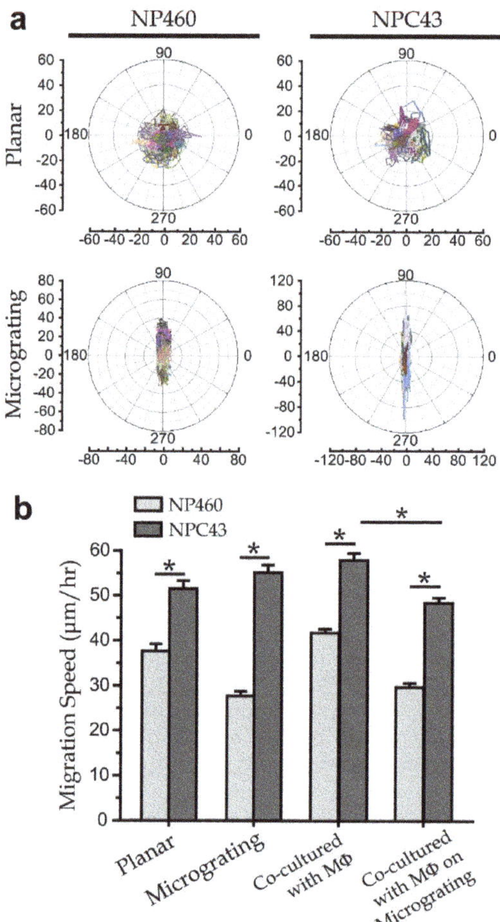

Figure 5. (**a**) Migration trajectories of single NP460 and NPC43 cells on flat/grating substrates. (**b**) Migration speed of single NP460, single NPC43 and those co-cultured with THP-1 derived macrophages on platforms with/without grating. $n > 35$ for all cases. Error bars represent the standard errors. Asterisk represents a *p*-value of <0.05.

Our results (Figure 5b) further indicate that NPC43 cells migrate faster than NP460 cells on both planar and micrograting surfaces. NPC43 cells can migrate even faster when they are cultured on micro-grating substrates than on planar substrates. On the other hand, NP460 cells migrate slower on microgratings than they do on planar substrates, which suggests the different responses of NP460 and NPC43 cells upon the microgratings topography. Furthermore, the co-cultures of macrophages with NP460/NPC43 cells on planar substrates induce faster cell migration, suggesting that molecular secretions of macrophages can promote cell migration. In fact, it has been reported that the MCP-1 secreted by macrophages can promote cell migration [36]. Furthermore, our results indicate that >5% of macrophages and NPC43 cells adhered together on microgratings without further migration, whereas the other cells appearing as single cells without noticeable cell-cell contact can still maintain at a faster migration speed. Interestingly, the cytokine and migration measurements exhibit that nasopharyngeal cancer cells can stay on the microgratings with suppressed IL-12 secretion of the contacting macrophages, supporting the higher tendency of nasopharyngeal cancer spreading to the grating-like pterygoid muscles [37]. Together, the microgratings can affect cell migration behaviors and possibly the intracellular interactions of nasopharyngeal cells

(NP460 and NPC43) and immune cells. For example, it is worthwhile to further examine the correlation between the direct NPC43/macrophage contact and the suppressed IL-12 secretion, and the underlying mechanism.

There are several limitations in the current study. For instance, the effects of interference components such as cell debris in detection samples would be eliminated by integrating a porous membrane filter [38] for sample pretreatments before cytokine detection. Integrating the current microfluidic immunoassay with non-washing cytokine detection strategy such as AlphaLISA [39] may further improve the detection sensitivity and specificity. Simultaneous monitoring of cell behaviors and highly multiplex cytokine detection during the coculture periods using the developed immunoassay would provide valuable insights into the comprehensive and dynamic immune status in solid tumors and during inflammatory states that result in heterogeneous tumor microenvironmental features for precision medicine. Moreover, standardized and automated fabrication setup such as cost-efficient and multilayer PDMS aligner [40] should be further developed for large-scale and high throughput fabrication of the developed microfluidic immunoassay.

4. Conclusions

In conclusion, we have reported a multifunctional microfluidic immunoassay by integrating microtopographic cell-culture substrates with a microbeads-based immunofluorescence assay that enables parallel detection of different immune biomarkers and intercellular behaviors in a rapid, sensitive, and easy-to-implement manner. The developed assay exhibits the advantages of the simultaneous investigation of different cytokines and cell migration behaviors on flat/grating ECM substrates, requiring a low-volume sample (0.5 μL) and short assay time (30 min) but a sensitive performance in a wide range of cytokine concentrations (5–5000 pg/mL). Secretions of TNF and IL-12p70 were successfully monitored throughout the co-culture period to evaluate the different immunological states of undifferentiated THP-1 monocytic cells or PMA-differentiated THP-1 macrophages cocultured with immortal cells NP460/NPC43 on flat and micrograting surfaces. We believe that the reported immunoassay is a promising approach to allow continuous, broad-range and precise on-chip characterization of cytokine and intercellular interactions on different topographical substrates, and thus provides clinical significance for early tumor diagnosis and treatment.

Supplementary Materials: The following supporting information can be downloaded at: https://www.mdpi.com/article/10.3390/bios12110963/s1, Figure S1: Fabrication process of the integrated microfluidic immunoassay; Figure S2. A microscope system integrated with a confining shield (a) and computer-controlled compressed air supply manifolds (b); Video S1: Device operation for extracting culture media from the cell culture chamber; Video S2: Cytokine detection procedures in one detection unit integrated with an active micromixer.

Author Contributions: Conceptualization, X.C. and R.H.W.L.; Data curation, X.C., L.L., J.L., Y.L. (Yi Liu) and D.H.; Formal analysis, L.L., J.L., D.H., R.Z., S.H., Z.J. and Y.W.; Funding acquisition, X.C. and R.H.W.L.; Investigation, X.C., S.W.P. and R.H.W.L.; Methodology, X.C., L.L. and S.W.P.; Project administration, S.W.P.; Resources, Y.L. (Yi Liu), Y.L. (Ya Liu), R.Z., S.H., Y.Q., S.W.P. and R.H.W.L.; Software, J.L. and Y.Q.; Supervision, R.H.W.L.; Validation, L.L., J.L., Y.L. (Yi Liu), Y.L. (Ya Liu) and Y.W.; Visualization, J.L., Y.L. (Ya Liu), Z.J. and Y.Q.; Writing—original draft, X.C., L.L. and J.L.; Writing—review & editing, X.C. and R.H.W.L. All authors have read and agreed to the published version of the manuscript.

Funding: This research was funded by the General Research Grant (11215619 and 11216220) of Hong Kong and Natural Science Foundation of Guangdong Province (2021A1515011167 and 2020A1515010332).

Institutional Review Board Statement: Not applicable.

Informed Consent Statement: Not applicable.

Data Availability Statement: Not applicable.

Conflicts of Interest: The authors declare no conflict of interest.

References

1. Low, H.B.; Png, C.W.; Li, C.; Wang, D.Y.; Wong, S.B.J.; Zhang, Y. Monocyte-derived factors including PLA2G7 induced by macrophage-nasopharyngeal carcinoma cell interaction promote tumor cell invasiveness. *Oncotarget* **2016**, *7*, 55473. [CrossRef] [PubMed]
2. Lee, A.W.M.; Ng, W.T.; Chan, Y.H.; Sze, H.; Chan, C.; Lam, T.H. The battle against nasopharyngeal cancer. *Radiother. Oncol.* **2012**, *104*, 272–278. [CrossRef] [PubMed]
3. Tsang, J.; Lee, V.H.F.; Kwong, D.L.W. Novel therapy for nasopharyngeal carcinoma—Where are we. *Oral Oncol.* **2014**, *50*, 798–801. [CrossRef] [PubMed]
4. Koon, H.-K.; Lo, K.-W.; Leung, K.-N.; Lung, M.L.; Chang, C.C.-K.; Wong, R.N.-S.; Leung, W.-N.; Mak, N.-K. Photodynamic therapy-mediated modulation of inflammatory cytokine production by Epstein–Barr virus-infected nasopharyngeal carcinoma cells. *Cell. Mol. Immunol.* **2010**, *7*, 323–326. [CrossRef] [PubMed]
5. Magalhães, L.M.; Viana, A.; Chiari, E.; Galvão, L.M.; Gollob, K.J.; Dutra, W.O. Differential activation of human monocytes and lymphocytes by distinct strains of Trypanosoma cruzi. *PLoS Negl. Trop. Dis.* **2015**, *9*, e0003816. [CrossRef]
6. Beutler, B.A. The role of tumor necrosis factor in health and disease. *J. Rheumatol. Suppl.* **1999**, *57*, 16–21. [PubMed]
7. Gaudreault, E.; Fiola, S.; Olivier, M.; Gosselin, J. Epstein-Barr virus induces MCP-1 secretion by human monocytes via TLR2. *J. Virol.* **2007**, *81*, 8016–8024. [CrossRef]
8. Chensue, S.W.; Warmington, K.S.; Ruth, J.H.; Sanghi, P.S.; Lincoln, P.; Kunkel, S.L. Role of monocyte chemoattractant protein-1 (MCP-1) in Th1 (mycobacterial) and Th2 (schistosomal) antigen-induced granuloma formation: Relationship to local inflammation, Th cell expression, and IL-12 production. *J. Immunol.* **1996**, *157*, 4602–4608.
9. Zergoun, A.A.; Draleau, K.S.; Chettibi, F.; Touil-Boukoffa, C.; Djennaoui, D.; Merghoub, T.; Bourouba, M. Plasma secretome analyses identify IL-8 and nitrites as predictors of poor prognosis in nasopharyngeal carcinoma patients. *Cytokine* **2022**, *153*, 155852. [CrossRef]
10. Picard, C.; Bobby Gaspar, H.; Al-Herz, W.; Bousfiha, A.; Casanova, J.L.; Chatila, T.; Crow, Y.J.; Cunningham-Rundles, C.; Etzioni, A.; Franco, J.L.; et al. International Union of Immunological Societies: 2017 Primary Immunodeficiency Diseases Committee Report on Inborn Errors of Immunity. *J. Clin. Immunol.* **2018**, *38*, 96–128. [CrossRef]
11. Al-Herz, W.; Bousfiha, A.; Casanova, J.L.; Chatila, T.; Conley, M.E.; Cunningham-Rundles, C.; Etzioni, A.; Franco, J.L.; Gaspar, H.B.; Holland, S.M.; et al. Primary immunodeficiency diseases: An update on the classification from the international union of immunological societies expert committee for primary immunodeficiency. *Front. Immunol.* **2014**, *5*, 162. [CrossRef] [PubMed]
12. Malhotra, R.; Patel, V.; Chikkaveeraiah, B.V.; Munge, B.S.; Cheong, S.C.; Zain, R.B.; Abraham, M.T.; Dey, D.K.; Gutkind, J.S.; Rusling, J.F. Ultrasensitive detection of cancer biomarkers in the clinic by use of a nanostructured microfluidic array. *Anal. Chem.* **2012**, *84*, 6249–6255. [CrossRef] [PubMed]
13. Han, D.; Park, J.K. Microarray-integrated optoelectrofluidic immunoassay system. *Biomicrofluidics* **2016**, *10*, 034106. [CrossRef] [PubMed]
14. Gao, X.; Jiang, L.; Su, X.; Qin, J.; Lin, B. Microvalves actuated sandwich immunoassay on an integrated microfluidic system. *Electrophoresis* **2009**, *30*, 2481–2487. [CrossRef] [PubMed]
15. Hung, L.Y.; Chang, J.C.; Tsai, Y.C.; Huang, C.C.; Chang, C.P.; Yeh, C.S.; Lee, G.B. Magnetic nanoparticle-based immunoassay for rapid detection of influenza infections by using an integrated microfluidic system. *Nanomedicine* **2014**, *10*, 819–829. [CrossRef]
16. Han, Q.; Bradshaw, E.M.; Nilsson, B.; Hafler, D.A.; Love, J.C. Multidimensional analysis of the frequencies and rates of cytokine secretion from single cells by quantitative microengraving. *Lab Chip* **2010**, *10*, 1391–1400. [CrossRef]
17. Baraket, A.; Lee, M.; Zine, N.; Sigaud, M.; Bausells, J.; Errachid, A. A fully integrated electrochemical biosensor platform fabrication process for cytokines detection. *Biosens. Bioelectron.* **2017**, *93*, 170–175. [CrossRef]
18. Salminen, T.; Juntunen, E.; Talha, S.M.; Pettersson, K. High-sensitivity lateral flow immunoassay with a fluorescent lanthanide nanoparticle label. *J. Immunol. Methods* **2019**, *465*, 39–44. [CrossRef]
19. Soares, R.R.G.; Santos, D.R.; Pinto, I.F.; Azevedo, A.M.; Aires-Barros, M.R.; Chu, V.; Conde, J.P. Multiplexed microfluidic fluorescence immunoassay with photodiode array signal acquisition for sub-minute and point-of-need detection of mycotoxins. *Lab Chip* **2018**, *18*, 1569–1580. [CrossRef]
20. Kim, J.; Staunton, J.R.; Tanner, K. Independent Control of Topography for 3D Patterning of the ECM Microenvironment. *Adv. Mater.* **2016**, *28*, 132–137. [CrossRef]
21. Jeon, H.; Tsui, J.H.; Jang, S.I.; Lee, J.H.; Park, S.; Mun, K.; Boo, Y.C.; Kim, D.H. Combined effects of substrate topography and stiffness on endothelial cytokine and chemokine secretion. *ACS Appl. Mater. Interfaces* **2015**, *7*, 4525–4532. [CrossRef] [PubMed]
22. Lam, B.P.; Cheung, S.K.C.; Lam, Y.W.; Pang, S.W. Microenvironmental topographic cues influence migration dynamics of nasopharyngeal carcinoma cells from tumour spheroids. *RSC Adv.* **2020**, *10*, 28975–28983. [CrossRef] [PubMed]
23. Liu, Y.; Ren, J.; Zhang, R.; Hu, S.; Pang, S.W.; Lam, R.H. Spreading and Migration of Nasopharyngeal Normal and Cancer Cells on Microgratings. *ACS Appl. Bio Mater.* **2021**, *4*, 3224–3231. [CrossRef] [PubMed]
24. Andersson, A.-S.; Bäckhed, F.; von Euler, A.; Richter-Dahlfors, A.; Sutherland, D.; Kasemo, B. Nanoscale features influence epithelial cell morphology and cytokine production. *Biomaterials* **2003**, *24*, 3427–3436. [CrossRef]
25. Cui, X.; Liu, Y.; Hu, D.; Qian, W.; Tin, C.; Sun, D.; Chen, W.; Lam, R.H.W. A fluorescent microbead-based microfluidic immunoassay chip for immune cell cytokine secretion quantification. *Lab Chip* **2018**, *18*, 522–531. [CrossRef]

26. Liu, Y.; Li, J.; Hu, D.; Lam, J.H.M.; Sun, D.; Pang, S.W.; Lam, R.H.W. Microfluidic implementation of functional cytometric microbeads for improved multiplexed cytokine quantification. *Biomicrofluidics* **2018**, *12*, 044112. [CrossRef]

27. Lin, W.; Yip, Y.L.; Jia, L.; Deng, W.; Zheng, H.; Dai, W.; Ko, J.M.Y.; Lo, K.W.; Chung, G.T.Y.; Yip, K.Y.; et al. Establishment and characterization of new tumor xenografts and cancer cell lines from EBV-positive nasopharyngeal carcinoma. *Nat. Commun.* **2018**, *9*, 4663. [CrossRef]

28. Priante, G.; Bordin, L.; Musacchio, E.; Clari, G.; Baggio, B. Fatty acids and cytokine mRNA expression in human osteoblastic cells: A specific effect of arachidonic acid. *Clin. Sci.* **2002**, *102*, 403–409. [CrossRef]

29. Yip, H.M.; Li, J.C.S.; Xie, K.; Cui, X.; Prasad, A.; Gao, Q.N.; Leung, C.C.; Lam, R.H.W. Automated Long-Term Monitoring of Parallel Microfluidic Operations Applying a Machine Vision-Assisted Positioning Method. *Sci. World J.* **2014**, *2014*, 608184. [CrossRef]

30. Liao, Q.; Zeng, Z.; Guo, X.; Li, X.; Wei, F.; Zhang, W.; Li, X.; Chen, P.; Liang, F.; Xiang, B.; et al. LPLUNC1 suppresses IL-6-induced nasopharyngeal carcinoma cell proliferation via inhibiting the Stat3 activation. *Oncogene* **2014**, *33*, 2098–2109. [CrossRef]

31. Sharma, D.; Jia, W.; Long, F.; Pati, S.; Chen, Q.; Qyang, Y.; Lee, B.; Choi, C.K.; Zhao, F. Polydopamine and collagen coated micro-grated polydimethylsiloxane for human mesenchymal stem cell culture. *Bioact. Mater.* **2019**, *4*, 142–150. [CrossRef] [PubMed]

32. McKee, T.J.; Perlman, G.; Morris, M.; Komarova, S.V. Extracellular matrix composition of connective tissues: A systematic review and meta-analysis. *Sci. Rep.* **2019**, *9*, 10542. [CrossRef] [PubMed]

33. Gosselin, J.; Menezes, J.; D'Addario, M.; Hiscott, J.; Flamand, L.; Lamoureux, G.; Oth, D. Inhibition of tumor necrosis factor-α transcription by Epstein-Barr virus. *Eur. J. Immunol.* **1991**, *21*, 203–208. [CrossRef] [PubMed]

34. Zhang, B.; Miao, T.; Shen, X.; Bao, L.; Zhang, C.; Yan, C.; Wei, W.; Chen, J.; Xiao, L.; Sun, C. EB virus-induced ATR activation accelerates nasopharyngeal carcinoma growth via M2-type macrophages polarization. *Cell Death Dis.* **2020**, *11*, 742. [CrossRef] [PubMed]

35. Zheng, X.-F.; Hong, Y.-X.; Feng, G.-J.; Zhang, G.-F.; Rogers, H.; Lewis, M.A.; Williams, D.W.; Xia, Z.-F.; Song, B.; Wei, X.-Q. Lipopolysaccharide-induced M2 to M1 macrophage transformation for IL-12p70 production is blocked by *Candida albicans* mediated up-regulation of EBI3 expression. *PLoS ONE* **2013**, *8*, e63967. [CrossRef]

36. Cheng, Z.; Zhu, W.; Li, H.; He, X. Macrophages promote the migration of neural stem cells into mouse spinal cord injury site. *Xi Bao Yu Fen Zi Mian Yi Xue Za Zhi Chin. J. Cell. Mol. Immunol.* **2016**, *32*, 1174–1177.

37. Dubrulle, F.; Souillard, R.; Hermans, R. Extension patterns of nasopharyngeal carcinoma. *Eur. Radiol.* **2007**, *17*, 2622–2630. [CrossRef]

38. Li, P.; Kaslan, M.; Lee, S.H.; Yao, J.; Gao, Z. Progress in exosome isolation techniques. *Theranostics* **2017**, *7*, 789. [CrossRef]

39. Chen, W.; Huang, N.T.; Oh, B.; Lam, R.H.; Fan, R.; Cornell, T.T.; Shanley, T.P.; Kurabayashi, K.; Fu, J. Surface-Micromachined Microfiltration Membranes for Efficient Isolation and Functional Immunophenotyping of Subpopulations of Immune Cells. *Adv. Healthc. Mater.* **2013**, *2*, 965–975. [CrossRef]

40. Nguyen, T.; Sarkar, T.; Tran, T.; Moinuddin, S.M.; Saha, D.; Ahsan, F. Multilayer Soft Photolithography Fabrication of Microfluidic Devices Using a Custom-Built Wafer-Scale PDMS Slab Aligner and Cost-Efficient Equipment. *Micromachines* **2022**, *13*, 1357. [CrossRef]

Review

Semiconducting Polymer Dots for Point-of-Care Biosensing and In Vivo Bioimaging: A Concise Review

Sile Deng, Lingfeng Li, Jiaxi Zhang, Yongjun Wang, Zhongchao Huang * and Haobin Chen *

Department of Biomedical Engineering, School of Basic Medical Sciences, Central South University, Changsha 410013, China
* Correspondence: bme_hzc@csu.edu.cn (Z.H.); chenhb@csu.edu.cn (H.C.)

Abstract: In recent years, semiconducting polymer dots (Pdots) have attracted much attention due to their excellent photophysical properties and applicability, such as large absorption cross section, high brightness, tunable fluorescence emission, excellent photostability, good biocompatibility, facile modification and regulation. Therefore, Pdots have been widely used in various types of sensing and imaging in biological medicine. More importantly, the recent development of Pdots for point-of-care biosensing and in vivo imaging has emerged as a promising class of optical diagnostic technologies for clinical applications. In this review, we briefly outline strategies for the preparation and modification of Pdots and summarize the recent progress in the development of Pdots-based optical probes for analytical detection and biomedical imaging. Finally, challenges and future developments of Pdots for biomedical applications are given.

Keywords: polymer dots; biosensors; fluorescent probes; semiconducting polymers; molecular imaging

Citation: Deng, S.; Li, L.; Zhang, J.; Wang, Y.; Huang, Z.; Chen, H. Semiconducting Polymer Dots for Point-of-Care Biosensing and In Vivo Bioimaging: A Concise Review. *Biosensors* **2023**, *13*, 137. https://doi.org/10.3390/bios13010137

Received: 16 December 2022
Revised: 11 January 2023
Accepted: 12 January 2023
Published: 14 January 2023

1. Introduction

Nanomedicine is the study of the application of nanoparticles in the field of biomedicine, and it has made progress in medical diagnosis and treatment, including biosensing, tissue engineering, medical imaging, cell tracking, drug transporting and cancer optical therapy [1–4]. Generally, biosensors can specifically detect analytes to provide physiological information in a fast and accurate way, and point-of-care testing has become a medical trend, as it greatly facilitates patient self-monitoring of health [5–8]. Apart from biosensing applications, biological imaging helps to visualize the internal structures or enables functional imaging for disease diagnosis, and multimodal imaging combines several imaging methods to integrate the respective signal containing different aspects of biological information for a more comprehensive diagnosis and accurate treatment [9,10]. With the development of materials and principles, biosensing and bioimaging technologies have received considerable attention due to their advantages of high resolution, real-time and non-invasiveness [11–13]. However, since the properties of the materials could exert influence on the sensitivity and accuracy of biosensing and optical applications, traditional small-molecule organic dyes suffer from inherent weakness such as short lifetime, poor photostability and low absorption, which limit further biomedical applications [14–16].

Nanomaterials with better properties and performance have been developed and widely used in the biological field [17–19]. Nanomaterials used to constitute biosensors have great properties and performance due to their unique nanoscale and easily modifiable characteristics, which benefit the energy transfer. On the other hand, nanomaterials for contrast agents contribute to better penetration depth and conversion efficiency, resulting in higher-quality imaging [20–22]. Typical luminescent nanomaterials include quantum dots (Qdots) [23], carbon dots [24], upconversion nanoparticles (UCNPs) [25], aggregation-induced emission (AIE) dots [26] and polymer dots (Pdots) [27]. In particular, Pdots have demonstrated the utilization of optical and biosensing applications in recent years, such as super-resolution imaging [28,29], fluorescence imaging [30,31] and disease-related marker detection [32,33]. Pdots are organic nanoparticles assembled from polymer chains with π-conjugated systems, and the nanoscale size endows Pdots with unique properties, which have attracted extensive attention. According to the definition given by Wu and Chiu [34], the Pdots, specifically considered as a part of semiconducting polymer nanoparticles (SPNs), are nanoparticles consisting of hydrophobic semiconducting polymers with a volume or weight fraction more than 50% and a diameter generally less than 50 nm, sometimes the particles size can be less than 30 nm. Pdots have shown characteristics of large absorption cross section, high brightness, good photostability, low toxicity and various forms of existence and modification, which are the basis of fluorescence probes for complex biological applications, typically for fluorescence-based biosensing and bioimaging [35–37].

This review focuses on the fundamental content and recent advances of Pdots in biosensing and bioimaging applications. The preparation and properties of Pdots are briefly introduced. Modification and functionalization are basic and crucial parts of practical applications, which are related to the attachment of functional groups to the surface of nanoparticles. Thus, several surface modification methods are also introduced. Many Pdots have been presented for in vitro biosensing applications and therapy applications, such as ion sensors [38], reactive oxygen species sensors [39], nucleic acid assays [40,41], enzymatic activity assays [42], photodynamic therapy [43], photothermal therapy [44], gene therapy [45] and chemotherapy [46], which are referred to in recent reviews [47–49]. Herein, we focus on the latest reported Pdots for point-of-care biosensing and in vivo imaging (Figure 1). Pdots point-of-care biosensors, applied to disease-related-metabolites assays, nicotinamide adenine dinucleotide (NAD) sensing, tumor markers quantification and cancer diagnostics, are detailed to demonstrate their great potential in biosensing and transducing techniques. Then, Pdots used as optical probes in bioimaging, such as fluorescence imaging, photoacoustic imaging (PAI), afterglow imaging, chemiluminescence imaging and multimodal imaging, are highlighted. The properties and biomedical applications of the Pdots summarized in this review are listed in Table 1. In the end, we share the challenges and perspective in this field.

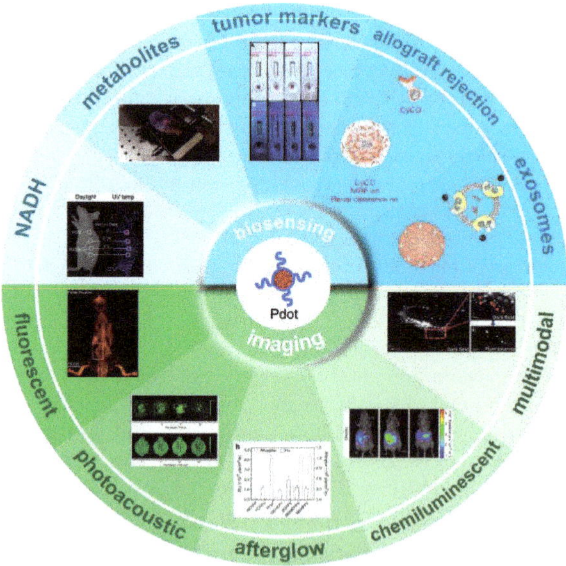

Figure 1. Semiconducting polymer dots for biosensing and imaging.

Table 1. Pdots for biosensing and in vivo imaging.

Pdots	λ_{max}^{abs} (nm)	λ_{max}^{em} (nm)	Φ (%)	Application	Ref.
DPA-CNPPV	294	627	10.8	NADH sensing	[50]
PD4Gx	380	425, 672	11.5	Glucose monitoring	[51]
PF-TC6FQ	~360	~670	N.A.	PSA detection	[52]
PFCN	~390	~450	N.A.	AFP detection	[52]
PFO	~350	~490	N.A.	CEA detection	[52]
APNs	700	720	N.A.	Cancer and allograft	[53]
ASPNC	~440	680	N.A.	Exosomes sensing	[54]
NIR MEH-PPV	504	776	N.A.	Fluorescent imaging	[55]
m-PBTQ4F	946	1123	3.2	Fluorescent imaging	[56]
RET$_2$IR NPs	503	778	0.18	Fluorescent imaging	[57]
Pdots-C6	745	1055	N.A.	Fluorescent imaging	[58]
SPN-PT	1064	N.A.	N.A.	Photoacoustic imaging	[59]
DPP-BTzTD	~1000	N.A.	N.A.	Photoacoustic imaging	[60]
SPNs	~490	780	N.A.	Afterglow imaging	[61]
SPPVN	500, 775	775	51.0	Afterglow imaging	[62]
SPN-NIR	452, 773	507, 775	2.12	Chemiluminescent imaging	[63]
SPNRs	450, 460, 580	520, 540, 700	2.7 $\pm$ 0.014	Chemiluminescent imaging	[64]
rSPN2	~680	840	N.A.	Multimodal imaging	[65]
PPE Gd-SPNs	388	440, 470	22.0	Multimodal imaging	[66]
Au-NP-Pdots	525	~440, 460	18.0	Multimodal imaging	[67]

N.A.: Not applicable.

2. Semiconducting Polymer Dots

This section may be divided by subheadings. It should provide a concise and precise description of the experimental results, their interpretation, as well as the experimental conclusions that can be drawn.

2.1. Methods of Preparation

As different preparation methods generate the Pdots with different sizes and performance, it is critical to choose the corresponding method to obtain suitable size and properties according to various application scenarios. The main preparation methods include the direct polymerization method, miniemulsion method, nanoprecipitation method and self-assembly method. Direct polymerization, referring to the preparation of Pdots from low molecular weight monomers by chemical reactions, offers a wide range of options for size and structure since it also applies to the polymers that are insoluble in organic solvents [68]. Miniemulsion and nanoprecipitation methods dissolve conjugated polymers in organic solvents and interact with water [69,70]. The self-assembly method requires stirring of the solution to mix conjugated polymers and reagents for functionalization. In this part, nanoprecipitation and miniemulsion methods are mainly illustrated (Figure 2).

During the preparation process of the miniemulsion method, the conjugated polymers or monomers to be polymerized are dissolved in a water-immiscible organic solvent [71]. Under vigorous sonication or stirring, it forms microemulsion droplets with aqueous solutions containing surfactants. Finally, stable and uniformly-dispersed Pdots are obtained by removing the organic solvent. In particular, the surfactants are used to avoid aggregation of microemulsion droplets. The concentration of polymers and surfactants in the mixed solution can affect the size of Pdots.

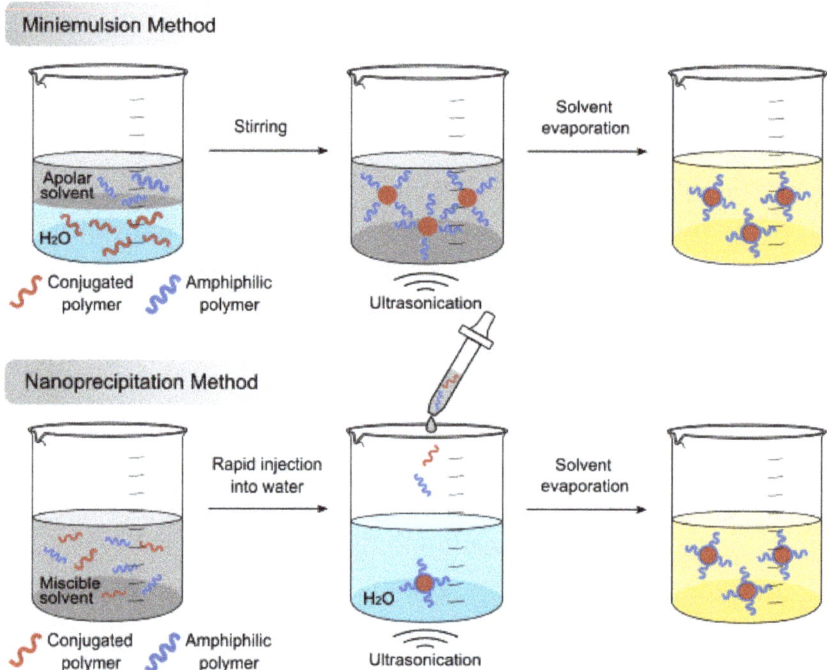

Figure 2. The miniemulsion and nanoprecipitation methods of Pdots.

In the nanoprecipitation method, conjugated polymers and amphiphilic polymers are dissolved in a water-miscible organic solvent. Then, the mixed solution is rapidly injected into water under vigorous sonication, and the nanoprecipitation occurs during this process. Pdots with great water dispersibility are obtained by removing the organic solvent. The biggest difference between the miniemulsion and nanoprecipitation methods is the solvent. The nanoprecipitation method uses a water-miscible organic solvent such as tetrahydrofuran (THF), while the miniemulsion method uses a water-immissible organic solvent such as chloroform. Typically, both methods use surfactants or amphiphilic polymers to increase the yield of nanoparticles. The size of the Pdots depends on the concentration of conjugated polymers in the organic solvents, which ranges from 5 to 50 nm, while the miniemulsion method often gives larger Pdots (larger than 40 nm). In addition, different kinds of amphiphilic polymers can realize different modifications for Pdots in the process of preparation. Liu's group fabricated uniform Pdots by a microfluidic-assisted nanoprecipitation process with a coaxial microfluidic glass capillary mixer [72]. Wu's group used the nanoprecipitation method to prepare functional Pdots with carboxyl groups on the surface for further bioconjugation [73]. Further, they combined photo-crosslinking technology to prepare Pdot-based nanocavities, nanoellipsoids, triangular nanorings and nanowires [74–76].

2.2. Properties and Performance

The critical factors to evaluate the quality of fluorescent probes are absorption cross section, quantum yield, emission rate and photostability. Absorption cross section is used to describe the ability of Pdots to absorb a photon of a particular wavelength and polarization. Studies have shown that the peak absorption cross section of single particles (15 nm in diameter) is about 10–100 times of CdSe Qdots [77]. Moreover, another key property, called quantum yield, is the ratio of the number of photons emitted to the number absorbed; the typical value is below 40% due to aggregation-induced self-quenching [78], and high quantum yield can reach 50–80%. It is generally considered that the brightness of fluorescent molecules depends on the product of absorption cross section and quantum yield. Photostability is assessed by the photobleaching quantum yield calculated from the ratio of photobleaching photons number to the photons absorbed number. Typical photobleaching quantum yield ranges from 10^{-4} to 10^{-6} [79]. Additionally, different kinds of Pdots have been proven to have low toxicity, thus Pdots are widely used in biological applications [80]. Several relevant research results about properties of Pdots are given in Figure 3.

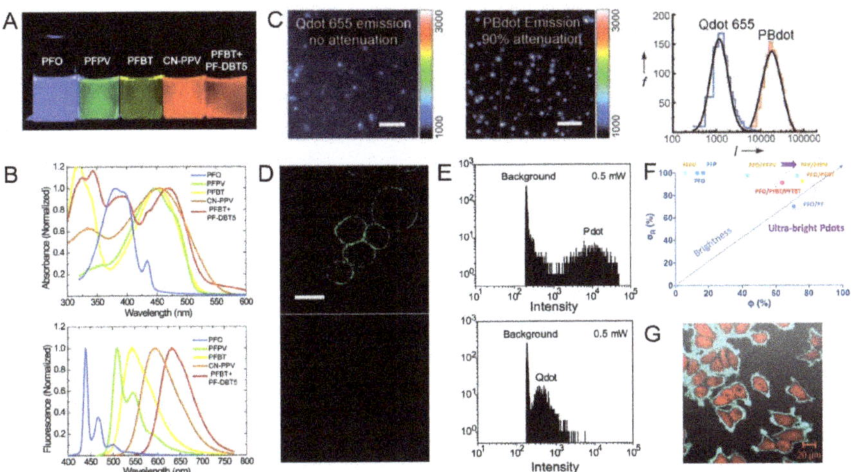

Figure 3. The properties of Pdots. (**A**) A photograph of various Pdots emitted by ultraviolet light. (**B**) Absorption and emission spectra of various Pdots. Reproduced from Ref. [34] with permission. (**C**) Single-particle images and intensity distributions of Qdot 655 and PBdots. Reproduced from Ref. [81] with permission. (**D**) Fluorescence imaging of MCF-7 cells incubated with anti-EpCAM primary antibody and Pdot-IgG conjugates. The bottom panels show the imaging of cells incubated with Pdot-IgG alone. (**E**) Fluorescence intensity distributions for Pdot-streptavidin-labeled MCF-7 cells and Qdot 565-streptavidin-labeled MCF-7 cells. Reproduced from Ref. [82] with permission. (**F**) Ultrabright FRET-based Pdots with simultaneously high absorption cross section and quantum yield. (**G**) Combined fluorescence microscopy images of MCF-7 cells labeled with PEP/PFPV Pdot-streptavidin and biotinylated primary antibody. Reproduced from Ref. [83] with permission.

The improvement of properties is a constant proposition in biological applications of Pdots. Recently, Zhang et al. reported fluorescence resonance energy transfer (FRET)-based Pdots with both large absorption cross section and high quantum yield [83]. By choosing acceptors that had a greater spectral overlap with donors or mixing different kinds of Pdots to create cascade FRET Pdots, they obtained ultrabright blue-, green- and red-emitting Pdots that were among the brightest Pdots reported in the visible region. In other examples, Kuo et al. found that the photostability of Pdots can be improved by adding 4-(2-hydroxyethyl)-1-piperazineethanesulfonic acid (HEPES) or 2-(N-morpholino)ethanesulfonic acid (MES) buffer to quench photoinduced radicals, which aided long-term cell tracking in biological imaging [84]. Chang et al. also designed low-toxic cycloplatinated Pdots, used as a photocatalyst to strengthen the photocatalytic efficiency and stability [85].

2.3. Surface Modification and Biological Functionalization

2.3.1. Encapsulation Method

Silica encapsulation is widely used for surface modification, as other functional groups could be easily attached to silica, which encapsulates particles in a 2–3 nm shell [86,87]. This method greatly promotes biological functionalization of Pdots. However, the silica shell is possible to hydrolyze in biological environments, and the amino groups used to stabilize silica-encapsulated Pdots (10–20 nm in diameter) also cause nonspecific adhesion between Pdots and the cell surface. Another way for surface modification is to encapsulate Pdots using poly (lactic-co-glycolic acid) (PLGA) [88,89] or phospholipids [90,91], which increases nanoparticle stability and reduces nonspecific adhesion. However, the size of nanoparticles modified by PLGA (230–270 nm) and phospholipids (80–100 nm) is too large to apply at the cellular and subcellular levels. Furthermore, the low concentration of fluorescent polymers eventually limits the brightness of nanoparticles, which causes the failure of the encapsulation method to take full advantages of the Pdots.

2.3.2. Amphiphilic Polymer Coprecipitation Method

Chiu's group developed some effective functionalization methods [34,82]. They pre-added amphiphilic polymers in organic solvents, and then prepared Pdots by nanoprecipitation. In this process, amphiphilic polymers covered the surface of the hydrophobic nanoparticles, and their hydrophobic ends were randomly bound to hydrophobic Pdots, while the hydrophilic ends were exposed to water. As a result, Pdots with hydrophilic groups were formed to covalently link biomolecules for biological conjugation and functionalization. For example, an amphiphilic polymer, polystyrene-polyethylene glycol-carboxyl (PS-PEG-COOH), was used for surface modification of Pdots [82]. The average diameter of the product was about 15 nm, and more than 80% of the constituents were significantly effective fluorophores. These research results indicated that this strategy can generate efficient nanoparticle probes, since neither the size nor fluorescent properties of Pdots were affected.

Wu et al. used another amphiphilic polymer, poly (styrene-co-maleic anhydride) (PSMA), to realize biological functionalization [73]. The hydrophilic ends were hydrolyzed in an aqueous environment to form Pdots with carboxyl groups, which facilitated further subsequent bioorthogonal labeling by click chemistry (Figure 4A). Dynamic light scattering (DLS) and transmission electron microscopy (TEM) measurements showed the typical image and hydrodynamic diameter (~15 nm) of functionalized Pdots. Among different functional groups on the Pdots surface, the carboxyl-functionalized Pdots had a significant increase in mobility in the gel electrophoresis (Figure 4B).

2.3.3. Direct Functionalization

In the above modification methods, functional molecules are non-covalently linked to Pdots, which is the main reason for functional molecules falling from the surface of Pdots, which ultimately affects the performance of functionalized Pdots. To overcome these drawbacks, Zhang et al. developed an alternative direct functionalization method, in which Pdots covalently link functional groups [92]. They synthesized conjugated polymers with different percentages of carboxyl groups and used them directly to prepare functionalized Pdots to avoid extra surface modification procedures. Moreover, they found that the degree of functionalization influences the stability and performance of Pdots. In addition, the low carboxylic acid group density (2.3%) brings the greatest properties, including fluorescence brightness, colloidal stability, non-specific absorption and compact internal structure. Yu et al. reported a cross-linking strategy to form functionalized Pdots with enhanced labeling efficiency and sensitivity for cellular assays (Figure 4C) [93]. In addition, Zhang and Chen et al. developed facile strategies with an optical stimulus to covalently link polyethylene glycol and/or carboxyl functional groups to the Pdots (Figure 4D) [94,95], and demonstrated effective bioconjugation of these functionalized photocross-linkable Pdots for specific cellular labeling.

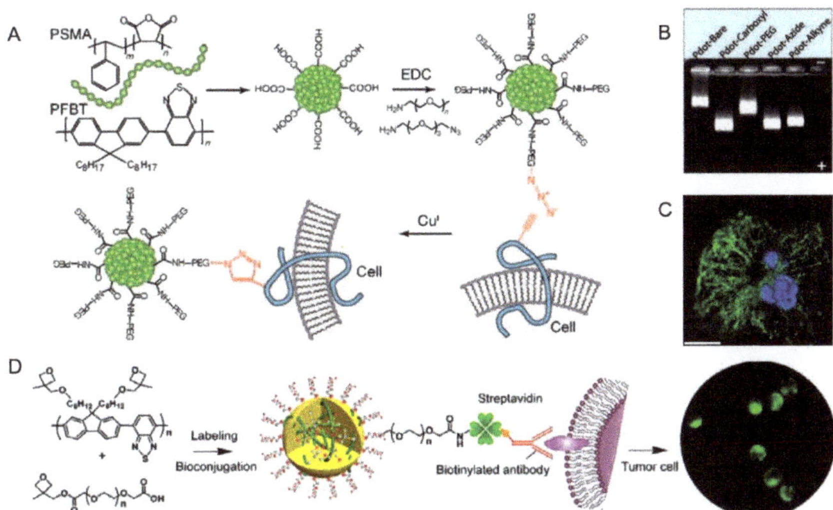

Figure 4. The functionalization methods of Pdots. (**A**) Pdots coprecipitated with amphiphilic polymer PSMA for bioorthogonal labeling via click chemistry. (**B**) Gel electrophoresis of Pdots with various surface functional groups using a 0.7% agarose gel. Reproduced from Ref. [73] with permission. (**C**) Fluorescence microscopy image of microtubules in HeLa cells labeled with cross-linked Pdot-streptavidin. Reproduced from Ref. [93] with permission. (**D**) PEGylated and carboxyl-functionalized Pdots for bioconjugation for specific cellular targeting. Reproduced from Ref. [95] with permission.

3. Application of Pdots Biosensors in Point-of-Care Diagnostics

Point-of-care diagnostics are analytical assays outside the laboratory in order to ensure the convenience of fast testing for target analytes in patients with the same accuracy as laboratory tests. Recently, Pdots-based biosensors have been used for point-of-care diagnostics due to their superior photophysical properties and efficient energy transfer or electron transfer.

FRET has facilitated tremendous advances in biosensing for point-of-care applications. FRET is a phenomenon in which energy is non-radiatively transferred from a donor fluorophore to an acceptor fluorophore, where the fluorescence of the acceptor is emitted while the fluorescence of the donor is quenched. Nanoscale Pdots enable efficient energy transfer, as the efficiency of FRET strongly depends on the distance between donor and acceptor [96]. Additionally, FRET in Pdots biosensors could enhance the brightness of Pdots and obtain high absorption cross section and great photostability [97]. Specific dyes are added into Pdots to realize corresponding FRET-based biosensing [98], which could be applied to detect various metabolites and physiological information [95], including reactive oxygen species [99–101], pH [102], temperature [103,104] and metal ions [105,106]. In this review, Pdots-based biosensors for biomolecule detection are mainly discussed.

3.1. Nicotinamide Adenine Dinucleotide (Oxidized Form: NAD⁺; Reduced Form: NADH)

NADH plays an extremely important role in redox reactions as a coenzyme in enzyme-catalyzed reactions associated with physiological processes [107]. The variation of NAD^+ and NADH concentration is one of the manifestations of diseases such as cancer, epilepsy and Parkinson's disease [108–110]. Biosensors, realizing the convenient and accurate measurement of NAD^+ and NADH concentrations in point-of-care, can promote research advances in the diagnosis of related diseases.

Chen et al. developed a series of Pdots for ratiometric NADH sensing [50], including PFO, PDHF, PFBT, PFBTTBT, PFTBT and DPA-CNPPV Pdots (Figure 5A), which were prepared by the nanoprecipitation method with PSMA. For DPA-CNPPV, the emission maximum was exhibited at 627 nm with excitation at 385 nm. In the presence of NADH, electron transfer from excited DPA-CNPPV to NADH caused fluorescence quenching of DPA-CNPPV Pdots, which was manifested as a decrease in emission intensity at 627 nm. Since NADH in solution had a blue emission (peak at 458 nm), while DPA-CNPPV hardly absorbed in the blue region, there was almost no energy transfer from NADH to DPA-CNPPV. In short, with the increasing concentration of NADH, the emission intensity decreased at 627 nm and increased at 485 nm in the physiological range (0–2 mM, Figure 5B). For other Pdots, for example, the absorption of PFBT in the blue region caused less color contrast, which was not favorable for ratiometric sensing. On the other hand, the NADH-sensitive DPA-CNPPV Pdots had shown great performance in terms of photostability, response time, selectivity (Figure 5C) and reversibility.

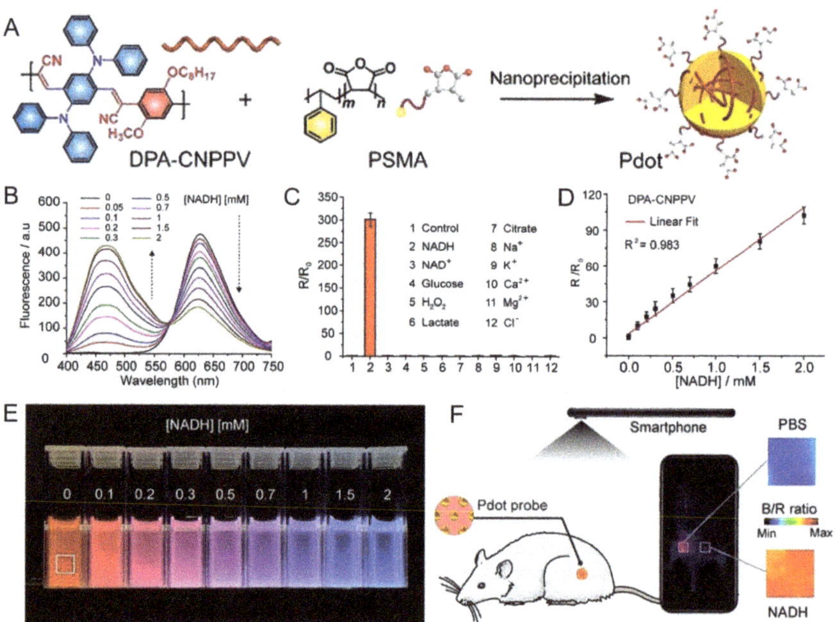

Figure 5. DPA-CNPPV Pdots for reversible ratiometric NADH sensing. (**A**) The preparation of DPA-CNPPV Pdots by nanoprecipitation. (**B**) Emission spectra of DPA-CNPPV Pdots with 0–2 mM NADH. (**C**) Selectivity of DPA-CNPPV to NADH versus 10 other analytes. (**D**) Calibration curve of DPA-CNPPV Pdots in the range of 0–2 mM NADH. $R = I_{458\,nm}/I_{627\,nm}$; $R_0 = R$ in the absence of NADH. (**E**) DPA-CNPPV Pdots at 0–2 mM NADH with excitation at 365 nm. (**F**) Ratiometric imaging taken by smartphone for NADH in vivo measurement. DPA-CNPPV were injected into two locations of mice with or without NADH (0.1 µmol). Heatmap images of blue/red-channel intensities ratios are shown on the right, with a high ratio (red) indicating high concentration of NADH. Reproduced from Ref. [50] with permission.

A DPA-CNPPV Pdots probe used for NADH measurement in vivo showed the potential for point-of-care testing. The emission proportion of the DPA-CNPPV Pdots probe changed with the increasing NADH concentration, and the true-color photographs were taken by smartphone for further analysis. The blue/red ratio, with a linear response to NADH concentration, was considered as a key parameter for NADH sensing, which was calculated by dividing the true-color photographs into blue and red channels (Figure 5D,E). Applying this protocol to mice (Figure 5F) enabled ratiometric NADH sensing.

3.2. Disease-Related Metabolites

Metabolites assays play a key role in early disease diagnosis and management for the reason that its blood levels are closely related to the disease or injuries; for example, phenylalanine in phenylketonuria, glucose in diabetes, tyrosinase in tyrosinemia, glutamate during ischemic strokes, galactose in galactosemia and leucine in maple syrup urine disease. However, it cannot reliably detect most metabolites, hindering disease diagnosis and management [111]. To address this issue, many researchers have developed new methods for stoichiometric-based metabolite detection using Pdots biosensors.

3.2.1. Glucose

According to the International Diabetes Federation, an estimated 8.8% of adults aged 20–79 had diabetes in 2015, and the proportion is expected to rise to 10.4% by 2040 [112]. Diabetes is a metabolic disease with increased concentration of blood glucose. Long-term high blood glucose can cause great damage to the heart, kidneys and nervous system of diabetic patients. Blood glucose management is a vital part of diabetes treatment, including instant sampling detection and dynamic real-time monitoring.

Recently, functionalized Pdots covalently linked to glucose oxidase (GOx) have been used as biosensors for real-time monitoring due to their excellent properties and outstanding detection results [113]. In the presence of analytes, internal oxygen is depleted to translate the oxygen concentration into a fluorescent (or phosphorescent) signal. Based on this principle, Wu's group developed an ultrasensitive Pdots transducer that enables wireless glucose monitoring [51]. The transducer was mainly composed of phosphorescent dyes, GOx and Pdots functionalized by PSMA (Figure 6A). PDHF was selected as the light-harvesting host. In addition, the oxygen-sensitive Pd(II) meso-tetra (pentafluorophenyl) porphine (PdTFPP, D4) was doped in PDHF Pdots. The glucose biosensor (PD4Gx) was formed by binding GOx to carboxyl groups on the surface of PDHF. FRET between PDHF and D4 caused changes in emission light in the presence of glucose, since GOx-catalyzed oxidation of glucose causes changes in oxygen levels. Upon ultraviolet (UV) excitation, the blue fluorescence from PDHF Pdots was quenched, and the glucose biosensors showed red phosphorescence from D4. The emission spectrum of solutions with different glucose concentrations are shown in Figure 6B. With increasing glucose concentration, the ratio of blue and red light changed, causing the sensors to emit different colors. During the experiment, they implanted PD4Gx sensors subcutaneously in mice, took photographs of the light-emitting part with a smartphone and found that the luminescent images were obviously altered after glucose infusion. Glucose concentrations were acquired by comparing the calibration curve (Figure 6C) with the emission ratio obtained from true-color image processing. Data from the continuous monitoring of the P4DGx glucose sensors were in good agreement with the discrete measurements from commercial blood glucometers (Figure 6D), indicating that this ultrabright Pdots transducer enabled dynamic real-time wireless monitoring of blood glucose in living mice.

However, there are some drawbacks that can be improved in the above research. Hydrogen peroxide was produced during the process of glucose oxidation, resulting in photobleaching that degraded glucose sensor performance (Figure 6E), reducing enzymatic activity and generating cytotoxicity. Thus, Sun et al. developed their research approach to construct Pdots-based glucose sensors with an enzymatic cascade system (Pdots-GOx/CAT) by adding catalase to the above sensors to rapidly decompose hydrogen peroxide and improve the photostability and biocompatibility of glucose sensors [114]. Figure 6F,G shows the performance of this Pdots-GOx/CAT glucose transducer, which indicates an excellent long-term sensing ability for monitoring glucose. However, the reaction of peroxide decomposition, physical activity and pathological factors changed oxygen levels in tissues, thereby affecting the accuracy of glucose concentration. Therefore, Sun et al. also proposed an assumption to prepare a second oxygen-sensitive Pdots sensor not coupled with GOx, for the purpose of measuring the altered oxygen levels outside the glucose biosensor and further correcting the changes to improve the accuracy of the results [115].

The feasibility of this method had been demonstrated by numerical simulations and in vivo experiments. Other problems arising from implantable Pdots sensors were the aggregation and migration of nanoparticles in subcutaneous tissue, which was possibly caused by the direct implantation of free Pdots, and affected the detection of luminous intensity. Liu et al. proposed an injectable hydrogel implant to disperse Pdots evenly in it. The hydrogel remained at the implant site for one month without migration. The findings suggest that this method can be used for long-term glucose monitoring [116].

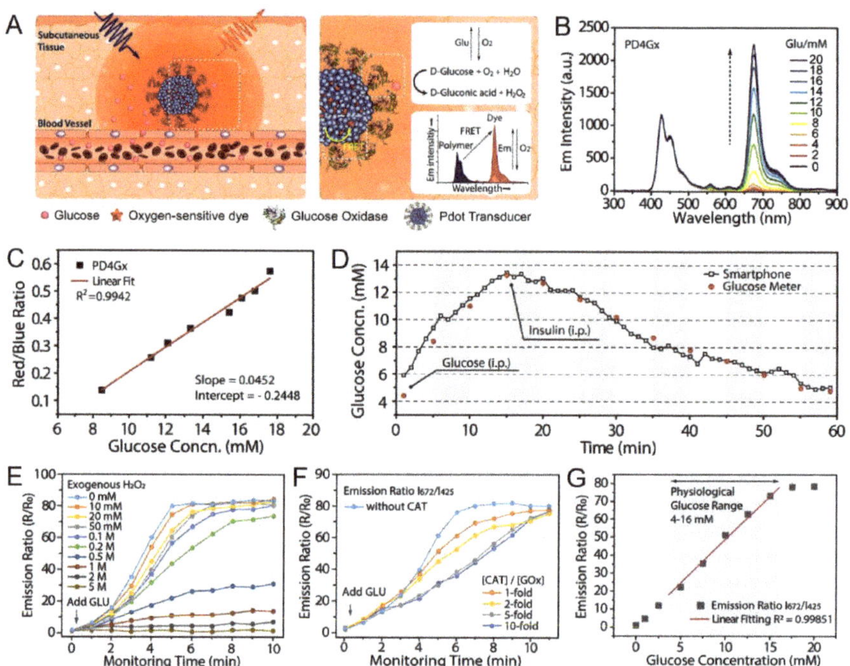

Figure 6. Pdots transducer used for wireless glucose monitoring. (**A**) The design and detection principle of PD4Gx. (**B**) Emission spectra of PD4Gx in a series of PBS solutions with different glucose concentrations. (**C**) Calibration curve by comparing red/blue ratio with standard glucose values. (**D**) Real-time detection for glucose concentration by PD4Gx transducer (black) and commercial glucose meter (red). Reproduced from Ref. [51] with permission. (**E**) Response curves of the Pdot–GOx transducer with different concentrations of exogenous hydrogen peroxide. (**F**) Response curves of the Pdot–GOx/CAT transducer with different CAT/GOx ratios. (**G**) Liner relationship between glucose concentration and emission ratio of Pdot-GOx/CAT with 5-fold CAT/GOx ratio. Reproduced from Ref. [114] with permission.

3.2.2. Phenylalanine

Phenylketonuria (PKU) is a metabolic genetic disorder that causes a defect in phenylalanine hydroxylase, preventing the conversion of phenylalanine to tyrosine and increasing blood phenylalanine levels [117]. This mechanism typically results in neurological damage in infants and children with PKU. PKU screening for newborns and management of blood phenylalanine levels in PKU patients have received much attention.

Chen et al. designed a metabolite biosensor consisting of NADH-sensitive Pdots and phenylalanine dehydrogenase (PheDH) [118]. On the basis of this biosensor, a paper-based point-of-care assay for blood phenylalanine levels was developed for PKU screening. Phenylalanine underwent oxidation catalyzed by PheDH to form phenylpyruvate and NADH. In the presence of NADH, red fluorescence was quenched and blue fluorescence was emitted. The emission intensities ratio at 458 nm and 627 nm showed an 18.9-fold change in concentration, from 0 to 2400 µM. In the process of the assay, plasma samples were added into test paper with lyophilized buffer containing modified Pdots, NAD$^+$ and PheDH. The blood phenylalanine concentrations were calculated from the emission intensity ratio of the blue and red channel. The performance of Pdots biosensors applied to a healthy person (60 µM) and a classic PKU patient (1200 µM) showed a significant difference in the ratio of blue and red channel intensity. Moreover, the error caused by endogenous NADH was corrected by analyzing samples both in the presence and absence of PheDH. The difference in concentrations obtained in the absence and presence of PheDH, respectively, was considered to be the exact blood phenylalanine concentration. The measurement result of plasma samples with phenylalanine were obtained by using a digital camera. Such a system can be promoted to the concentration measurement of any other metabolite oxidized by NAD$^+$ or reduced by NADH.

3.3. Tumor Markers

Immunochromatographic test strips (ICTS) have become an important tool for point-of-care testing (POCT) of tumor markers. Especially, ICTS based on multicolor Pdots with excellent fluorescent properties can be used for multiplex detection of target analytes. Chan's group developed a series of Pdots-based ICTS for the detection of multiple tumor markers [52,119,120]. For the simultaneous detection of prostate-specific antigen (PSA), α-fetoprotein (AFP) and carcinoembryonic antigen (CEA) in a single test strip, Fang et al. developed ICTS based on PF-TC6FQ/PFCN/PFO Pdots, which emit red, green and blue fluorescence, respectively [52]. The test strip consists of an absorbent pad, conjugate pad, sample pad and nitrocellulose membrane. Pdot–antibody conjugates, prepared by conjugation of Pdots functionalized with carboxyl groups and PSA/AFP/CEA antibody, respectively, were added into the conjugate pad of the test strip (Figure 7A). The control lines and test lines in the nitrocellulose membranes were modified with a caption antibody and bare lgG, respectively. The absorbent pad, conjugate pad and sample pad were then sequentially adhered to the nitrocellulose membrane to assemble the test strips. Samples dropped on the sample pad would move due to the capillary force. Regardless of the presence of target tumor markers, Pdot–antibody conjugates connected with the bare lgG in the control line to emit fluorescence to indicate the validity of test strip. In addition, in the presence of target tumor markers, Pdot–antibody conjugates were specifically captured by the corresponding caption antibody in the test line and also emitted the specific fluorescence, which gave the positive results. In the absence of target analytes, no emission can be observed in the test line, which gives the negative results. The multiplexed detection of PSA/AFP/CEA was obtained by direct observation with the naked eye under ultraviolet light, as shown in Figure 7B.

Quantitative detection should be calculated according to the fluorescence intensity ratio of the test line to the control line (T/C). With the increasing of tumor markers concentration, the fluorescence brightness of the test line enhanced, while that of the control line decreased. The reason for this phenomenon was that the number of Pdot–antibody conjugates was certain in a test strip; the more Pdot–antibody conjugates were bound to the test line in the presence of tumor markers, the less were bond to the control line, resulting in the comparison of fluorescence intensity of these two lines. According to the study results, the fluorescence ratio of T/C was linearly related to the log [PSA/AFP/CEA] concentration in the range of 3–15 ng/mL. The limit of detection was two orders of magnitude lower than that of conventional detection methods, which indicated that Pdot-based ICTS was beneficial for early diagnosis, rapid screening or regular monitoring of cancer.

Other types of test strips have been developed to improve the detection performance and expand the range of applications. You et al. used Pdot–Au hybrid nanocomposites, formed from Pdots and Au nanorods, with dual colorimetric and fluorometric readout abilities, for rapid screening (colorimetry) and accurate detection (fluorometry) of PSA [119]. The calibration curve of quantitative performance for PSA of this Au650@Pdot immunoassay platform is given in Figure 7C. A drop of whole blood could realize the detection of

PSA, since the plasma was captured by the sample pad, whereas conventional testing requires pretreatment of whole blood. Yang et al. reported another dual-modal Au@Pdot-based immunoassay for the detection of CEA and cytokeratin 19 fragments (CYFRA21-1) in the blood of non-small-cell lung cancer patients [120]. The test line was simultaneously modified by the two corresponding caption antibodies, causing four different types of luminous modals with two luminous lines, representing four detection results (Figure 7D,E). Similarly, this Au@Pdot-based ICTS had good linearity in the quantitative analysis for CYFRA21-1 and CEA detection.

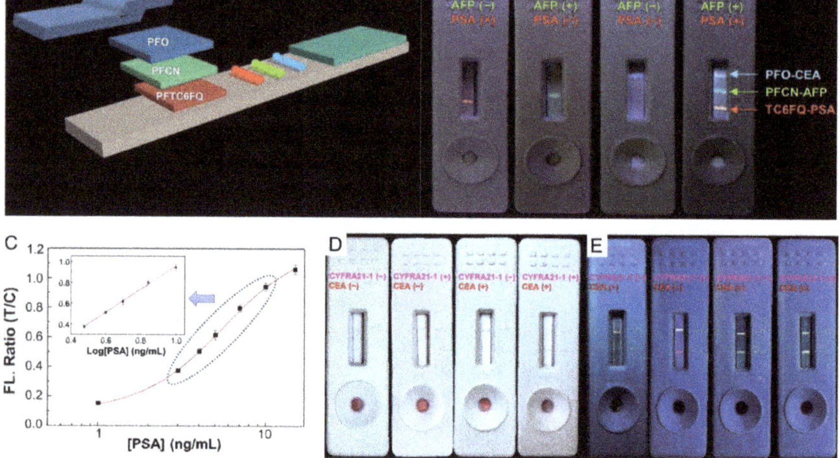

Figure 7. Multiplexed ICTS based on Pdots for detection of CEA/AFP/PSA. (**A**) The structure and principle of multiplexed ICTS for detection of multiplexed tumor markers. (**B**) Photographs of detection results with multiplexed tumor markers concentration (0/0/5, 0/5/0, 5/0/0, 5/5/5 ng/mL). Reproduced from Ref. [52] with permission. (**C**) The calibration curve of the Au650@Pdot immunoassay platform with 3–10 ng/mL PSA. Reproduced from Ref. [119] with permission. (**D,E**) Photographs of Au@Pdot-based test strips with samples containing different concentrations of CYFRA21-1 and CEA obtained under ambient light (**D**) and UV light (**E**). Reproduced from Ref. [120] with permission.

3.4. Cancer and Allograft Rejection

Polymer nanoparticles can be used for detecting biomarkers related to cancer and allograft rejection in early diagnosis. An advanced method for in vivo imaging and therapy utilizes the interconversion between nanoparticles and small molecules, since it has lots of advantages such as deeper penetration and broader biodistribution due to the small size of molecules, longer retention at the disease site of nanoparticles formed by the biomarker-activated conversion, fast clearance and specific sensitivity [121–123]. Pu's group recently reported the activatable polyfluorophore nanosensors (APNs) with biomarker-activated renal clearance and fluorescence response for bioimaging and urinalysis, which consisted of protease-reactive peptide brushes, cascaded self-immolative linker and caged fluorophore units [53]. Cathepsin B and granzyme B, corresponding with tumor status and lymphocyte activation in allograft rejection, were chosen to be the biomarkers of APNs. In the intrinsic state, APNs were non-fluorescent and accumulated at the disease site. Then, in the activated state, which was caused by the presence of disease-related biomarkers, the protease released the renal-clearable fluorophore fragments. These fragments were further cleared though the kidneys for fluorescent urine analysis. The high renal clearance efficacy and specificity testing of biomarkers made APNs-based urinalysis superior to other detection methods.

3.5. Exosomes

Exosomes, extracellular vesicles containing protein, DNA and RNA from the cells that secrete them, take part in cell communication and are involved in pathogenesis of several diseases, including cancer, neurodegeneration and infections [124]. Therefore, they have recently been used in new assays for disease-related biomarker identification and therapeutic response monitoring. Development of biosensors for exosome detection is significant for early cancer diagnosis. Lyu et al. reported the first near-infrared (NIR) afterglow nanosensor for multiplex differentiation of cancer-related exosomes, which consisted of a complex (ASPNC) formed by an NIR semiconducting polyelectrolyte with a quencher-tagged aptamer [54]. Poly(phenylenevinylene)-based (PPV-based) Pdots were used as the backbone for afterglow luminescence, while an NIR $^1O_2$ photosensitizer, tetraphenylporphyrin (TPP), was added into the PPV backbone for red-shift emission and afterglow signal amplification. The side chains on PPV, cationic quaternary ammonium groups, enabled the formation of ASNPC with the black hole quencher 2 (BHQ-2)-tagged aptamer. The fluorescence and afterglow signals of ASPNC were both quenched due to the electron transfer between PPV and BHQ-2. However, in the presence of exosomes, the specific binding between exosomes and the designed aptamer hampers the electron transfer and activates the fluorescence and afterglow signals. In their studies, a comparison of the limit of detection (LOD) and of afterglow signal with fluorescence signal indicated that afterglow detection enabled the minimization of background interference and achieved an LOD two orders of magnitude (~93-fold) lower than fluorescence detection. By orthogonally assaying of five kinds of exosomes at the expression levels of four biomarkers, they showed the recognition ability of this afterglow nanosensor and suggested that the different BHQ-2 targeted aptamers mediated the specific binding of ASPNC for potential orthogonal analysis of multiplex exosomes.

Additionally, Jiang et al. developed a method using exosomes labeled with Pdots for superresolution mapping with location error less than 5 nm and excellent optical adjustable duty cycles [125]. In their studies, the switch-on frequency of Pdots were tuned to obtain the structure of a large number of exosomes within a few minutes. A combination of two Pdots and one fluorophore that conjugated antibodies against three different tetraspanins on a seminal exosome were used for multicolor superresolution mapping to simultaneously achieve high throughput and high imaging quality of three tetraspanins. This method can also be applied to understand the structure of other similar biological vesicles, which promotes the application of vesicles in disease diagnosis.

4. Application of Pdots Optical Probes in Bioimaging

Optical imaging plays a critical role in disease diagnosis, which includes fluorescence imaging, PAI, afterglow imaging, chemiluminescent imaging, bioluminescence imaging, multiphoton imaging and harmonic imaging. Traditional materials for bioimaging have limitations such as low brightness, photobleaching and toxicity, while Pdots are considered as potential bioimaging materials due to their high brightness and good biocompatibility [126–128].

4.1. Pdots for Fluorescence Imaging

Fluorescence imaging enables tumor imaging, cell labeling and targeting, vascular structure imaging, etc. This imaging technology requires no ionizing radiation and enables real-time imaging with targetability and high spatial resolution to provide the exact location and silhouette of the targeted object. In particular, NIR fluorescence imaging allows in vivo fluorescence imaging with deeper penetration in biological tissue and less background fluorescence than visible-light fluorescence imaging. An ideal NIR fluorescent agent applied for targeting optical probes should work in the NIR window (NIR-I at 700–900 nm and NIR-II at 1000–1700 nm) and have great photophysical properties, such as brightness, quantum yield and photostability and sufficient strokes shift [125,126]. Indocyanine green (ICG) and methylene blue (MB) are commonly used as NIR fluorescent agents. However, ICG and

MB lack sufficient targetability against tumors and the ability of specific conjugation, with low quantum yield and poor photostability [129–132]. Thus, NIR fluorescent Pdots have developed as fluorescence probes in fluorescence imaging due to their good targetability and low background fluorescence and have been applied in lymph node localization [57], tumor imaging [58] and cancer cell tracking and imaging [133].

NIR fluorescent Pdots can be used for cell labeling to track cell migration non-invasively. Xiong et al. prepared NIR fluorescent Pdots for long-term tumor cell tracking in vivo [55]. They doped the NIR dye, silicon 2,3-naphthalocyaninebis(trihexylsilyloxide) (NIR775), into poly [2-methoxy-5-(2-ethylhexyloxy)-1,4-phenylenevinylene] (MEH-PPV) to prepare the fluorescent nanoprobes. FRET between MEH-PPV as a donor and NIR775 as an acceptor achieved NIR emission, with the absorption peak at 504 nm dominated by MEH-PPV and the emission peak at 776 nm dominated by NIR775. Additionally, the optimal weight ratio of NIR775 to MEH-PPV (0.012:1) decreased the self-quenching and provided the highest efficiency. In their study, 20 μg of NIR Pdots were used to treat 2×10^5 HeLa cells, and then cells were injected into nude mice. The fluorescence of NIR Pdots remained at 75% after 7 days and 28% after 23 days. According to their research results, NIR Pdots proved to have proper strokes shift for reducing errors caused by background fluorescence, weak cytotoxicity evaluated by CCK-8 assays, long-term labeling ability and photostability, which indicated the potential of NIR Pdots-based nanoprobes in the field of in vivo tumor imaging. Furthermore, Feng et al. developed an ultrasmall Pdots with excellent specificity and fast clearance for targeted tumor cell imaging, suppressing nonspecific cell uptake that limited the targetability and sensitivity of nanoprobes [134].

Based on the application of NIR fluorescent Pdots for labeling and tracking of targeted cells, in vivo tumor imaging technology has expanded and developed new methods by several improved strategies. Liu et al. developed a fluorination method of Pdots for high brightness in the NIR-II window, which promoted the application of NIR-II probes in brain tumor imaging [56]. They fluoridated PBTQ (m-PBTQ4F) to improve its photophysical properties by using semiconducting polymer synthesized with benzodithiophene (BDT) as a donor and triazole [4,5-g]-quinoxaline (TQ) as an acceptor and changing the quantity and location of the fluorine substituent on the TQ acceptor. The m-PBTQ4F showed excellent fluorescence intensity and photostability compared to ICG and IR26 NIR fluorophores (Figure 8A–C) due to the nanoscale fluorous effect. To assess the quality of m-PBTQ4F used in in vivo fluorescence imaging, they used m-PBTQ4F Pdots for tail-vein injection to show the whole-body vasculature structure of C57BL/6 mice in both prone and supine positions (Figure 8D). In addition, in the prone position, blood vessels in the back were observed clearly, which indicated that m-PBTQ4F Pdots could successfully display in vivo vasculature for tumor fluorescence imaging, since microvascular proliferation and pleomorphic vessels are one of the typical characterizations of malignant brain tumors [135]. The blood vessels were evenly distributed in the normal brain, whereas they were unevenly and chaotically distributed in brain tumors. Therefore, the brain tumors were revealed by vascular structure images obtained from NIR-II fluorescent imaging using m-PBTQ4F Pdots.

4.2. Pdots for Photoacoustic Imaging

PAI is a non-invasive biomedical imaging technique that involves the energy conversion from biological tissue. The photoacoustic (PA) contrast agents in the biological tissue absorb energy when they are irradiated by pulsed laser, and then generate ultrasound signals due to transient thermoelastic expansion, which is also referred to PA signals. PA images with high selectivity and penetration depth can be reconstructed by detecting the PA signals containing information about light absorption characteristics. The intensity of PA signal depends on the internal competition of the PA contrast agents between fluorescence emission and non-radioactive heat dissipation. Pdots have been used as PA contrast agents for PAI since Pdots have high photothermal conversion efficiencies and PA intensity [136–140].

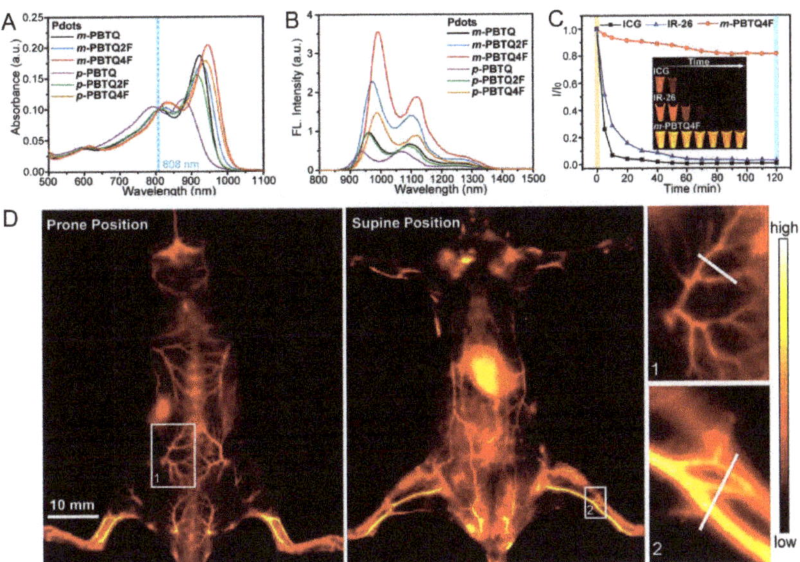

Figure 8. The performance of m-PBTQ4F Pdots for vasculature imaging. (**A**) Absorption spectra of the various Pdots. (**B**) Emission spectra of the various Pdots. (**C**) The photostability under excitation at 808 nm of ICG, IR-26 and m-PBTQ4F in corresponding solvent. (**D**) The NIR-II fluorescent images of C57BL/6 mice injected with 100 mL of m-PBTQ4F into the tail vein in prone and supine positions. Reproduced from Ref. [56] with permission.

4.2.1. Amplification of PA Signal from Pdots

To reduce the toxicity, many design strategies have been developed to amplify the PA signal, which aids to decrease the dosage of PA contrast agents formed by Pdots [141,142]. Adapting the structure of Pdots is one of the design strategies to amplify the PA signal. Guo et al. designed a series of Pdots by using different electron acceptors and planar electron donors, which demonstrated the high photothermal conversion efficiencies and signal-to-background ratio (SBR) of 47 in PAI for tumors in vivo at a depth of 3.2 mm [143]. Dong et al. also prepared PTIGSVS nanoparticles with a high photothermal conversion efficiency of 74% that proved to be a superior PA contrast agent. Another way to amplify the PA signal is to enhance the non-radioactive heat dissipation by strengthening the fluorescence quenching [144]. Lyu et al. developed an intraparticle molecular orbital engineering approach to induce electron transfer with light irradiation, causing enhanced heat production, which increased the PA signal intensity of Pdots by 2.6-fold and maximum photothermal temperature by 1.3-fold [145]. For some Pdots with original faint fluorescence, accelerating the heat dissipation can also enhance the PA signal intensity. Zhen et al. reported that Pdots with a silica layer on the surface simulatively improved the fluorescence and PA brightness due to the higher interfacial thermal conductance between the silica layer and water [146]. Duan et al. also developed CP-IO nanocomposites in which the additional heat production and faster heat dissipation caused by IO nanoparticles resulted in amplification of the PA signal. The above research progress indicates that Pdots for PA contrast agents have potential development space and application prospects for PAI [147].

4.2.2. Pdots for NIR-II PAI

NIR-II PAI usually has higher SBR and penetration depth compared to NIR-I imaging due to reduced light attenuation and lower absorption by biological tissues in the NIR-II window. Recently, Numerous research studies have demonstrated that the Pdots-based NIR-II PA contrast agents have increased SBR and penetration depth, as well as a higher maximum permissible exposure and image resolution [148–151]. In addition, metabolizable ability related to biotoxicity is another key property of NIR-II PA contrast agents for in vivo PAI, which is an issue to be researched.

Jiang et al. designed a group of metabolizable NIR-II Pdots contrast agents that were easily degraded by relevant enzymes in phagocytes to ultrasmall metabolites, which were cleared out by hepatobiliary and renal excretions after PAI [59]. Three different Pdots (SPN-OT, SPN-PT and SPN-DT) were obtained from three corresponding semiconducting polymers encapsulated into poly(ethyleneglycol)-methyl ether-block-poly(lactide-co-glycolide) (PLGA–PEG) by nanoprecipitation. The obtained Pdots had peak absorption at about 1079 nm and an average hydrodynamic diameter of about 30 nm. The PA spectrum of obtained Pdots ranged from 680 nm to 1064 nm, while the PA amplitude of blood proved to significantly decrease at 1064 nm, indicating these Pdots had increased SBR in PAI due to the background noise from blood, which decreased at 1064 nm in the NIR-II window. SPN-PT was detected for the metabolism and clearance processes as a representative Pdot. Myeloperoxidase (MPO) in phagocytes efficiently degraded semiconducting particles, and lipase in phagocytes catalyzed the hydrolysis of ester linkages of PLGA-PEG. After combined treatment of MPO and lipase, the hydrodynamic diameter of SPN-PT reduced from 30 nm to about 1 nm. The degradation product was fluorescent and then expelled from phagocytes, as shown in Figure 9A. The NIR-II fluorescence signal of degraded products of SPN-PT in the cytoplasm of cells was detected using a confocal laser scanning microscopy (Figure 9B), and the fluorescence was obviously observed after 48 h of incubation, which indicated the degradation in phagocytes was efficient and successful. They then delivered SPN-PT into living mice to study the clearance pathways inside the organism. The fluorescence signal of blood reached a maximum on the first day, then, that of urine dominated by renal clearance also reached a maximum on the second day, as shown in Figure 9C. Subsequently, the signal from the liver reached the top on the fourth day (Figure 9D), and, finally, feces on the fifth day. Such a sequence demonstrated that SPN-PT was successfully degraded in living mice and cleared out though the renal pathway in the beginning and hepatobiliary pathway later. To investigate the NIR-II PA ability of SPN-PT, they used a superficial tumor modal and a deep transcranial brain vasculature modal via vein injection of SPN-PT. The PA amplitude of tumor region and brain vasculature both increased (Figure 9E), making it easier to observe the structure of superficial tumor and brain blood vessels. All the results indicated that these Pdots were efficient NIR-II contrast agents for PAI in vivo and had great biosafety ensured by renal and hepatobiliary clearance.

Men et al. also reported metabolizable highly absorbing NIR-II Pdots for PAI-guided photothermal therapy (PTT), with ultrasmall size at 4 nm, good biocompatibility and photostability, high photothermal conversion efficiency and tumor ablation capability [60]. These metabolizable and ultrasmall Pdots were prepared by a conjugated polymer, Poly([2,5-bis(2-decyltetradecyl)-2,5-dihydropyrrolo [3,4-c]pyrrole-1,4-dione-3,6-dithienyl]-co-[6-(2-ethylhexyl)-[1,2,5]thiadiazolo [3,4-f]benzotriazole-4,8-diyl]) (DPP-BTzTD), and PSMA though the nanoprecipitation method. The particle size revealed by DLS of the obtained DPP-BTzTD S-Pdots was 4 nm, and L-Pdots were 25 nm. The NIR spectra of obtained S-Pdots and L-Pdots showed high absorption in the NIR-II region.

DPP-BTzTD S-Pdots exhibited good photothermal properties. The temperature variation of S-Pdots at different power densities integrated the laser-power-dependency of the photothermal effect. In addition, the rapid temperature increases of S-Pdots at different concentrations also showed the concentration dependency of the photothermal heating effect. The photothermal stability of S-Pdots was assessed by five repeated laser on/off cycles with NIR-II laser irradiation, while the highest temperature was detected within the same range. The photothermal conversion efficiency of S-Pdots was calculated to be 53.1%. Furthermore, the PAI capability of S-Pdots was demonstrated by in vitro phantom tests. The PA signal intensities of S-Pdots were detected at different concentrations from 50 to 200 μg mL^{-1} to show the linear relationship between PA signal intensity and concentration of S-Pdots. The in vivo PAI performance of S-Pdots was inspected by monitoring the PA intensities at the tumor site at designed time points of a 4T1 tumor-bearing nude mice modal. The PA intensity peaked 2 h post injection and stabilized at 24 h post injection, which also revealed the increased metabolizable ability of the S-Pdots. DPP-BTzTD S-Pdots showed the potential utility for NIR-II contract agents with ultrasmall particle size, high photothermal conversion efficiency, rapid excretion and strong PA signal.

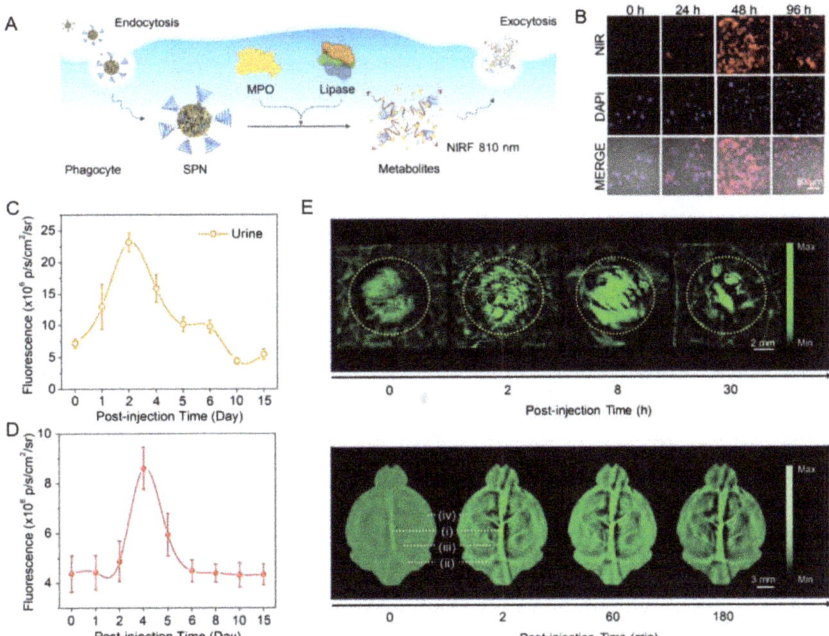

Figure 9. Metabolizable Pdots NIR-II contrast agents for in vivo PAI. (**A**) The degradation and metabolism process of Pdots in phagocytes. (**B**) Confocal images of cells incubated by SPN-PT at different time points. (**C**) Variation of fluorescence intensity with time in the urine of living mice after vein injection of SPN-PT. (**D**) Variation of fluorescence intensity with time in the liver region. (**E**) PA images of a superficial tumor and brain vasculature of injected mice at special time points under irradiation at 1064 nm. Reproduced from Ref. [59] with permission.

4.3. Pdots for Afterglow Imaging

In addition to the imaging methods mentioned above, other luminescence imaging including afterglow imaging [152], chemiluminescent imaging [153] and bioluminescence imaging [154] provides different imaging characteristics and advantages. Such methods realize imaging decrease background interference caused by irradiation light in fluorescence imaging, causing higher SBR of images. Pu and co-workers have developed a series of polymer nanoparticles for afterglow imaging. Afterglow materials emit light long after ceasing to provide excitation light. Thus, the fluorescence from afterglow materials and biological tissues can be separated to obtain better images. Miao et al. presented Pdots with a diameter less than 40 nm, emitting luminescence at 780 nm with a half-life of about 6 min. Such afterglow Pdots were demonstrated to be more than 100-fold brighter than inorganic afterglow materials in afterglow intensity, and 127-fold higher than NIR fluorescence imaging in SBR of living mice tumor imaging [61]. Xie et al. reported self-assemble poly(p-phenylenevinylene) derivatives for metastatic tumor afterglow imaging in living mice, which detected the xenograft tumors with volumes of 1 mm³ and tiny peritoneal metastatic tumors [62]. These afterglow nanoparticles also showed the potential utility for oxygen partial pressure imaging due to the oxygen-sensitive afterglow property.

4.4. Pdots for Chemiluminescent Imaging

Electrochemiluminescent and chemiluminescent Pdots for imaging generate light without light irradiation, which significantly reduces the background interference and photodamage [155]. Zhen et al. developed Pdots doped with naphthalocyanine dye for detection of hydrogen peroxide with intraparticle chemiluminescence resonance energy transfer, realizing the ultrasensitive imaging for hydrogen peroxide in several mouse modals [63]. Moreover, Cui et al. firstly reported Pdots activated by superoxide anion to generate the chemiluminescent signal for in vivo imaging in cancer immunotherapy [64]. Pdots accumulated into tumors and produced the chemiluminescent signal corresponding with the concentration of superoxide anion.

4.5. Pdots for Multimodal Imaging

Pdots with dual imaging capabilities have been developed since the multimodal imaging attracts much interest in biological imaging. In relevant studies, the designed Pdots served as fluorescent probes and gave other forms of imaging signals, such as signals in PAI, magnetic resonance imaging (MRI), computed tomography (CT) imaging and singlephoton emission computed tomography (SPECT) imaging to offer the location or physiological information of the detection area for better diagnosis and treatment effects [156–160]. Lyu et al. reported reaction-based Pdots (rSPNs) with a sulfenic acid reactive group (1,3-cyclohexanedione) on the surface [65]. Figure 10A,B shows the PA maximum intensity projection images and fluorescence images of tumors in living mice injected with rSPNs via the tail vein. rSPNs with intense NIR absorption and fluorescence made it possible to achieve fluorescence imaging and real-time PAI for protein sulfenic acids in tumors.

Multimodal Pdots for fluorescence and MRI have also been developed. Hashim et al. reported bifunctional Gd-Pdots prepared by semiconducting polymers, amphiphilic phospholipids and gadolinium-containing lipids, and measured their fluorescence quantum yield, extinction coefficient and MRI T_1-weighted relaxation times in water [66]. Figure 10C shows the photograph, IVIS image and IVIS processed image of living mice injected with Gd-Pdots, the red fluorescence from Gd-Pdots and green fluorescence from mice were distinguished from each other. The linear correlation between relaxation rate values (R_1) and gadolinium concentration of Gd-Pdots at 3T and 7T was calculated and plotted in Figure 10D. The fluorescence and magnetic resonance properties of Pdots indicated their potential utility for dual modal fluorescence imaging and MRI.

Additionally, Sun et al. reported small Pdot nanocomposites containing gold nanoparticles to demonstrate the practicality of Pdot-Au nanoparticles for dual-modal imaging involving fluorescence from Pdots and scattering from Au [67]. They used Au passivated by hydrophobic molecules and semiconducting polymers in tetrahydrofuran to prepare Au-NP-Pdots by nanoprecipitation. Au-NP-Pdots inside mammalian cells were imaged by a fluorescence microscope equipped with dark field optics (Figure 10E). Au nanoparticles were useful in long-term tracking and imaging in dark field mode since there was no photobleach. However, the nanoscale cellular organelles strongly scattering light made it difficult to distinguish from Au nanoparticles. In the presence of Pdots, the bright spots in fluorescence mode had corresponding bright spots in dark field mode so that the Au-NP-Pdots that generated both the fluorescence and scattering signals were differentiated from other cellular features.

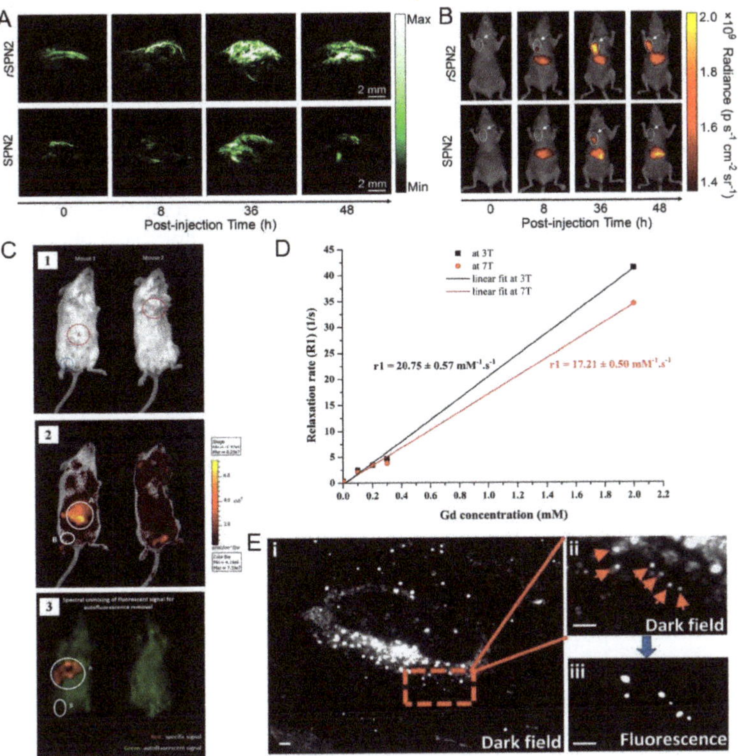

Figure 10. Pdots for multimodal imaging. (**A**) PA images and fluorescence images (**B**) of tumor in living mice injected with rSPNs though the tail vein. Reproduced from Ref. [65] with permission. (**C**) The photograph (1), IVIS image (2) and IVIS processed image (3) of living mice after injection of Gd-SPNs. (**D**) The linear correlation between relaxation rate values (R1) and gadolinium concentration of Gd-SPNs. Reproduced from Ref. [66] with permission. (**E**) Dark field and fluorescence images of Au-NP-Pdots inside mammalian cells. Reproduced from Ref. [67] with permission.

5. Conclusions and Outlooks

Pdots have become an important material for point-of-care testing and in vivo imaging, which greatly assist the diagnosis and treatment of diseases. In this review, several preparation methods are described to demonstrate ways to obtain Pdots with different ranges of particle diameter. In addition, Pdots have excellent photophysical properties and performance, including large absorption cross section, high quantum yield, good photostability and low toxicity. Moreover, recent studies have reported different approaches to further improve these properties. In addition to the size and performance, Pdots can be adapted to special application aims and environments though several modification and functionalization methods. All these methods mentioned in this review indicate that the size, properties and function of Pdots can be tuned for practical applications, suggesting greater utility compared to the traditional materials. For biosensing application, Pdots biosensors based on FRET or electron transfer have been used for the detection of glucose, phenylalanine, NAD and tumor markers, all of which have shown good assay results with convenient, accurate and rapid characteristics. For in vivo bioimaging applications, Pdots can be used as optical probes for fluorescence imaging or as contrast agents for NIR-II PAI, which not only have deeper penetration depth, higher brightness, better resolution and higher contrast efficiency, but also exhibit rapid metabolic capacity relevant for biosafety. Such a large range of applications and outstanding results have proven the significance of Pdots for point-of-care diagnostics and in vivo bioimaging.

Considering the highly homogeneous product landscape, Pdots-based diagnostic techniques have great potential in the sensing and imaging markets. However, Pdots for disease diagnosis and treatment are mainly in the research stage, which still have some distance from clinical application. In this period of rapid development in the materials sector, relevant research on Pdots needs to constantly combine new techniques and explore more applications. In our opinion, future work in this field mainly includes the following aspects: (1) To develop a large-scale, eco-friendly and low-cost method for the preparation of Pdots. (2) Optimizing biofunctionalization strategies to obtain smart Pdots-based probes with stimuli-responsive targeting. (3) Pdots with ultra-small particles (<10 nm) and uniform particle size that are particularly suitable for the development of optical transducers. Furthermore, for in vivo applications, the ultra-small-size Pdots can further enhance biodistribution and rapid metabolism. The biological metabolism of nanoparticles is currently the biggest problem limiting their clinical application. (4) The exploration of NIR-II Pdots can improve the penetration depth of light in biological tissues. Although NIR-II imaging and therapy have received considerable attention, biosensing applications of the NIR-II window are largely unexplored. (5) Emerging hybrid nanocomposites composed of organic and inorganic nanomaterials are expected to endow the nanosystems of Pdots with complementary multimodal imaging modalities and synergistic therapeutics. (6) Computational simulations can also help us design functionalized Pdots for specific biomedical applications. With the rapid development of artificial intelligence, it will be more effective to design rational Pdots for clinical translation.

Author Contributions: Conceptualization, S.D., Z.H. and H.C.; writing—original draft preparation, S.D.; writing—review and editing, L.L., J.Z., Y.W., Z.H. and H.C.; visualization, S.D., Z.H. and H.C.; supervision, Z.H. and H.C.; funding acquisition, H.C. All authors have read and agreed to the published version of the manuscript.

Funding: We thank the National Natural Science Foundation of China (Grant No. 82202296) and Central South University for the financial support.

Institutional Review Board Statement: Not applicable.

Informed Consent Statement: Not applicable.

Data Availability Statement: Not applicable.

Conflicts of Interest: The authors declare no conflict of interest.

References

1. Bhatia, S.N.; Chen, X.; Dobrovolskaia, M.A.; Lammers, T. Cancer nanomedicine. *Nat. Rev. Cancer* **2022**, *22*, 550–556. [CrossRef] [PubMed]
2. De Lázaro, I.; Mooney, D.J. Obstacles and opportunities in a forward vision for cancer nanomedicine. *Nat. Mater.* **2021**, *20*, 1469–1479. [CrossRef] [PubMed]
3. Chen, G.; Roy, I.; Yang, C.; Prasad, P.N. Nanochemistry and nanomedicine for nanoparticle-based diagnostics and therapy. *Chem. Rev.* **2016**, *116*, 2826–2885. [CrossRef] [PubMed]
4. Barreto, J.A.; O'Malley, W.; Kubeil, M.; Graham, B.; Stephan, H.; Spiccia, L. Nanomaterials: Applications in cancer imaging and therapy. *Adv. Mater.* **2011**, *23*, H18–H40. [CrossRef] [PubMed]
5. Giljohann, D.A.; Mirkin, C.A. Drivers of biodiagnostic development. *Nature* **2009**, *462*, 461–464. [CrossRef]
6. Chen, Y.T.; Lee, Y.C.; Lai, Y.H.; Lim, J.C.; Huang, N.T.; Lin, C.T.; Huang, J.J. Review of integrated optical biosensors for point-of-care applications. *Biosensors* **2020**, *10*, 209. [CrossRef]
7. Holzinger, M.; Le Goff, A.; Cosnier, S. Nanomaterials for biosensing applications: A review. *Front. Chem.* **2014**, *2*, 63. [CrossRef]
8. Cavalcante, F.T.; Falcão, I.R.d.A.; da Souza, J.E.S.; Rocha, T.G.; de Sousa, I.G.; Cavalcante, A.L.; de Oliveira, A.L.; de Sousa, M.C.; dos Santos, J.C. Designing of nanomaterials-based enzymatic biosensors: Synthesis, properties, and applications. *Electrochem* **2021**, *2*, 149–184. [CrossRef]
9. Duan, Y.; Liu, B. Recent advances of optical imaging in the second near-infrared window. *Adv. Mater.* **2018**, *30*, 1802394.
10. Haupt, K.; Medina Rangel, P.X.; Bui, B.T.S. Molecularly imprinted polymers: Antibody mimics for bioimaging and therapy. *Chem. Rev.* **2020**, *120*, 9554–9582. [CrossRef]
11. Zhang, Y.; Zhang, G.; Zeng, Z.; Pu, K. Activatable molecular probes for fluorescence-guided surgery, endoscopy and tissue biopsy. *Chem. Soc. Rev.* **2022**, *51*, 566–593. [CrossRef]

12. Zhao, L.; Zhao, C.; Zhou, J.; Ji, H.; Qin, Y.; Li, G.; Wu, L.; Zhou, X. Conjugated polymers-based luminescent probes for ratiometric detection of biomolecules. *J. Mater. Chem. B* **2022**, *10*, 7309–7327. [CrossRef]
13. Qiu, X.; Xu, J.; Cardoso dos Santos, M.; Hildebrandt, N. Multiplexed biosensing and bioimaging using lanthanide-based time-gated forster resonance energy transfer. *Acc. Chem. Res.* **2022**, *55*, 551–564. [CrossRef]
14. Resch-Genger, U.; Grabolle, M.; Cavaliere-Jaricot, S.; Nitschke, R.; Nann, T. Quantum dots versus organic dyes as fluorescent labels. *Nat. Methods* **2008**, *5*, 763–775. [CrossRef]
15. Riahin, C.; Meares, A.; Esemoto, N.N.; Ptaszek, M.; LaScola, M.; Pandala, N.; Lavik, E.; Yang, M.; Stacey, G.; Hu, D.; et al. Hydroporphyrin-doped near-infrared-emitting polymer dots for cellular fluorescence imaging. *ACS Appl. Mater. Interfaces* **2022**, *14*, 20790–20801. [CrossRef]
16. Zhang, F.; Tang, B.Z. Near-infrared luminescent probes for bioimaging and biosensing. *Chem. Sci.* **2021**, *12*, 3377–3378. [CrossRef]
17. Wang, X.; Zhong, X.; Li, J.; Liu, Z.; Cheng, L. Inorganic nanomaterials with rapid clearance for biomedical applications. *Chem. Soc. Rev.* **2021**, *50*, 8669–8742. [CrossRef]
18. Pei, P.; Chen, Y.; Sun, C.; Fan, Y.; Yang, Y.; Liu, X.; Lu, L.; Zhao, M.; Zhang, H.; Zhao, D.; et al. X-ray-activated persistent luminescence nanomaterials for NIR-II imaging. *Nat. Nanotechnol.* **2021**, *16*, 1011–1018. [CrossRef]
19. Zhou, J.; Chizhik, A.I.; Chu, S.; Jin, D. Single-particle spectroscopy for functional nanomaterials. *Nature* **2020**, *579*, 41–50. [CrossRef]
20. Chung, Y.H.; Cai, H.; Steinmetz, N.F. Viral nanoparticles for drug delivery, imaging, immunotherapy, and theranostic applications. *Adv. Drug Deliv. Rev.* **2020**, *156*, 214–235. [CrossRef]
21. Wu, Y.; Ali, M.R.; Chen, K.; Fang, N.; El-Sayed, M.A. Gold nanoparticles in biological optical imaging. *Nano Today* **2019**, *24*, 120–140. [CrossRef]
22. Meng, Z.; Hou, W.; Zhou, H.; Zhou, L.; Chen, H.; Wu, C. Therapeutic considerations and conjugated polymer-based photosensitizers for photodynamic therapy. *Macromol. Rapid Commun.* **2018**, *39*, 1700614. [CrossRef] [PubMed]
23. García de Arquer, F.P.; Talapin, D.V.; Klimov, V.I.; Arakawa, Y.; Bayer, M.; Sargent, E.H. Semiconductor quantum dots: Technological progress and future challenges. *Science* **2021**, *373*, eaaz8541. [CrossRef] [PubMed]
24. Liu, J.; Li, R.; Yang, B. Carbon dots: A new type of carbon-based nanomaterial with wide applications. *ACS Cent. Sci.* **2020**, *6*, 2179–2195. [CrossRef] [PubMed]
25. Chen, B.; Wang, F. Emerging frontiers of upconversion nanoparticles. *Trends Chem.* **2020**, *2*, 427–439. [CrossRef]
26. Peng, Q.; Shuai, Z. Molecular mechanism of aggregation-induced emission. *Aggregate* **2021**, *2*, e91. [CrossRef]
27. Mayder, D.M.; Tonge, C.M.; Nguyen, G.D.; Tran, M.V.; Tom, G.; Darwish, G.H.; Gupta, R.; Lix, K.; Kamal, S.; Algar, W.R.; et al. Polymer dots with enhanced photostability, quantum yield, and two-photon cross-section using structurally constrained deep-blue fluorophores. *J. Am. Chem. Soc.* **2021**, *143*, 16976–16992. [CrossRef]
28. Jiang, Y.; Hu, Q.; Chen, H.; Zhang, J.; Chiu, D.T.; McNeill, J. Dual-mode superresolution imaging using charge transfer dynamics in semiconducting polymer dots. *Angew. Chem. Int. Ed.* **2020**, *59*, 16173–16180. [CrossRef]
29. Chen, X.; Li, R.; Liu, Z.; Sun, K.; Sun, Z.; Chen, D.; Xu, G.; Xi, P.; Wu, C.; Sun, Y.; et al. Small photoblinking semiconductor polymer dots for fluorescence nanoscopy. *Adv. Mater.* **2017**, *29*, 1604850. [CrossRef]
30. Wang, Y.; Feng, L.; Wang, S. Conjugated polymer nanoparticles for imaging, cell activity regulation, and therapy. *Adv. Funct. Mater.* **2019**, *29*, 1806818. [CrossRef]
31. Peng, H.S.; Chiu, D.T. Soft fluorescent nanomaterials for biological and biomedical imaging. *Chem. Soc. Rev.* **2015**, *44*, 4699–4722. [CrossRef]
32. Solhi, E.; Hasanzadeh, M. Recent advances on the biosensing and bioimaging based on polymer dots as advanced nanomaterial: Analytical approaches. *TrAC Trends Anal. Chem.* **2019**, *118*, 840–852. [CrossRef]
33. Sun, J.; Mei, H.; Gao, F. Ratiometric detection of copper ions and alkaline phosphatase activity based on semiconducting polymer dots assembled with rhodamine B hydrazide. *Biosens. Bioelectron.* **2017**, *91*, 70–75. [CrossRef]
34. Wu, C.; Chiu, D.T. Highly fluorescent semiconducting polymer dots for biology and medicine. *Angew. Chem. Int. Ed.* **2013**, *52*, 3086–3109. [CrossRef]
35. Massey, M.; Wu, M.; Conroy, E.M.; Algar, W.R. Mind your P's and Q's: The coming of age of semiconducting polymer dots and semiconductor quantum dots in biological applications. *Curr. Opin. Biotechnol.* **2015**, *34*, 30–40. [CrossRef]
36. Chabok, A.; Shamsipur, M.; Yeganeh-Faal, A.; Molaabasi, F.; Molaei, K.; Sarparast, M. A highly selective semiconducting polymer dots-based "off–on" fluorescent nanoprobe for iron, copper and histidine detection and imaging in living cells. *Talanta* **2019**, *194*, 752–762. [CrossRef]
37. Shi, X.M.; Mei, L.P.; Wang, Q.; Zhao, W.W.; Xu, J.J.; Chen, H.Y. Energy transfer between semiconducting polymer dots and gold nanoparticles in a photoelectrochemical system: A case application for cathodic bioanalysis. *Anal. Chem.* **2018**, *90*, 4277–4281. [CrossRef]
38. Kuo, S.Y.; Li, H.H.; Wu, P.J.; Chen, C.P.; Huang, Y.C.; Chan, Y.H. Dual colorimetric and fluorescent sensor based on semiconducting polymer dots for ratiometric detection of lead ions in living cells. *Anal. Chem.* **2015**, *87*, 4765–4771. [CrossRef]
39. Cai, L.; Deng, L.; Huang, X.; Ren, J. Catalytic chemiluminescence polymer dots for ultrasensitive in vivo imaging of intrinsic reactive oxygen species in mice. *Anal. Chem.* **2018**, *90*, 6929–6935. [CrossRef]
40. Bao, B.; Ma, M.; Zai, H.; Zhang, L.; Fu, N.; Huang, W.; Wang, L. Conjugated polymer nanoparticles for label-free and bioconjugate-Recognized DNA sensing in serum. *Adv. Sci.* **2015**, *2*, 1400009. [CrossRef]

41. Wang, N.; Chen, L.; Chen, W.; Ju, H. Potential-and color-resolved electrochemiluminescence of polymer dots for array imaging of multiplex microRNAs. *Anal. Chem.* **2021**, *93*, 5327–5333. [CrossRef] [PubMed]
42. Li, Q.; Wang, Y.; Yu, G.; Liu, Y.; Tang, K.; Ding, C.; Chen, H.; Yu, S. Fluorescent polymer dots and graphene oxide based nanocomplexes for "off-on" detection of metalloproteinase-9. *Nanoscale* **2019**, *11*, 20903–20909. [CrossRef] [PubMed]
43. Yuan, Y.; Hou, W.; Sun, Z.; Liu, J.; Ma, N.; Li, X.; Yin, S.; Qin, W.; Wu, C. Measuring cellular uptake of polymer dots for quantitative imaging and photodynamic therapy. *Anal. Chem.* **2021**, *93*, 7071–7078. [CrossRef] [PubMed]
44. Yin, C.; Tai, X.; Li, X.; Tan, J.; Lee, C.S.; Sun, P.; Fan, Q.; Huang, W. Side chain engineering of semiconducting polymers for improved NIR-II fluorescence imaging and photothermal therapy. *Chem. Eng. J.* **2022**, *428*, 132098. [CrossRef]
45. Wang, Y.; Xiong, X.; Zhu, Y.; Song, X.; Li, Q.; Zhang, S. A pH-responsive nanoplatform based on fluorescent conjugated polymer dots for imaging-guided multitherapeutics delivery and combination cancer therapy. *ACS Biomater. Sci. Eng.* **2021**, *8*, 161–169. [CrossRef]
46. Men, X.; Chen, H.; Sun, C.; Liu, Y.; Wang, R.; Zhang, X.; Wu, C.; Yuan, Z. Thermosensitive polymer dot nanocomposites for trimodal computed tomography/photoacoustic/fluorescence imaging-guided synergistic chemo-photothermal therapy. *ACS Appl. Mater. Interfaces* **2020**, *12*, 51174–51184. [CrossRef]
47. Bai, X.; Wang, K.; Chen, L.; Zhou, J.; Wang, J. Semiconducting polymer dots as fluorescent probes for in vitro biosensing. *J. Mater. Chem. B* **2022**, *10*, 6248–6262. [CrossRef]
48. Yuan, Y.; Hou, W.; Qin, W.; Wu, C. Recent advances in semiconducting polymer dots as optical probes for biosensing. *Biomater. Sci.* **2021**, *9*, 328–346. [CrossRef]
49. Li, J.; Pu, K. Semiconducting polymer nanomaterials as near-infrared photoactivatable protherapeutics for cancer. *Acc. Chem. Res.* **2020**, *53*, 752–762. [CrossRef]
50. Chen, H.; Yu, J.; Men, X.; Zhang, J.; Ding, Z.; Jiang, Y.; Wu, C.; Chiu, D.T. Reversible ratiometric NADH sensing using semiconducting polymer dots. *Angew. Chem. Int. Ed.* **2021**, *60*, 12007–12012. [CrossRef]
51. Sun, K.; Yang, Y.; Zhou, H.; Yin, S.; Qin, W.; Yu, J.; Chiu, D.T.; Yuan, Z.; Zhang, X.; Wu, C. Ultrabright polymer-dot transducer enabled wireless glucose monitoring via a smartphone. *ACS Nano* **2018**, *12*, 5176–5184. [CrossRef]
52. Fang, C.C.; Chou, C.C.; Yang, Y.Q.; Wei-Kai, T.; Wang, Y.T.; Chan, Y.H. Multiplexed detection of tumor markers with multicolor polymer dot-based immunochromatography test strip. *Anal. Chem.* **2018**, *90*, 2134–2140. [CrossRef]
53. Huang, J.; Chen, X.; Jiang, Y.; Zhang, C.; He, S.; Wang, H.; Pu, K. Renal clearable polyfluorophore nanosensors for early diagnosis of cancer and allograft rejection. *Nat. Mater.* **2022**, *21*, 598–607. [CrossRef]
54. Lyu, Y.; Cui, D.; Huang, J.; Fan, W.; Miao, Y.; Pu, K. Near-nfrared afterglow semiconducting nano-polycomplexes for the multiplex differentiation of cancer exosomes. *Angew. Chem. Int. Ed.* **2019**, *58*, 4983–4987. [CrossRef]
55. Xiong, L.; Guo, Y.; Zhang, Y.; Cao, F. Highly luminescent and photostable near-infrared fluorescent polymer dots for long-term tumor cell tracking in vivo. *J. Mater. Chem. B* **2016**, *4*, 202–206. [CrossRef]
56. Liu, Y.; Liu, J.; Chen, D.; Wang, X.; Zhang, Z.; Yang, Y.; Jiang, L.; Qi, W.; Ye, Z.; He, S.; et al. Fluorination enhances NIR-II fluorescence of polymer dots for quantitative brain tumor imaging. *Angew. Chem. Int. Ed.* **2020**, *59*, 21049–21057. [CrossRef]
57. Xiong, L.; Shuhendler, A.J.; Rao, J. Self-luminescing BRET-FRET near-infrared dots for in vivo lymph-node mapping and tumour imaging. *Nat. Commun.* **2012**, *3*, 1193. [CrossRef]
58. Men, X.; Geng, X.; Zhang, Z.; Chen, H.; Du, M.; Chen, Z.; Liu, G.; Wu, C.; Yuan, Z. Biomimetic semiconducting polymer dots for highly specific NIR-II fluorescence imaging of glioma. *Mater. Today Bio* **2022**, *16*, 100383. [CrossRef]
59. Jiang, Y.; Upputuri, P.K.; Xie, C.; Zeng, Z.; Sharma, A.; Zhen, X.; Li, J.; Huang, J.; Pramanik, M.; Pu, K.; et al. Metabolizable semiconducting polymer nanoparticles for second near-infrared photoacoustic imaging. *Adv. Mater.* **2019**, *31*, 1808166. [CrossRef]
60. Men, X.; Wang, F.; Chen, H.; Liu, Y.; Men, X.; Yuan, Y.; Zhang, Z.; Gao, D.; Wu, C.; Yuan, Z.; et al. Ultrasmall semiconducting polymer dots with rapid clearance for second near-infrared photoacoustic imaging and photothermal cancer therapy. *Adv. Funct. Mater.* **2020**, *30*, 1909673. [CrossRef]
61. Miao, Q.; Xie, C.; Zhen, X.; Lyu, Y.; Duan, H.; Liu, X.; Jokerst, J.V.; Pu, K. Molecular afterglow imaging with bright, biodegradable polymer nanoparticles. *Nat. Biotechnol.* **2017**, *35*, 1102–1110. [CrossRef]
62. Xie, C.; Zhen, X.; Miao, Q.; Lyu, Y.; Pu, K. Self-assembled semiconducting polymer nanoparticles for ultrasensitive near-infrared afterglow imaging of metastatic tumors. *Adv. Mater.* **2018**, *30*, 1801331. [CrossRef] [PubMed]
63. Zhen, X.; Zhang, C.; Xie, C.; Miao, Q.; Lim, K.L.; Pu, K. Intraparticle energy level alignment of semiconducting polymer nanoparticles to amplify chemiluminescence for ultrasensitive in vivo imaging of reactive oxygen species. *ACS Nano* **2016**, *10*, 6400–6409. [CrossRef] [PubMed]
64. Cui, D.; Li, J.; Zhao, X.; Pu, K.; Zhang, R. Semiconducting polymer nanoreporters for near-infrared chemiluminescence imaging of immunoactivation. *Adv. Mater.* **2020**, *32*, 1906314. [CrossRef] [PubMed]
65. Lyu, Y.; Zhen, X.; Miao, Y.; Pu, K. Reaction-based semiconducting polymer nanoprobes for photoacoustic imaging of protein sulfenic acids. *ACS Nano* **2017**, *11*, 358–367. [CrossRef] [PubMed]
66. Hashim, Z.; Green, M.; Chung, P.H.; Suhling, K.; Protti, A.; Phinikaridou, A.; Botnar, R.; Khanbeigi, R.A.; Thanou, M.; Dailey, L.A.; et al. Gd-containing conjugated polymer nanoparticles: Bimodal nanoparticles for fluorescence and MRI imaging. *Nanoscale* **2014**, *6*, 8376–8386. [CrossRef]
67. Sun, W.; Hayden, S.; Jin, Y.; Rong, Y.; Yu, J.; Ye, F.; Chan, Y.H.; Zeigler, M.; Wu, C.; Chiu, D.T.; et al. A versatile method for generating semiconducting polymer dot nanocomposites. *Nanoscale* **2012**, *4*, 7246–7249. [CrossRef]

68. Pecher, J.; Mecking, S. Nanoparticles of conjugated polymers. *Chem. Rev.* **2010**, *110*, 6260–6279. [CrossRef]
69. Gharieh, A.; Khoee, S.; Mahdavian, A.R. Emulsion and miniemulsion techniques in preparation of polymer nanoparticles with versatile characteristics. *Adv. Colloid Interface Sci.* **2019**, *269*, 152–186. [CrossRef]
70. He, Y.; Fan, X.; Sun, J.; Liu, R.; Fan, Z.; Zhang, Z.; Chang, X.; Wang, B.; Gao, F.; Wang, L.; et al. Flash nanoprecipitation of ultra-small semiconducting polymer dots with size tunability. *Chem. Commun.* **2020**, *56*, 2594–2597. [CrossRef]
71. Yu, M.; Zhang, P.; Krishnan, B.P.; Wang, H.; Gao, Y.; Chen, S.; Zeng, R.; Cui, J.; Chen, J. From a molecular toolbox to a toolbox for photoswitchable fluorescent polymeric nanoparticles. *Adv. Funct. Mater.* **2018**, *28*, 1804759. [CrossRef]
72. Guo, B.; Feng, Z.; Hu, D.; Xu, S.; Middha, E.; Pan, Y.; Liu, C.; Zheng, H.; Qian, J.; Sheng, Z.; et al. Precise deciphering of brain vasculatures and microscopic tumors with dual NIR-II fluorescence and photoacoustic imaging. *Adv. Mater.* **2019**, *31*, 1902504. [CrossRef] [PubMed]
73. Wu, C.; Jin, Y.; Schneider, T.; Burnham, D.R.; Smith, P.B.; Chiu, D.T. Ultrabright and bioorthogonal labeling of cellular targets using semiconducting polymer dots and click chemistry. *Angew. Chem.* **2010**, *49*, 9436–9440. [CrossRef] [PubMed]
74. Chen, H.; Fang, X.; Jin, Y.; Hu, X.; Yin, M.; Men, X.; Chen, N.; Fan, C.; Chiu, D.T.; Wan, Y.; et al. Semiconducting polymer nanocavities: Porogenic synthesis, tunable host–guest interactions, and enhanced drug/siRNA delivery. *Small* **2018**, *14*, 1800239. [CrossRef] [PubMed]
75. Jiang, Y.; Chen, H.; Men, X.; Sun, Z.; Yuan, Z.; Zhang, X.; Chiu, D.T.; Wu, C.; McNeill, J. Multimode time-resolved superresolution microscopy revealing chain packing and anisotropic single carrier transport in conjugated polymer nanowires. *Nano Lett.* **2021**, *21*, 4255–4261. [CrossRef]
76. Men, X.; Fang, X.; Liu, Z.; Zhang, Z.; Wu, C.; Chen, H. Anisotropic assembly and fluorescence enhancement of conjugated polymer nanostructures. *View* **2022**, *3*, 20220020. [CrossRef]
77. Wu, C.; Bull, B.; Szymanski, C.; Christensen, K.; McNeill, J. Multicolor conjugated polymer dots for biological fluorescence imaging. *ACS Nano* **2008**, *2*, 2415–2423. [CrossRef]
78. Wu, C.; Szymanski, C.; McNeill, J. Preparation and encapsulation of highly fluorescent conjugated polymer nanoparticles. *Langmuir* **2006**, *22*, 2956–2960. [CrossRef]
79. Eggeling, C.; Widengren, J.; Rigler, R.; Seidel, C. Photobleaching of fluorescent dyes under conditions used for single-molecule detection: Evidence of two-step photolysis. *Anal. Chem.* **1998**, *70*, 2651–2659. [CrossRef]
80. Fernando, L.P.; Kandel, P.K.; Yu, J.; McNeill, J.; Ackroyd, P.C.; Christensen, K.A. Mechanism of cellular uptake of highly fluorescent conjugated polymer nanoparticles. *Biomacromolecules* **2010**, *11*, 2675–2682. [CrossRef]
81. Wu, C.; Hansen, S.J.; Hou, Q.; Yu, J.; Zeigler, M.; Jin, Y.; Burnham, D.R.; McNeill, J.D.; Olson, J.M.; Chiu, D.T. Design of highly emissive polymer dot bioconjugates for in vivo tumor targeting. *Angew. Chem. Int. Ed.* **2011**, *50*, 3430–3434. [CrossRef]
82. Wu, C.; Schneider, T.; Zeigler, M.; Yu, J.; Schiro, P.G.; Burnham, D.R.; McNeill, J.D.; Chiu, D.T. Bioconjugation of ultrabright semiconducting polymer dots for specific cellular targeting. *J. Am. Chem. Soc.* **2010**, *132*, 15410–15417. [CrossRef]
83. Zhang, J.; Yu, J.; Jiang, Y.; Chiu, D.T. Ultrabright pdots with a large absorbance cross section and high quantum yield. *ACS Appl. Mater. Interfaces* **2022**, *14*, 13631–13637. [CrossRef]
84. Kuo, C.T.; Wu, I.C.; Chen, L.; Yu, J.; Wu, L.; Chiu, D.T. Improving the photostability of semiconducting polymer dots using buffers. *Anal. Chem.* **2018**, *90*, 11785–11790. [CrossRef]
85. Chang, C.L.; Lin, W.C.; Jia, C.Y.; Ting, L.Y.; Jayakumar, J.; Elsayed, M.H.; Yang, Y.Q.; Chan, Y.H.; Wang, W.S.; Lu, C.Y.; et al. Low-toxic cycloplatinated polymer dots with rational design of acceptor co-monomers for enhanced photocatalytic efficiency and stability. *Appl. Catal. B Environ.* **2020**, *268*, 118436. [CrossRef]
86. Ow, H.; Larson, D.R.; Srivastava, M.; Baird, B.A.; Webb, W.W.; Wiesner, U. Bright and stable core−shell fluorescent silica nanoparticles. *Nano Lett.* **2005**, *5*, 113–117. [CrossRef]
87. Wang, L.; Yang, C.; Tan, W. Dual-luminophore-doped silica nanoparticles for multiplexed signaling. *Nano Lett.* **2005**, *5*, 37–43. [CrossRef]
88. Li, K.; Liu, Y.; Pu, K.Y.; Feng, S.S.; Zhan, R.; Liu, B. Polyhedral oligomeric silsesquioxanes-containing conjugated polymer loaded PLGA nanoparticles with trastuzumab (herceptin) functionalization for HER2-positive cancer cell detection. *Adv. Funct. Mater.* **2011**, *21*, 287–294. [CrossRef]
89. Li, K.; Pan, J.; Feng, S.S.; Wu, A.W.; Pu, K.Y.; Liu, Y.; Liu, B. Generic strategy of preparing fluorescent conjugated-polymer-loaded poly (dl-lactide-co-glycolide) nanoparticles for targeted cell imaging. *Adv. Funct. Mater.* **2009**, *19*, 3535–3542. [CrossRef]
90. Howes, P.; Green, M.; Bowers, A.; Parker, D.; Varma, G.; Kallumadil, M.; Hughes, M.; Warley, A.; Brain, A.; Botnar, R.; et al. Magnetic conjugated polymer nanoparticles as bimodal imaging agents. *J. Am. Chem. Soc.* **2010**, *132*, 9833–9842. [CrossRef]
91. Howes, P.; Green, M.; Levitt, J.; Suhling, K.; Hughes, M. Phospholipid encapsulated semiconducting polymer nanoparticles: Their use in cell imaging and protein attachment. *J. Am. Chem. Soc.* **2010**, *132*, 3989–3996. [CrossRef] [PubMed]
92. Zhang, X.; Yu, J.; Wu, C.; Jin, Y.; Rong, Y.; Ye, F.; Chiu, D.T. Importance of having low-density functional groups for generating high-performance semiconducting polymer dots. *ACS Nano* **2012**, *6*, 5429–5439. [CrossRef] [PubMed]
93. Yu, J.; Wu, C.; Zhang, X.; Ye, F.; Gallina, M.E.; Rong, Y.; Wu, I.C.; Sun, W.; Chan, Y.H.; Chiu, D.T.; et al. Stable functionalization of small semiconducting polymer dots via covalent cross-linking and their application for specific cellular imaging. *Adv. Mater.* **2012**, *24*, 3498–3504. [CrossRef] [PubMed]
94. Zhang, Y.; Ye, F.; Sun, W.; Yu, J.; Wu, I.C.; Rong, Y.; Zhang, Y.; Chiu, D.T. Light-induced crosslinkable semiconducting polymer dots. *Chem. Sci.* **2015**, *6*, 2102–2109. [CrossRef]

95. Chen, H.; Zhou, H.; Men, X.; Sun, K.; Sun, Z.; Fang, X.; Wu, C. Light-induced PEGylation and functionalization of semiconductor polymer dots. *ChemNanoMat* **2017**, *3*, 755–759. [CrossRef]
96. Jiang, Y.; McNeill, J. Light-harvesting and amplified energy transfer in conjugated polymer nanoparticles. *Chem. Rev.* **2017**, *117*, 838–859. [CrossRef]
97. Lyu, J.; Wang, C.; Zhang, X. Rational construction of a mitochondria-targeted reversible fluorescent probe with intramolecular fret for ratiometric monitoring sulfur dioxide and formaldehyde. *Biosensors* **2022**, *12*, 715. [CrossRef]
98. Liu, S.; Liu, Y.; Zhang, Z.; Wang, X.; Yang, Y.; Sun, K.; Yu, J.; Chiu, D.T.; Wu, C. Near-infrared optical transducer for dynamic imaging of cerebrospinal fluid glucose in brain tumor. *Anal. Chem.* **2022**, *94*, 14265–14272. [CrossRef]
99. Hou, W.; Yuan, Y.; Sun, Z.; Guo, S.; Dong, H.; Wu, C. Ratiometric fluorescent detection of intracellular singlet oxygen by semiconducting polymer dots. *Anal. Chem.* **2018**, *90*, 14629–14634. [CrossRef]
100. Wu, L.; Wu, I.C.; DuFort, C.C.; Carlson, M.A.; Wu, X.; Chen, L.; Kuo, C.T.; Qin, Y.; Yu, J.; Hingorani, S.R.; et al. Photostable ratiometric pdot probe for in vitro and in vivo imaging of hypochlorous acid. *J. Am. Chem. Soc.* **2017**, *139*, 6911–6918. [CrossRef]
101. Pu, K.; Shuhendler, A.J.; Rao, J. Semiconducting polymer nanoprobe for in vivo imaging of reactive oxygen and nitrogen species. *Angew. Chem. Int. Ed.* **2013**, *52*, 10325–10329. [CrossRef]
102. Chen, L.; Wu, L.; Yu, J.; Kuo, C.T.; Jian, T.; Wu, I.C.; Rong, Y.; Chiu, D. Highly photostable wide-dynamic-range pH sensitive semiconducting polymer dots enabled by dendronizing the near-IR emitters. *Chem. Sci.* **2017**, *8*, 7236–7245. [CrossRef]
103. Ye, F.; Wu, C.; Jin, Y.; Chan, Y.H.; Zhang, X.; Chiu, D.T. Ratiometric temperature sensing with semiconducting polymer dots. *J. Am. Chem. Soc.* **2011**, *133*, 8146–8149. [CrossRef]
104. Zhen, X.; Xie, C.; Pu, K. Temperature-correlated afterglow of a semiconducting polymer nanococktail for imaging-guided photothermal therapy. *Angew. Chem. Int. Ed.* **2018**, *57*, 3938–3942. [CrossRef]
105. Feng, L.; Guo, L.; Wang, X. Preparation, properties and applications in cell imaging and ions detection of conjugated polymer nanoparticles with alcoxyl bonding fluorene core. *Biosens. Bioelectron.* **2017**, *87*, 514–521. [CrossRef]
106. Chan, Y.H.; Jin, Y.; Wu, C.; Chiu, D.T. Copper (II) and iron (II) ion sensing with semiconducting polymer dots. *Chem. Commun.* **2011**, *47*, 2820–2822. [CrossRef]
107. Yang, L.; Canaveras, J.C.G.; Chen, Z.; Wang, L.; Liang, L.; Jang, C.; Mayr, J.A.; Zhang, Z.; Ghergurovich, J.M.; Zhan, L.; et al. Serine catabolism feeds NADH when respiration is impaired. *Cell Metab.* **2020**, *31*, 809–821.e6. [CrossRef]
108. Covarrubias, A.J.; Perrone, R.; Grozio, A.; Verdin, E. NAD$^+$ metabolism and its roles in cellular processes during ageing. *Nat. Rev. Mol. Cell Biol.* **2021**, *22*, 119–141. [CrossRef]
109. Luongo, T.S.; Eller, J.M.; Lu, M.J.; Niere, M.; Raith, F.; Perry, C.; Bornstein, M.R.; Oliphint, P.; Wang, L.; McReynolds, M.R.; et al. SLC25A51 is a mammalian mitochondrial NAD$^+$ transporter. *Nature* **2020**, *588*, 174–179. [CrossRef]
110. Zaremba, M.; Dakineviciene, D.; Golovinas, E.; Zagorskaitė, E.; Stankunas, E.; Lopatina, A.; Sorek, R.; Manakova, E.; Ruksenaite, A.; Silanskas, A. Short prokaryotic Argonautes provide defence against incoming mobile genetic elements through NAD$^+$ depletion. *Nat. Microbiol.* **2022**, *7*, 1857–1869. [CrossRef]
111. Yu, Q.; Xue, L.; Hiblot, J.; Griss, R.; Fabritz, S.; Roux, C.; Binz, P.A.; Haas, D.; Okun, J.G.; Johnsson, K. Semisynthetic sensor proteins enable metabolic assays at the point of care. *Science* **2018**, *361*, 1122–1126. [CrossRef] [PubMed]
112. Ogurtsova, K.; da Rocha Fernandes, J.; Huang, Y.; Linnenkamp, U.; Guariguata, L.; Cho, N.H.; Cavan, D.; Shaw, J.; Makaroff, L. IDF diabetes atlas: Global estimates for the prevalence of diabetes for 2015 and 2040. *Diabetes Res. Clin. Pract.* **2017**, *128*, 40–50. [CrossRef] [PubMed]
113. Sun, K.; Tang, Y.; Li, Q.; Yin, S.; Qin, W.; Yu, J.; Chiu, D.T.; Liu, Y.; Yuan, Z.; Zhang, X.; et al. In vivo dynamic monitoring of small molecules with implantable polymer-dot transducer. *ACS Nano* **2016**, *10*, 6769–6781. [CrossRef] [PubMed]
114. Sun, K.; Ding, Z.; Zhang, J.; Chen, H.; Qin, Y.; Xu, S.; Wu, C.; Yu, J.; Chiu, D.T. Enhancing the long-term stability of a polymer dot glucose transducer by using an enzymatic cascade reaction system. *Adv. Healthc. Mater.* **2021**, *10*, 2001019. [CrossRef] [PubMed]
115. Sun, K.; Liu, S.; Liu, J.; Ding, Z.; Jiang, Y.; Zhang, J.; Chen, H.; Yu, J.; Wu, C.; Chiu, D.T.; et al. Improving the accuracy of Pdot-based continuous glucose monitoring by using external ratiometric calibration. *Anal. Chem.* **2021**, *93*, 2359–2366. [CrossRef] [PubMed]
116. Liu, J.; Fang, X.; Zhang, Z.; Liu, Z.; Liu, J.; Sun, K.; Yuan, Z.; Yu, J.; Chiu, D.T.; Wu, C.; et al. Long-term in vivo glucose monitoring by polymer-dot transducer in an injectable hydrogel implant. *Anal. Chem.* **2022**, *94*, 2195–2203. [CrossRef] [PubMed]
117. Van Spronsen, F.J.; Blau, N.; Harding, C.; Burlina, A.; Longo, N.; Bosch, A.M. Phenylketonuria. *Nat. Rev. Dis. Primers* **2021**, *7*, 36. [CrossRef]
118. Chen, H.; Yu, J.; Zhang, J.; Sun, K.; Ding, Z.; Jiang, Y.; Hu, Q.; Wu, C.; Chiu, D.T. Monitoring metabolites using an NAD (P) H-sensitive polymer dot and a metabolite-specific enzyme. *Angew. Chem. Int. Ed.* **2021**, *60*, 19331–19336. [CrossRef]
119. You, P.Y.; Li, F.C.; Liu, M.H.; Chan, Y.H. Colorimetric and fluorescent dual-mode immunoassay based on plasmon-enhanced fluorescence of polymer dots for detection of PSA in whole blood. *ACS Appl. Mater. Interfaces* **2019**, *11*, 9841–9849. [CrossRef]
120. Yang, Y.C.; Liu, M.H.; Yang, S.M.; Chan, Y.H. Bimodal multiplexed detection of tumor markers in non-small cell lung cancer with polymer dot-based immunoassay. *ACS Sens.* **2021**, *6*, 4255–4264. [CrossRef]
121. Xie, C.; Zhen, X.; Lyu, Y.; Pu, K. Nanoparticle regrowth enhances photoacoustic signals of semiconducting macromolecular probe for in vivo imaging. *Adv. Mater.* **2017**, *29*, 1703693. [CrossRef]

122. Huynh, E.; Leung, B.Y.; Helfield, B.L.; Shakiba, M.; Gandier, J.A.; Jin, C.S.; Master, E.R.; Wilson, B.C.; Goertz, D.E.; Zheng, G.; et al. In situ conversion of porphyrin microbubbles to nanoparticles for multimodality imaging. *Nat. Nanotechnol.* **2015**, *10*, 325–332. [CrossRef]

123. Kwon, E.J.; Dudani, J.S.; Bhatia, S.N. Ultrasensitive tumour-penetrating nanosensors of protease activity. *Nat. Biomed. Eng.* **2017**, *1*, 0054. [CrossRef]

124. Kalluri, R.; LeBleu, V.S. The biology, function, and biomedical applications of exosomes. *Science* **2020**, *367*, eaau6977. [CrossRef]

125. Jiang, Y.; Andronico, L.A.; Jung, S.R.; Chen, H.; Fujimoto, B.; Vojtech, L.; Chiu, D.T. High-throughput counting and superresolution mapping of tetraspanins on exosomes using a single-molecule sensitive flow technique and transistor-like semiconducting polymer dots. *Angew. Chem.* **2021**, *60*, 13470–13475. [CrossRef]

126. Gupta, N.; Chan, Y.H.; Saha, S.; Liu, M.H. Near-infrared-II semiconducting polymer dots for deep-tissue fluorescence imaging. *Chem. Asian J.* **2021**, *16*, 175–184. [CrossRef]

127. Wu, Y.; Shi, C.; Wang, G.; Sun, H.; Yin, S. Recent advances in the development and applications of conjugated polymer dots. *J. Mater. Chem. B* **2022**, *10*, 2995–3015. [CrossRef]

128. Yang, Z.; Dai, Y.; Shan, L.; Shen, Z.; Wang, Z.; Yung, B.C.; Jacobson, O.; Liu, Y.; Tang, W.; Wang, S.; et al. Tumour microenvironment-responsive semiconducting polymer-based self-assembling nanotheranostics. *Nanoscale Horiz.* **2019**, *4*, 426–433. [CrossRef]

129. Zhang, R.R.; Schroeder, A.B.; Grudzinski, J.J.; Rosenthal, E.L.; Warram, J.M.; Pinchuk, A.N.; Eliceiri, K.W.; Kuo, J.S.; Weichert, J.P. Beyond the margins: Real-time detection of cancer using targeted fluorophores. *Nat. Rev. Clin. Oncol.* **2017**, *14*, 347–364. [CrossRef]

130. Desmettre, T.; Devoisselle, J.; Mordon, S. Fluorescence properties and metabolic features of indocyanine green (ICG) as related to angiography. *Surv. Ophthalmol.* **2000**, *45*, 15–27. [CrossRef]

131. Levitus, M.; Ranjit, S. Cyanine dyes in biophysical research: The photophysics of polymethine fluorescent dyes in biomolecular environments. *Q. Rev. Biophys.* **2011**, *44*, 123–151. [CrossRef] [PubMed]

132. Altınoglu, E.I.; Russin, T.J.; Kaiser, J.M.; Barth, B.M.; Eklund, P.C.; Kester, M.; Adair, J.H. Near-infrared emitting fluorophore-doped calcium phosphate nanoparticles for in vivo imaging of human breast cancer. *ACS Nano* **2008**, *2*, 2075–2084. [CrossRef] [PubMed]

133. Zhang, Z.; Yuan, Y.; Liu, Z.; Chen, H.; Chen, D.; Fang, X.; Zheng, J.; Qin, W.; Wu, C. Brightness enhancement of near-infrared semiconducting polymer dots for in vivo whole-body cell tracking in deep organs. *ACS Appl. Mater. Interfaces* **2018**, *10*, 26928–26935. [CrossRef] [PubMed]

134. Feng, G.; Liu, J.; Liu, R.; Mao, D.; Tomczak, N.; Liu, B. Ultrasmall conjugated polymer nanoparticles with high specificity for targeted cancer cell imaging. *Adv. Sci.* **2017**, *4*, 1600407. [CrossRef] [PubMed]

135. Louis, D.N.; Ohgaki, H.; Wiestler, O.D.; Cavenee, W.K.; Burger, P.C.; Jouvet, A.; Scheithauer, B.W.; Kleihues, P. The 2007 Who classification of tumours of the central nervous system. *Acta Neuropathol.* **2007**, *114*, 97–109. [CrossRef]

136. Miao, Q.; Pu, K. Organic semiconducting agents for deep-tissue molecular imaging: Second near-infrared fluorescence, self-luminescence, and photoacoustics. *Adv. Mater.* **2018**, *30*, 1801778. [CrossRef]

137. Li, J.; Rao, J.; Pu, K. Recent progress on semiconducting polymer nanoparticles for molecular imaging and cancer phototherapy. *Biomaterials* **2018**, *155*, 217–235. [CrossRef]

138. Jiang, Y.; Pu, K. Molecular fluorescence and photoacoustic imaging in the second near-infrared optical window using organic contrast agents. *Adv. Biosyst.* **2018**, *2*, 1700262. [CrossRef]

139. Pu, K.; Shuhendler, A.J.; Jokerst, J.V.; Mei, J.; Gambhir, S.S.; Bao, Z.; Rao, J. Semiconducting polymer nanoparticles as photoacoustic molecular imaging probes in living mice. *Nat. Nanotechnol.* **2014**, *9*, 233–239. [CrossRef]

140. Chen, H.; Yuan, Z.; Wu, C. Nanoparticle probes for structural and functional photoacoustic molecular tomography. *BioMed Res. Int.* **2015**, *2015*, 757101. [CrossRef]

141. Zhen, X.; Pu, K.; Jiang, X. Photoacoustic imaging and photothermal therapy of semiconducting polymer nanoparticles: Signal amplification and second near-infrared construction. *Small* **2021**, *17*, 2004723. [CrossRef]

142. Chen, H.; Zhang, J.; Chang, K.; Men, X.; Fang, X.; Zhou, L.; Li, D.; Gao, D.; Yin, S.; Zhang, X.; et al. Highly absorbing multispectral near-infrared polymer nanoparticles from one conjugated backbone for photoacoustic imaging and photothermal therapy. *Biomaterials* **2017**, *144*, 42–52. [CrossRef]

143. Guo, B.; Sheng, Z.; Hu, D.; Li, A.; Xu, S.; Manghnani, P.N.; Liu, C.; Guo, L.; Zheng, H.; Liu, B.; et al. Molecular engineering of conjugated polymers for biocompatible organic nanoparticles with highly efficient photoacoustic and photothermal performance in cancer theranostics. *ACS Nano* **2017**, *11*, 10124–10134. [CrossRef]

144. Dong, T.; Wen, K.; Chen, J.; Xie, J.; Fan, W.; Ma, H.; Yang, L.; Wu, X.; Xu, F.; Peng, A.; et al. Significant enhancement of photothermal and photoacoustic efficiencies for semiconducting polymer nanoparticles through simply molecular engineering. *Adv. Funct. Mater.* **2018**, *28*, 1800135. [CrossRef]

145. Lyu, Y.; Fang, Y.; Miao, Q.; Zhen, X.; Ding, D.; Pu, K. Intraparticle molecular orbital engineering of semiconducting polymer nanoparticles as amplified theranostics for in vivo photoacoustic imaging and photothermal therapy. *ACS Nano* **2016**, *10*, 4472–4481. [CrossRef]

146. Zhen, X.; Feng, X.; Xie, C.; Zheng, Y.; Pu, K. Surface engineering of semiconducting polymer nanoparticles for amplified photoacoustic imaging. *Biomaterials* **2017**, *127*, 97–106. [CrossRef]

147. Duan, Y.; Xu, Y.; Mao, D.; Liew, W.H.; Guo, B.; Wang, S.; Cai, X.; Thakor, N.; Yao, K.; Zhang, C.J. Photoacoustic and magnetic resonance imaging bimodal contrast agent displaying amplified photoacoustic signal. *Small* **2018**, *14*, 1800652. [CrossRef]
148. Jiang, Y.; Upputuri, P.K.; Xie, C.; Lyu, Y.; Zhang, L.; Xiong, Q.; Pramanik, M.; Pu, K. Broadband absorbing semiconducting polymer nanoparticles for photoacoustic imaging in second near-infrared window. *Nano Lett.* **2017**, *17*, 4964–4969. [CrossRef]
149. Wu, J.; You, L.; Lan, L.; Lee, H.J.; Chaudhry, S.T.; Li, R.; Cheng, J.X.; Mei, J. Semiconducting polymer nanoparticles for centimeters-deep photoacoustic imaging in the second near-infrared window. *Adv. Mater.* **2017**, *29*, 1703403. [CrossRef]
150. Guo, B.; Sheng, Z.; Hu, D.; Lin, X.; Xu, S.; Liu, C.; Zheng, H.; Liu, B. Biocompatible conjugated polymer nanoparticles for highly efficient photoacoustic imaging of orthotopic brain tumors in the second near-infrared window. *Mater. Horiz.* **2017**, *4*, 1151–1156. [CrossRef]
151. Guo, B.; Chen, J.; Chen, N.; Middha, E.; Xu, S.; Pan, Y.; Wu, M.; Li, K.; Liu, C.; Liu, B.; et al. High-resolution 3D NIR-II photoacoustic imaging of cerebral and tumor vasculatures using conjugated polymer nanoparticles as contrast agent. *Adv. Mater.* **2019**, *31*, 1808355. [CrossRef] [PubMed]
152. Jiang, Y.; Huang, J.; Zhen, X.; Zeng, Z.; Li, J.; Xie, C.; Miao, Q.; Chen, J.; Chen, P.; Pu, K.; et al. A generic approach towards afterglow luminescent nanoparticles for ultrasensitive in vivo imaging. *Nat. Commun.* **2019**, *10*, 2064. [CrossRef]
153. Wang, Y.; Shi, L.; Ye, Z.; Guan, K.; Teng, L.; Wu, J.; Yin, X.; Song, G.; Zhang, X.B. Reactive oxygen correlated chemiluminescent imaging of a semiconducting polymer nanoplatform for monitoring chemodynamic therapy. *Nano Lett.* **2020**, *20*, 176–183. [CrossRef] [PubMed]
154. Su, Y.; Walker, J.R.; Park, Y.; Smith, T.P.; Liu, L.X.; Hall, M.P.; Labanieh, L.; Hurst, R.; Wang, D.C.; Encell, L.P.; et al. Novel NanoLuc substrates enable bright two-population bioluminescence imaging in animals. *Nat. Methods* **2020**, *17*, 852–860. [CrossRef] [PubMed]
155. Li, M.; Huang, X.; Ren, J. Multicolor chemiluminescent resonance energy-transfer system for in vivo high-contrast and targeted imaging. *Anal. Chem.* **2021**, *93*, 3042–3051. [CrossRef]
156. Men, X.; Yuan, Z. Multifunctional conjugated polymer nanoparticles for photoacoustic-based multimodal imaging and cancer photothermal therapy. *J. Innov. Opt. Health Sci.* **2019**, *12*, 1930001. [CrossRef]
157. Men, X.; Yuan, Z. Polymer dots for precision photothermal therapy of brain tumors in the second near-infrared window: A mini-review. *ACS Appl. Polym. Mater.* **2020**, *2*, 4319–4330. [CrossRef]
158. Yu, N.; Zhao, L.; Cheng, D.; Ding, M.; Lyu, Y.; Zhao, J.; Li, J. Radioactive organic semiconducting polymer nanoparticles for multimodal cancer theranostics. *J. Colloid Interface Sci.* **2022**, *619*, 219–228. [CrossRef]
159. Xie, C.; Zhou, W.; Zeng, Z.; Fan, Q.; Pu, K. Grafted semiconducting polymer amphiphiles for multimodal optical imaging and combination phototherapy. *Chem. Sci.* **2020**, *11*, 10553–10570. [CrossRef]
160. Jiang, Y.; Pu, K. Multimodal biophotonics of semiconducting polymer nanoparticles. *Acc. Chem. Res.* **2018**, *51*, 1840–1849. [CrossRef]

biosensors

Review

Metal-Organic Frameworks-Based Optical Nanosensors for Analytical and Bioanalytical Applications

Cong Wen [1], Rongsheng Li [2], Xiaoxia Chang [3] and Na Li [1,*]

[1] Beijing National Laboratory for Molecular Sciences (BNLMS), Key Laboratory of Bioorganic Chemistry and Molecular Engineering of Ministry of Education, College of Chemistry and Molecular Engineering, Peking University, Beijing 100871, China

[2] National Demonstration Center for Experimental Chemistry and Chemical Engineering Education (Yunnan University), School of Chemical Science and Engineering, Yunnan University, Kunming 650091, China

[3] College of Chemistry and Molecular Engineering, Peking University, Beijing 100871, China

* Correspondence: lina@pku.edu.cn

Abstract: Metal-organic frameworks (MOFs)-based optical nanoprobes for luminescence and surface-enhanced Raman spectroscopy (SERS) applications have been receiving tremendous attention. Every element in the MOF structure, including the metal nodes, the organic linkers, and the guest molecules, can be used as a source to build single/multi-emission signals for the intended analytical purposes. For SERS applications, the MOF can not only be used directly as a SERS substrate, but can also improve the stability and reproducibility of the metal-based substrates. Additionally, the porosity and large specific surface area give MOF a sieving effect and target molecule enrichment ability, both of which are helpful for improving detection selectivity and sensitivity. This mini-review summarizes the advances of MOF-based optical detection methods, including luminescence and SERS, and also provides perspectives on future efforts.

Keywords: metal–organic frameworks; luminescence; surface-enhanced Raman spectroscopy; multiplexed detection

Citation: Wen, C.; Li, R.; Chang, X.; Li, N. Metal-Organic Frameworks-Based Optical Nanosensors for Analytical and Bioanalytical Applications. *Biosensors* **2023**, *13*, 128. https://doi.org/10.3390/bios13010128

Received: 13 December 2022
Revised: 9 January 2023
Accepted: 10 January 2023
Published: 12 January 2023

1. Introduction

Nanomaterial-based optical methods are advantageous for measuring quantitatively biomolecules of interest. With optical nanosensors, an extremely high photoluminescence intensity or enhancement effect in surface-enhanced Raman spectroscopy (SERS) can be achieved, and diversified signals, including wavelength/wavenumber, intensity, and excited-state lifetime, can be tuned to cater to specific applications. Most importantly, multiple signals can be integrated in a single optical nanoprobe to readily achieve ratiometric, multiplexed, and multimodality measurements. These detection modes are highly desirable for quantitative measurements in complex sample matrices [1,2].

Metal–organic frameworks (MOFs) are a kind of hybrid porous material consisting of inorganic metal ion or cluster nodes, and linkers including organic ligands and metal–organic complexes [3–7]. Due to the large specific surface area, the ultrahigh porosity, the adjustable internal surface property, the extraordinarily diversified structure, and the reasonable biocompatibility, MOFs are widely employed in storage and separation [8], catalysis, [9,10] drug delivery [5,11–15], and biomedicine [12,16–18], as well as in chemical sensors and biosensors [12,19–28]. In terms of optical measurements, MOFs are promising optical sensing materials because emission centers of MOFs can be constructed by "multiple photonic units" originating from inorganic metal ion or cluster nodes, linkers, or their combination to exemplify the features of structural diversity through combining inorganic and organic chemistry [29]. This unique property, together with tunable functional sites, imparts MOFs with highly designable and diversified luminescence that can be used for customized applications. On one hand, the "multiple photonic units" can be engineered or tailored with rational design to achieve the aforementioned diversified luminescence signals, all of which can be used for applications with specific needs; on the other hand, aside from attaching recognition entities for target-specific interaction, the functional sites can also be used to provide auxiliary interaction with emission centers to further tailor luminescence properties. Furthermore, taking the advantage of ultra-high porosity and modifiable internal surface property to encapsulate luminescent guest molecules in the porous structure, one more dimension of luminescence can be added to grant MOFs with ratiometric, multiplexing, and multimodality measurement capabilities [30,31]. In addition to luminescence, MOFs can serve as promising materials for SERS methods in which MOFs either act as substrates for signal enhancement or form composite substrates by encapsulating metallic nanomaterials. In addition to the above, the high surface area and controllable pore size allow a high capacity for adsorbing and concentrating analytes to achieve a low limit of detection (LOD) and a unique sieve effect, thus improving the selectivity [6,32].

There are quite a few review articles on MOFs and their applications in the optical measurement field [33–35]. This mini-review tries to highlight MOFs-based optical nanosensors, particularly luminescence and SERS, for applications in analytical chemistry. First, the application of MOFs in luminescence detection is summarized and discussed based on the evolution from a single luminophore, including MOFs itself or the encapsulated guest, to multiple signal sources for ratiometric and multimodality measurements. Second, the application of MOFs in SERS measurements is summarized and discussed based on the MOF itself as both a SERS substrate and a MOF-metal nanomaterial composite substrate. Finally, perspectives on future efforts to develop MOF-based nanosensors for analytical and bioanalytical applications are provided.

2. MOFs-Based Nanosensors for Luminescence Applications

Luminescence sensing has been recognized as an important tool in food safety detection, disease diagnosis, and environmental monitoring [36,37]. For these applications, it is required that the method has sufficient sensitivity and selectivity towards the analyte of interest. For analysis of targets in complex sample matrices, it is also desirable that the sensor or nanosensor system can self-calibrate to provide reproducible and quantitative results (Table 1).

Compared to conventional organic and inorganic luminophores, the luminescent MOF materials have excellent host and sensing features that can meet the aforementioned requirements for luminescence sensing [32,35,38]. The photoluminescence of MOFs may originate from metal centers, linkers, or guest molecules. Each emission unit can be designed and tuned for diversified application needs. Moreover, the integration of emissive units into MOFs can be adopted for the purpose of developing multi-emissive probes to realize self-calibrating measurements, multiplexed measurements, and multimodality measurements. The following discussion is based on the signal evolution in terms of the single luminophore signal, the ratiometric signal, and the multimodality signals. A perspective on multiplexed measurements for fluorescence applications will be discussed in the Conclusions and Future Perspectives section.

2.1. Single Luminophore Signal

The luminescence of MOFs may originate from the metal nodes [39], organic linkers [6], and guest molecules [40]. The luminescence may also originate from second building units (SBUs), including metal clusters [30,41,42] and metal–organic complex linkers [6,43], which are also used to build MOFs with extended porous networks and luminescence centers. All of the above can be utilized as luminescent probes to indicate the presence and even quantity of a tentative analyte [44]. The analyte can alter the luminescence of MOF sensors through one of the following mechanisms: intermolecular charge transfer, ligand-to-metal charge transfer (LMCT), photo-induced electron transfer (PET), Förster resonance energy transfer (FRET), dynamic quenching, or static quenching [45,46].

Lanthanide-based MOFs (Ln-MOFs), as metal node-based luminescence MOFs, present distinctive luminescence properties originating from abundant f-orbital configurations, long luminescence lifetimes, and "antenna effects" of linkers [47,48]. Qian and colleagues [39] used H_3TATAB (4,4′,4″-s-triazine-1,3,5-triyltri-*p*-aminobenzoic acid) as an organic linker to synthesize a series of isomorphic Ln-MOFs. For example, TbTATAB exhibited good stability in water and a high fluorescence quantum yield (77.48%). The large amount of N atoms afforded it a high ligand-Hg^{2+} affinity, which was able to block the antenna effect of the linker. As a result, the fluorescence of TbTATAB was quenched through either a dynamic or a static mechanism. The TbTATAB-based sensor showed excellent stability, reproducibility, and sensitivity, such that it could be practically employed to detect Hg^{2+} in environmental samples.

In addition to the metal nodes, the organic linker is another source of luminescence of MOFs. A luminescent MOF using aggregation-induced emission (AIE) molecules as an organic linker can confine the intramolecular motions of the linker molecule to give a bright emission [49]. Tang and colleagues [6] used an AIE linker, tetrakis(4-carboxyphenyl)ethylene (TCPE), to prepare ZnMOF and CoMOF with an unique $[M^+–L–M^-–L–M]_\infty$ (M = metal clusters, L = linker) configuration (Figure 1). During the sensing of HCl vapor, ZnMOF exhibited a blue-to-yellow-greenish transition of fluorescence due to the adsorption, rather than coordination, of HCl vapor. The adsorbed strong dipole HCl molecules were able to decrease the energy of the lowest singlet excited state via dipole–dipole interaction, leading to a red-shifted and weakened emission. For CoMOF, luminescence was quenched by cobalt ions. Introducing histidine could cause the collapse of the CoMOF framework and subsequent aggregation of TCPE to recover the blue emission of the TCPE aggregate.

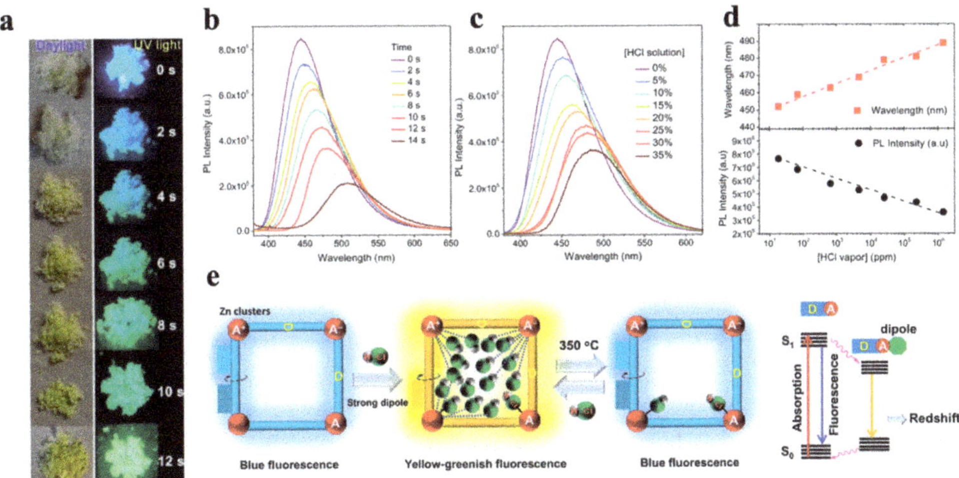

Figure 1. HCl vapor sensing by ZnMOF. (**a**) Photographs of ZnMOF fumigated with HCl vapor from 12% HCl solution. (**b**) Fumigation time dependence. (**c**) HCl solution concentration dependence (fumigation time = 5 s) of solid-state fluorescence spectra of HCl-vapor fumigated ZnMOF. (**d**) Emission peak and intensity responses to HCl vapor concentration. (**e**) Schematics showing the solid-state fluorescence change of ZnMOF with adsorption and desorption of HCl vapor (the blue dotted line for host–guest interaction, and the black solid line for coordination), and the energy diagram of ZnMOF showing the influence of dipole–dipole interaction [6]. Copyright 2021, John Wiley and Sons.

Guest encapsulation into MOFs offers advantages over traditional synthesis of luminescent materials, such as easiness and cost effectiveness, as well as the possibility of tuning the emission properties by selection of guest molecules. [50] To date, various luminescent molecules have been used as guests, including but not limited to Lanthanide (Ln) ions [47], quantum dots (QDs) [51,52], carbon dots (CDs) [40], upconversion nanoparticles (UCNPs) [3,11,16,53–59], noble metal nanoclusters [5,8,38,58,60–76], and organic dyes [20,77–82]. Encapsulation can prevent the aggregation-caused quenching (ACQ) of organic dyes and maintain the signal stability, the photostability, and the reasonable shelf-life of the fluorescent nanoparticles [30,38,77]. Wang and colleagues [40] reported a nanoscale complex based on CDs and MOFs (abbreviated as CDs@ZIF-8) for enhanced chemical sensing of quercetin. Quercetin can form a complex with CD in CDs@ZIF-8 via the electrostatic interaction between hydroxyl groups of quercetin and basic groups on the surface of CDs; thus, the fluorescence of CDs@ZIF-8 was quenched. Moreover, ZIF-8 endowed CDs@ZIF-8 with a high binding affinity to quercetin by π-π staking, as well as by improving the detection sensitivity and selectivity. The aforementioned characteristics of CDs@ZIF-8 guaranteed a LOD of 3.5 nM and its suitability for practical application in real samples for sensing of quercetin. In addition, Yan and colleagues [47] also used Ln ions as guest molecules to decorate MOF through post-synthetic modification in order to detect diphenyl phosphate in human urinary samples.

2.2. Ratiometric Signal

Although single emission-based luminescent materials have been widely used, they suffer from inherent defects, such as signal fluctuations, variations of probe concentrations, light scattering from the matrix, signal fluctuations due to the complex matrices and sometimes sample pretreatments, etc. [30,77,79,83–86]. The ratiometric fluorescence sensing, based on the intensity ratio of two or more well-resolved emission bands, has the self-calibrating capability to eliminate the aforementioned problems, and, thus, to enable more accurate measurements [80,84,87,88]. Due to the tunable multiple emission centers, the large absorption cross sections, and the tailorable skeletons [78,80], MOFs have been demonstrated to be a potential candidate for ratiometric sensing applications [78,79,83,84]. A ratiometric signal can be achieved between MOF and guest molecules [30,78–80,89], between guest and guest molecules [77], and even by MOF itself [32,42,84].

The tailor-made skeletons of MOFs offer specific host–guest interaction sites for intended recognition events. The guest molecules can be attached on the surface of MOFs through covalent linking [78] and electrostatic adsorption [40], or being encapsulated into the MOF channels [30,38,89]. Zhang and colleagues [78] synthesized nanoscale MOF (NMOF) for ratiometric peroxynitrite ($ONOO^-$) sensing based on FRET (Figure 2). Poly(vinyl alcohol) (PVA) was used to attach the energy acceptor (ABt or BDP) on the surface of the energy donor (MOF). With 340-nm excitation of the donor, the nanosensor presented the emissions of the acceptors: 540 nm for ABt and 610 nm for BDP. The presence of $ONOO^-$ disabled FRET by detaching the acceptor from the donor. The quantification of $ONOO^-$ was realized based on the donor-to-acceptor intensity ratio. The fast response and high selectivity made the nanosensor suitable for imaging of $ONOO^-$ in living cells.

Yin and colleagues [89] reported a turn-on ratiometric fluorescent sensor, Ru@MIL-NH$_2$, for water quantification. Ru(bpy)$_3{}^{2+}$ was trapped in the channels of MIL-101(Al)-NH$_2$ via a simple one-pot method. With the water content increasing from 0% to 100%, MIL-NH$_2$ emission at 465 nm was intensified while the Ru(bpy)$_3{}^{2+}$ emission remained stable at 615 nm with 300 nm excitation. It was revealed that the protonation of the nitrogen atom of the MIL-NH$_2$, the π-conjugation system, and the stable fluorescence of Ru(bpy)$_3{}^{2+}$ together facilitated the sensitive ratiometric measurements. This turn-on ratiometric fluorescence sensor showed low LOD (0.02%), fast response (less than 1 min), large dynamic range (0−100%), and good sensor reusability. The turn-on response is much simpler and more straightforward, and even more sensitive, than the quenching process [77].

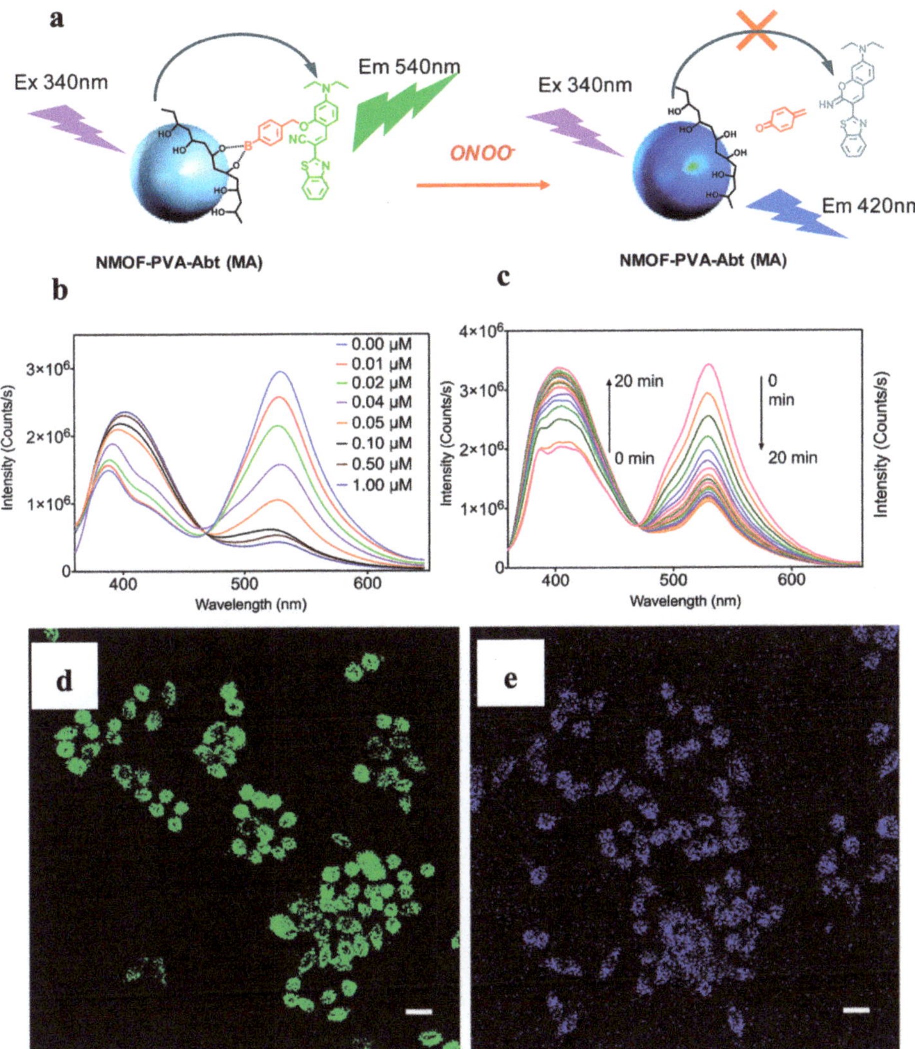

Figure 2. The sensing and cell imaging of ONOO⁻ with NMOF-Dye FRET pair. (**a**) Schematics of the sensing mechanism. (**b**) Fluorescence spectra of MA (1 mg L^{-1}) as a function of ONOO⁻ concentration in PBS buffer (pH 7.4) for 30 min; (**c**) fluorescence spectra of MA (1 mg L^{-1}) with the addition of 1.00 mM ONOO⁻ over time (0 to 20 min). (**d,e**) Fluorescence microscopic images of HeLa cells for exogenous ONOO⁻. (**d**) The cells were stained with MA (10 mg L^{-1}) for 1 h and then washed with PBS before imaging; (**e**) the cells were pretreated as shown in (**d**) and then treated with SIN-1(1 mM) for 30 min. Scale bar: 25 mm [78]. Copyright 2017, Royal Society Chemistry.

A ratiometric signal can be achieved from the guest molecules and metal nodes of MOFs. Chen and colleagues [80] developed a MOF for real-time ratiometric fluorescent monitoring of food freshness by covalently coupling fluorescein 5-isothiocyanate (5-FITC) with NH$_2$-rich EuMOF in a post-synthetic modification manner. Histamine, a biogenic amine produced by spoiled food, increased the emission of FITC at 525 nm and decreased the emission of Eu^{3+} at 611 nm. By doping the EuMOF-FITC probe on a flexible substrate (glass fiber), the complex was able to be integrated with a smartphone-based portable platform for on-site visual inspection of the freshness of raw fish samples.

MOFs with multi-emission centers are able to provide the ratiometric signal themselves. The metal-to-ligand and metal-to-metal energy transfers empower diversified luminescence responses. Shi and colleagues [32] synthesized a luminescent Eu-ZnMOF-*n* by a structure engineering strategy, rendering the material enhanced slope sensitivity within the "optimized useful detection window" (Figure 3). Therefore, this biosensor enabled the discrimination of small concentration variations of urinary vanillylmandelic acid (VMA), an early pathological signature of pheochromocytoma. Upon the addition of VMA, emissions from organic linker at 433 nm became conspicuous with 330-nm excitation, while emissions from Eu^{3+} at 615 nm decreased. The organic linker emission change was attributed to the formation of an exciplex between the linker and VMA, which held a lower-lying excited-state energy level; the emission change in Eu^{3+} was due to the static quenching by VMA. This structure engineering strategy provided a facile approach to detect the biomarker change within a small concentration range, making the biosensor more suitable for clinical applications.

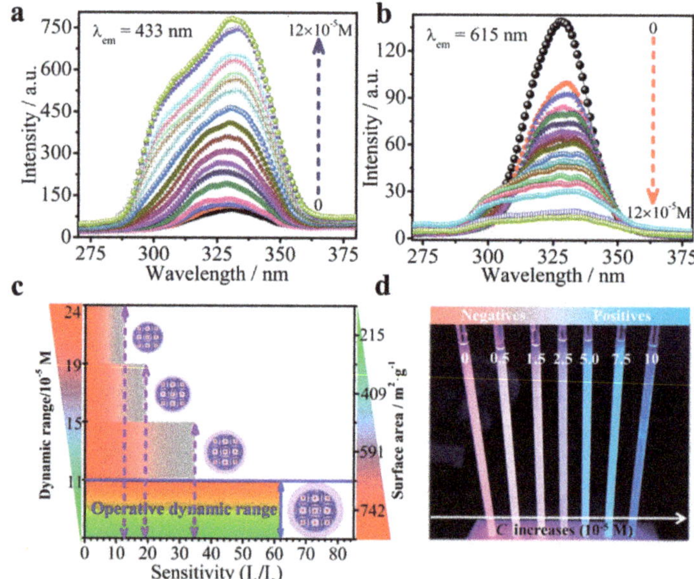

Figure 3. Ratiometric luminescence sensing of VMA by Eu-ZnMOF-*n*. (**a,b**) The widening of the excitation peak of Eu-ZnMOF-*n*, monitored at both 433 and 615 nm emissions with incremental addition of VMA. (**c**) The structure–property relationship of the probe's surface area to its dynamic range and sensitivity. (**d**) Color-coded digital array of Eu-ZnMOF-*n* probe after treatment with increased VMA concentrations under irradiation with a UV lamp (λ_{ex} = 254 nm). [32] Copyright 2020, John Wiley and Sons.

The similar atomic radii and chemical properties of Ln ions make it feasible to prepare multi-emissive Ln-MOFs [87,90]. Previous studies have shown that a Tb-to-Eu energy transfer can occur [91]; therefore, the Tb/Eu mixed-Ln MOF can serve as the ratiometric sensor. Chen and colleagues [84] developed a mixed Ln-MOF luminescence thermometer using 2,5-dimethoxy-1,4-benzenedicarboxylate (DMBDC) as the linker. The Tb-to-Eu luminescent intensity ratio in Tb-DMBDC and Eu-DMBDC decreased as the temperature increased due to the thermal activation of non-radiative decay pathways. In the mixed-Ln-MOF, $Eu_{0.0069}Tb_{0.9931}$-DMBDC, the increment of temperature led to a decrease in Tb emissions and an increase in Eu emissions, thus forming a ratiometric fluorescence nanothermometer that presented a wide temperature dynamic range. The unique phenomenon can be attributed not only to the efficient antenna effect between DMBDC and Ln ions, but also to the Tb-to-Eu energy transfer.

Some non-luminescent MOFs, such as ZIF-8, can serve as host matrices to encapsulate guest organic dyes for ratiometric sensing. The confinement and isolation of the organic dyes can effectively inhibit the intramolecular torsional motion and increase the conformational rigidity to produce a high quantum yield [92]. Qian and colleagues [77] reported a luminescent nanothermometer by encapsulating luminescent dyes, 4-methylumbelliferone (4-Mu), and fluorescein (Flu) in the pores of ZIF-8. The developed nanothermometer can response to temperature changes based on the Flu-to-4-MU emission ratio ($I_{Flu}/I_{4\text{-}MU}$) as well as the emission peak wavelength of 4-MU. These two kinds of readouts can self-calibrate to ensure the accuracy of the detection. Furthermore, the nanosized property of ZIF-8 and the excellent luminescence properties of dyes impart the sensor with a large dynamic range and a high spatial resolution, which are important for temperature mapping [93].

In addition to fluorescence, phosphorescence can also be tuned by rationally designing the structure of metal nodes or organic linkers of MOFs [42,79]. MOFs can integrate both fluorescence and phosphorescence in one nanosensor. For example, the emission and lifetime of phosphorescence, but not fluorescence, are easily quenched by triplet oxygen, which can be utilized as a ratiometric signal based on the fluorescence-to-phosphorescence ratio. Lin and colleagues [79] designed the mixed-linker nanoscale UiO MOF and decorated the structure with Rhodamine-B isothiocyanate (RITC) to form R-UiO MOF for ratiometric sensing of intracellular O_2 (Figure 4). In the nanostructure, the phosphorescent Pt-5,15-di(p-benzoato)porphyrin (DBP-Pt) linker acted as an O_2-sensitive probe, and the O_2-insensitive fluorescent RITC served as a reference. With 514-nm laser excitation, emissions at 630 nm from DBP-Pt and 570 nm from RITC were observed. With the increase of O_2 pressure, the DBP-Pt phosphorescence decreased significantly while the RITC fluorescence remained unchanged. The intracellular O_2 of CT26 cells at 4, 32, and 160 mmHg was detected based on ratiometric signals using confocal laser scanning microscopy.

Zang and colleagues [42] reported a fluorescence–phosphorescence dual-emissive oxygen sensing MOF: ([Ag$_{12}$(SBu^t)$_8$(CF$_3$COO)$_4$(bpy-NH$_2$)$_4$]$_n$ (abbreviated as Ag$_{12}$bpy-NH$_2$)), based on silver–chalcogenolate-cluster and bipyridine (bpy) linkers. The introduction of the amino group enhanced the spin–orbit coupling and increased the intersystem crossing efficiency to boost triplet excitons and prolong the lifetime of phosphorescence at 556 nm in vacuum. As a result, oxygen molecules quenched the phosphorescence at 556 nm, while the fluorescence emission at 456 nm remained nearly invariant. This ratiometric quantification manner ensured a LOD as low as 0.1 ppm. The introduction of other substitutional groups, such as methyl or F^- groups, extended the dynamic range of the ratiometric sensing. Therefore, tailoring the linker was deemed to be a powerful method for modulating luminescent sensing functionality.

Zhou and colleagues [94] used Prussian Blue (PB) and UCNPs to develop a nanoprobe (UC-PB) for the purpose of detecting and eliminating H_2S with a linear range of 0–150 μM and an LOD of 50 nM. The Er-doped UCNP, NaLuF$_4$:Yb,Er,Tm@NaLuF$_4$, presented multiple emission peaks at 550 nm, 650 nm, and 800 nm, which were quenched by adding the PB shell. H_2S triggered the decomposition of the PB shells to recover the strong upconversion luminescence (UCL) signal and the near infrared-to-green (N/G) ratio due to the cooperation of both redox and combination reactions. With the help of DL-propargylglycine (DL-PAG), the UC-PB was able to realize the *in vivo* near-infrared region ratiometric imaging, eliminate and inhibit the production of H_2S, which is meaningful for clinical acute pancreatitis treatment.

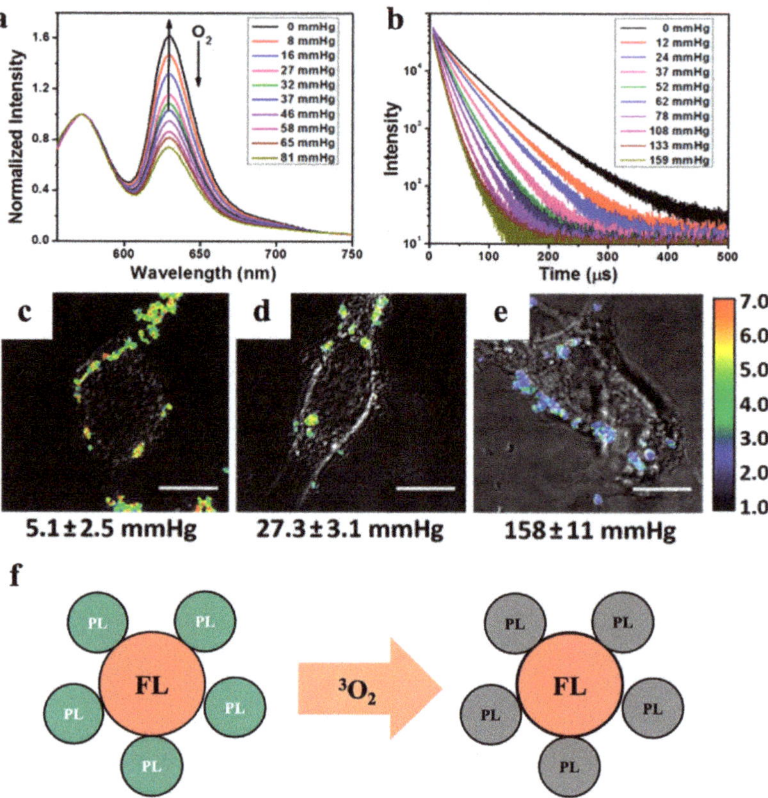

Figure 4. Oxygen sensing based on the phosphorescence-to-fluorescence ratio of R-UiO-1. (**a**) Emission spectra (λ_{ex} = 514 nm) and (**b**) phosphorescent decays (λ_{ex} = 405 nm) of R-UiO-1 in HBSS buffer under varied oxygen partial pressures. Ratiometric luminescence imaging (λ_{ex} = 514 nm) of CT26 cells after incubation with R-UiO-2 under (**c**) hypoxia, (**d**) normoxia, and (**e**) aerated conditions. Scale bar: 10 µm. (**f**) Schematics of fluorescence (Fl)–phosphorescence (Ph) dual emission centers, with Ph center quenched (Qu) in response to oxygen [79]. Copyright 2016, American Chemical Society.

2.3. Multi-Modal Signal

Fluorescence imaging (FLI), magnetic resonance imaging (MRI), and photoacoustic imaging (PAI) are widely used molecular imaging technologies which can realize non-invasive disease diagnosis and real-time *in vivo* lesion imaging [95,96]. Compared to single-modality imaging, which has the limitation of low penetration depth and low spatial resolution, multi-modality imaging integrates two or more modalities into one nanocomplex. Thus, it can provide more efficient and comprehensive information, which is desirable in the biomedical field [97].

Kuang and colleagues [53] developed an ultrasensitive and selective method for H_2O_2 detection based on a UCNP@ZIF-8/NiSx chiral complex. The NiSx moiety is a chiral nanoparticle with circular dichroism (CD) signals at 440 and 530 nm. The presence of the NiSx can quench the UCL signal of the UCNPs core at 540 nm, with the UCL signal at 660 nm remaining unchanged. The introduction of H_2O_2 caused NiSx to degrade, accompanied by the recovery of the UCL signal and the disappearance of the CD signal. This dual-mode signal of CD and fluorescence changes opens a new avenue for developing a toolbox for biomedical and biological analyses.

Yin and colleagues [98] reported a MnO_2-coated, hollow mixed metal (Mn/Cu/Zn) MOF that could allow the photosensitizer, indocyanine green (ICG), to have a high loading efficiency. The coexistence of Cu^+ and Cu^{2+} in the MOF, as verified by X-ray photoelectron spectroscopy, endows the complex with a glutathione-responsive "turn on" MRI ability. With the laser irradiation, the ICG can serve not only as the fluorescence and photothermal imaging agent, but also as the photodynamic therapy (PDT) agent. Therefore, this hollow MOF was used as a trimodality imaging-guided tumor therapy agent to highlight the efficiency of mixed-metal and mixed-valence strategies in tumor theranostic capacities.

For small functional molecule encapsulation, the risk of leakage and burst release is always a challenge [99]. To solve the aforementioned problem, Yang and colleagues [100] used the one-pot approach to prepare a Fe-based MOF (MIL-53) with defect structure due to the introduction of near-infrared dye (cypate). Further decoration of PEG and transferrin on the surface of nanoparticles (denoted as CMNP-Tf) was able to accelerate the passive and active targeting to the tumor region. The presence of cypate endowed the nanoparticles with excellent PAI and near-infrared fluorescence (NIRF) imaging properties, as well as reactive oxygen species (ROS) generation and photothermal therapy abilities. Furthermore, Fe also possesses a T_1-weighted MRI contrast property. Therefore, the CMNP-Tf can realize the NIRF-, PAI-, and MRI-guided tumor targeting imaging-guided photothermal/photodynamic performance.

Table 1. Summary of MOF based fluorescence applications.

MOF	Synthesis Method	Luminescence Center	$\lambda_{ex/em}$ (nm)	Target Molecules	LOD	Ref.
ZnMOF CoMOF	solvothermal	TCPE		HCl vapor histidine	2.63 ppm 2×10^{-6} M	[6]
Eu-ZnMOF	solvothermal	Eu & BPDC	330/433 & 615	vanillylmandelic acid		[32]
TbTATAB	solvothermal	Tb & linker		Hg^{2+}	4.4 nM	[39]
CDs@ZIF-8	one-pot room temperature crystal growth	CD	365/480	quercetin	3.5 nM	[40]
Ag_{12}bpy-NH_2		bpy-NH_2	370	O_2	11.4 mPa	[42]
Tb(III)@Cd-MOF	solvothermal	Tb^{3+}	325/544	diphenyl phosphate	0.022 mg/mL	[47]
MIL-101(Cr)	hydrothermal	Cy3	525/570	tetrodotoxin	0.006 ng/mL	[60]
ZIF-8⊃4-MU & Flu	solvothermal	4-MU & Flu	360/380–450 & 500–570	temperature		[77]
Zr-MOF	Suzuki coupling	Zr-MOF-ABt Zr-MOF-BDP	340/403 & 530 340/403 & 610	peroxynitrite		[78]
R-UiO	Suzuki coupling	DBP-Pt/RITC	514/570 & 630	O_2		[79]
EuMOF-FITC	hydrothermal	Eu^{3+} & FITC	380/525 & 611	biogenic amine	1.11 mg/L	[80]
Eu^{3+}/Tb^{3+} MOFs	solvothermal	Eu & Tb	280/547 & 491, 616 & 592	Fe^{3+}	3.86 μM	[83]
$Eu_{0.0069}Tb_{0.9931}$-DMBDC	solvothermal	Eu & Tb	355/613 & 545	temperature		[84]
Ru@MIL-101(Al)-NH_2	one-pot	Ru & linker	300/465 & 615	water	0.02% *v/v*	[89]
UC-PB	ligand-exchange and controllable complexation	UCNP	980/540 & 654	H_2S	50 nM	[94]
ZnMOF	solvothermal	Linker	370	Fe^{3+} Pb^{2+} $Cr_2O_7^{2-}$ CrO_4^{2-}	28 μM 600 μM 43 μM 45 μM	[101]
CdMOF				Fe^{3+}; Pb^{2+}; $Cr_2O_7^{2-}$; CrO_4^{2-}	57 μM; 370 μM; 71 μM; 31 μM	
ZJU-168(Tb or Eu)	solvothermal	Tb &linker Eu & linker	340/430 & 544 340/430 & 614	Glutamic acid	3.6 μM 4.3 μM	[102]
F-UiO	solvothermal	FITC	488 & 435/520	pH		[103]
BSA + KFP@ZIF-8/HP +primer + MB	one-pot room temperature crystal growth	Cy5		survivin mRNA	2.3 pM	[104]

3. MOFs-Based Nanosensors in SERS Applications

Surface-enhanced Raman spectroscopy is a hypersensitive technique that enhances Raman scattering of the analyte in proximity to a nanostructured substrate via electromagnetic or chemical enhancement [4,105]. It can provide structural fingerprint information on the low-concentration analyte in real time [106,107]. Owing to its high sensitivity and high selectivity, SERS has a breadth of applications in pharmaceutical and environmental analysis [69,108], food science [109], life sciences [5], clinical diagnosis [64,69], and other fields [106]. However, for metal substrates, problems remain regarding the application of real-world sample detection. To name a few, the selective adsorption of analytes onto the metal substrate is required to assure better sensitivity and specificity [110]; the binding strength between the analyte and plasmonic surfaces needs to be improved to facilitate better chemical enhancement; and the stability and reproducibility of metal substrates need to be improved to assure the robustness of the analytical methods [105].

As a new class of porous polymeric materials, MOFs present ultra-high porosity, a large surface area, and designable binding sites for multiple functionalization, which provide concentrating effects and multiple selectivity for the analytes [22,23,63]. In addition to directly serving as the SERS substrate, the MOF-metal composite SERS substrate can further improve both the sensitivity and the selectivity of the analytical measurements; in such a way, the stability and uniformity of the enhancement substrate can be improved [32,75]. Therefore, MOF materials have found increasing applications in SERS measurements (Scheme 1, Table 2).

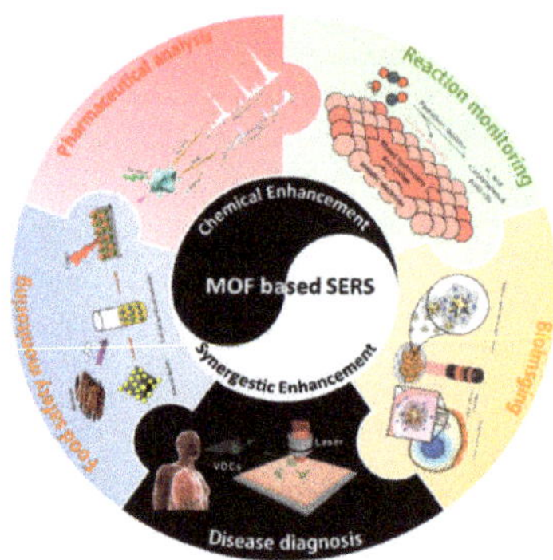

Scheme 1. Applications of MOF-based surface-enhanced Raman spectroscopy (SERS).

Table 2. Summary of MOF and MOF-plasmonic system-based SERS applications.

SERS Substrate	Target Molecules	Enhancement Mechanism	LOD	Ref
UiO-67	2,4,6-Trinitrophenol (TNP)	CM [a]		[23]
Au/MOF-199	acetamiprid	EM [b] & CM	10^{-8} M	[42]
Au@MIL-101 (Cr)	tetrodotoxin	EM & CM	0.008 ng/mL	[60]
GSPs@ZIF-8	aldehyde VOCs	EM & CM	10^{-9} M	[64]
Ag@MIL-101 (Cr)	nitrofurantion		10^{-7} M	[65]
APTES@ZIF-67	benzaldehyde	EM & CM	10^{-2} M	[65]
ZIF-8	methyl orange		10^{-4} M	[65]
Au-L/D-AlaZnCl	pseudoephedrine	EM & CM	10^{-12} M	[72]
$Cu_2O@SiO_2@ZIF-8@Ag$	phenol red	EM & CM	10^{-12} M	[73]
Au/MOF-74	4-nitrothiophenol	EM & CM	69 nmol·L^{-1}	[74]

Table 2. *Cont.*

SERS Substrate	Target Molecules	Enhancement Mechanism	LOD	Ref
Au@ZIF-8	4-nitrobenzenethiol	EM & CM	0.1 nM	[75]
Ag@MIL-101 (Fe)	dopamine	EM & CM	0.32 p M	[76]
Au/MIL-101	rhodamine 6G; benzadine	EM & CM	41.75 fmol; 0.54 fmol	[110]
Mo-MOF	crystal violet	CM	10^{-6} M	[111]
MIL-100 (Fe)	toluene	CM	2.5 ppm	[112]
	acetone		20 ppm	
	chloroform		20 ppm	
	isopropanol		100 ppm	
	4-ethylbenzaldehyde		26 ppm	
MIL-100 (Fe-Zr)	isopropanol		50 ppm	[112]
Au@MIL-101 (Fe)	toluene	EM & CM	0.48 ppb	[112]
Au@NU-901	4′-mercaptobiphenylcarbonitrile	EM & CM		[113]

[a]: Chemical enhancement; [b]: electromagnetic enhancement.

3.1. SERS Substrates

To date, many MOFs have been directly used as SERS substrates; examples include ZIF-67, ZIF-8, Mo-MOF, MIL-100(Fe), etc. [91,112,114]. The enhancement by MOF can be attributed to chemical enhancement (CM) [63,105], which has been proposed as the primary enhancement mechanism for plasmon-free substrates such as MOF, semiconductors, and other metal oxides (Cu_2O [73], WO_3 [115], TiO_2 [116], VO_2 [117–119]), etc. Particularly, charge transfer transitions may be major contributors to SERS [120]. The charge transfers between the highest occupied molecular orbital (HOMO) of the analyte and the conduction band (CB) edge of the substrate material, or between the valence band (VB) of the substrate material and the lowest unoccupied molecular orbital (LUMO) of the molecules, are crucial for the enhancement effects. Other than the resonance Raman enhancement due to electronic transition between HOMO and LUMO, the electronic transition between VB and CB of the substrate material can also contribute to the enhancement of the Raman signal due to the resonance process [121–123].

The first MOF-enhanced Raman spectroscopic study was conducted by Tsung-Han Yu et al. [121] using methyl orange (MO) adsorbed in MIL-100 and MIL-101 as the model system. The study suggested that the Raman intensity enhancement was due to the charge transfer between the metal oxide clusters in MOFs and the adsorbed MO molecules. The study also suggested that the SERS effect was also orientation-dependent, which is in agreement with the basic understanding of Raman spectroscopy.

MIL-100(Fe) has been demonstrated, for the first time, by Li and colleagues to act as a SERS-active substrate to detect volatile organic compounds (VOCs) that usually possess low Raman cross-sections [112], and a LOD of 2.5 ppm was achieved for toluene. Based on special adsorption energy and density functional theory (DFT) calculations, the charge transfer enhancement mechanism was suggested for high SERS activity. Selective enhancement was attributed to resonance between laser energy and the photo-induced charge transfer energy, as well as the different dispersive energy between the ligand of the MOFs and the analyte (Figure 5). Through the ion exchange strategy, the MIL-100(Fe-Zr) complex improved LOD for isopropanol (50 ppm) compared to the pristine MIL-100(Fe) (100 ppm). This MIL-100(Fe)-based sensing platform was successfully used to monitor gaseous indicators, including 4-ethylbenzaldehyde, acetone, and isopropanol, for early diagnosis of lung cancer.

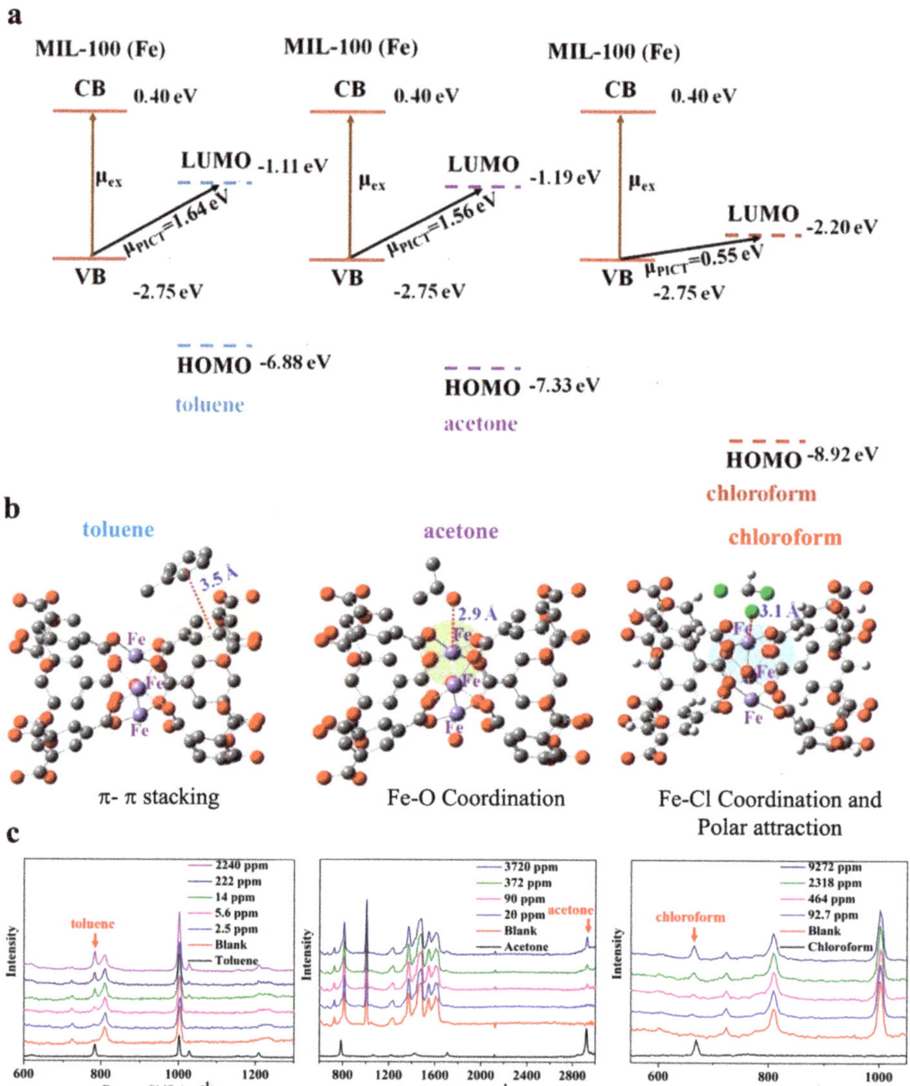

Figure 5. SERS of gaseous small molecules by MIL-100 (Fe). (**a**) Energy-level diagrams of toluene, acetone, and chloroform relative to MIL-100 (Fe) with respect to the vacuum level. (**b**) Adsorption affinity and binding energy of toluene, acetone, and chloroform on MIL-100(Fe). (**c**) SERS spectra of toluene, acetone, and chloroform vapor, respectively, on MIL-100(Fe), with concentrations as indicated [112]. Copyright 2020, John Wiley and Sons.

MOFs have a high degree of tailorability, with the ability to choose nodes and linkers as well as to tune the framework topologies. Consequently, the electronic energy band structure can be tuned to match the molecular orbital energy level of the analyte in order to facilitate controllable combinations of several resonances, such as the charge transfer, inter-band, and molecule resonances, in addition to the ground-state charge transfer interactions; thus, a significantly enhanced Raman signal can be achieved. Encouragingly, based on the pore-structure optimization and surface modification strategy, Zhao and colleagues synthesized a series of MOFs by using metal clusters $M_2(COO)_4$ (M = Zn, Co, and Cu) as the nodes and Tetrakis(4-carboxyphenyl)porphyrin (TCPP) or 2-methylimidazole as the linker (Figure 6) [114]. The electronic band structure of the MOFs can be tuned to match that of the target analyte, such that the enhancement factor (EF) of ZIF-67 can reach as high as 1.9×10^6 with the LOD as low as 10^{-8} M. The high flexibility of the MOF structure can provide high levels of variety for SERS applications.

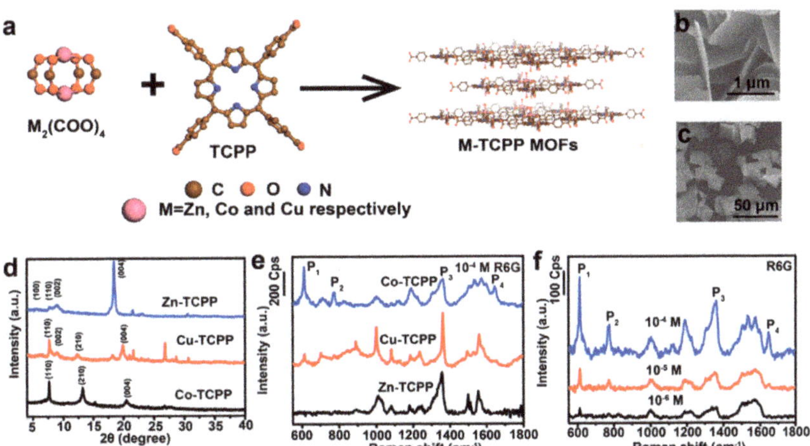

Figure 6. Structure and SERS effect of M-TCPP MOF. (**a**) Illustration of the process of synthesizing the M-TCPP (M = Zn, Co, and Cu) MOFs. (**b**,**c**) SEM images of the Cu-TCPP MOFs and Zn-TCPP MOFs. (**d**) PXRD patterns of the Co-TCPP MOFs, Zn-TCPP MOFs, and Cu-TCPP MOFs. (**e**) SERS spectra of R6G (10^{-4} M) on the Co-TCPP, Zn-TCPP, and Cu-TCPP substrates. (**f**) SERS spectra of R6G on the Co-TCPP substrate at three different concentrations: 10^{-4}, 10^{-5}, and 10^{-6} M [114]. Copyright 2019, American Chemical Society.

Using MOFs as the SERS-active substrate can realize selective enhancement of the target molecule for the Raman signal. However, precise tuning of the band energy is required to facilitate resonance with laser energy and charge transfer energies, which makes it difficult to realize SERS for a breadth of molecules of interest using one MOF substrate. Furthermore, there is still room to improve the sensitivity which can be achieved by the combination of MOFs with plasmonic substrates.

Electromagnetic enhancement (EM) is a physical enhancement process attributed to the localized surface plasmon resonance (LSPR) of the noble metal [4]. When the laser impinges on the metal nanostructure, e.g., a metal nanoparticle, the electromagnetic wave causes the collective oscillation of the delocalized conduction electrons. LSPR occurs when the frequency of the light matches with the oscillation frequency. The coupling of incident light with the metal nanostructures results in a huge enhancement of the local electromagnetic field, which is the major mechanism accounting for SERS. The combination of MOF with noble metal SERS substrates may be able to realize the synergetic enhancement by integrating CM and EM into one system, thus achieving maximum sensitivity [23,63]. Furthermore, such a MOF-SERS combination may be but one solution for problems associated with metal substrates, such as the low concentration and high mobility of the analyte at the enhancement spot, complicated matrix interferences, and modest substrate stability [63,124]. Specifically, the adjustable pore size of MOFs enables them to serve as molecular sieves to filter the analyte based on molecular size [71,75,104]; MOFs' shells can protect the metal substrates from oxidation and reactive species in the complex matrix, thus improving the stability [125].

MOFs coatings have the advantage of controlling the hotspot distribution to improve the SERS performance [71–73]. In the SERS technique, the Raman signal can be significantly enhanced only when the analyte is confined to the proximal distance to the plasmonic surface (<3–5 nm), because enhancement depends on the exponential decay of the plasmonic field at the metal–medium interface [63]. In the work by Wang and colleagues, gold superparticles (GSPs) were used to provide high-density hotspots, and the coating of ZIF-8 over GSPs provided a further enhanced SERS effect [64] (Figure 7). The finite-difference time-domain (FDTD) calculation revealed that the MOF coating was able to enhance the intensity of the electromagnetic field around the metal surface, and the high dielectric constant of MOF prevents the decay of the electromagnetic field along the radial direction. Consequently, a very intense SERS effect was observed [64,75].

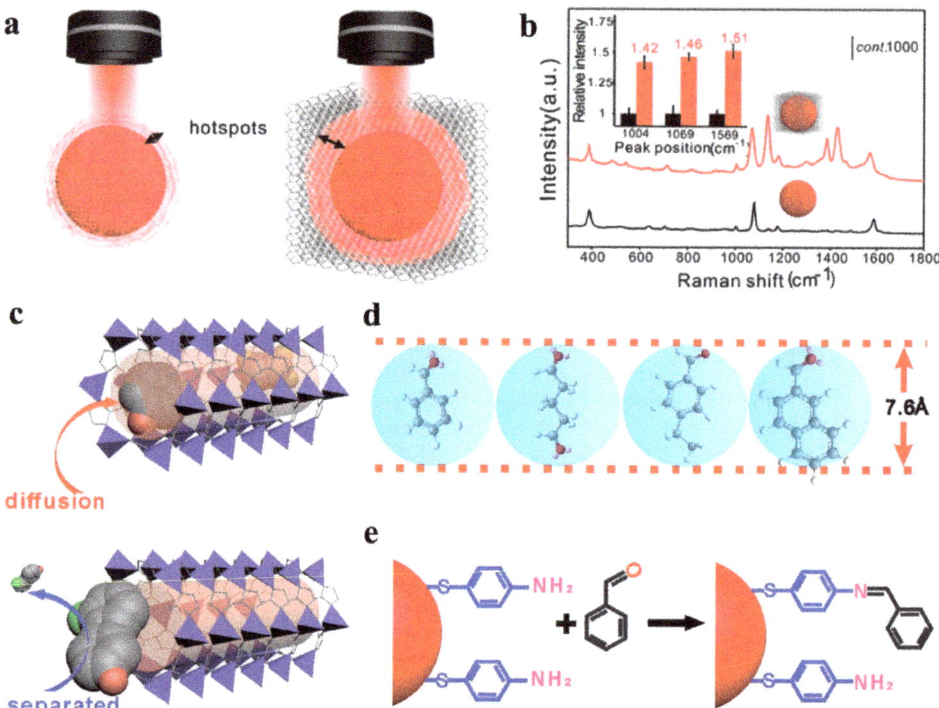

Figure 7. GSPs@ZIF-8 and the mechanism for selective SERS detection. (**a**) Hotspots in GSPs (left) and GSPs@ZIF-8 (right) around the edges of the plasmon. (**b**) SERS spectra of 4-ATP on GSPs (black) and GSPs@ZIF-8 (red), with the inset showing the relative intensity of the Raman spectra. (**c**) An illustration of the principle for achieving the selectivity. (**d**) The maximum diameter of four main analytes that were allowed to pass through the pores of ZIF-8. (**e**) The reaction used to capture aldehyde vapors via covalent linkage to the GSPs [64]. Copyright 2018, John Wiley and Sons.

The synergistic SERS effect was achieved by Wang and colleagues, via site-selective deposition of ZIF-8 on gold nanobipyramids (Au NBPs) [71]. Deposition of ZIF-8 was observed around the distal end, waist, or the surface of the Au NBPs (Figure 8). When ZIF-8 was located at the distal ends (Au BNPs@end-ZIF 8), the largest electric field enhancement was achieved, and the Raman signals on Au BNPs@end-ZIF 8 were at least twice those on the Au NBPs. Li and colleagues used ZIF-8-coated cuprous oxide/silica core–shell nanostructure ($Cu_2O@SiO_2@ZIF$-8) as a template to precisely control the growth of Ag NPs of varying sizes (2–29 nm) [73]. The results reflected that when the size of the Ag NPs matched well with the pores of $Cu_2O@SiO_2@ZIF$-8, the strongest electromagnetic field was generated. The $Cu_2O@SiO_2@ZIF$-8 provided abundant and uniformly distributed hotspots, thus resulting in an LOD of 5.76×10^{-12} mol·L^{-1} and a limit of quantification (LOQ) of 1.92×10^{-12} mol·L^{-1}, respectively, in the detection of phenol red.

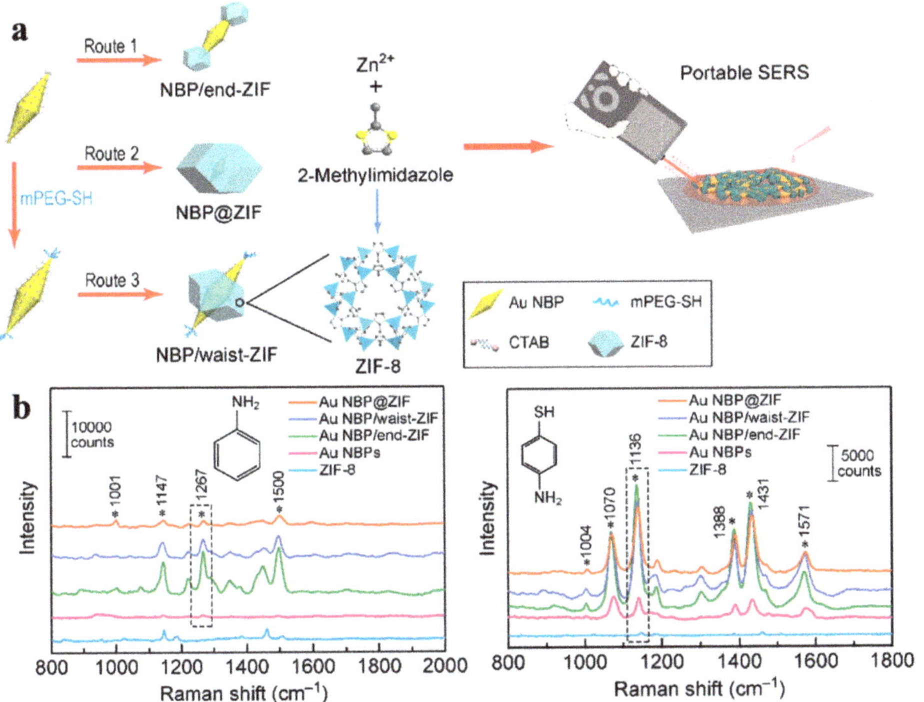

Figure 8. SERS of NBP@ZIF. (**a**) Schematics showing the routes for selective deposition of ZIF-8 on the Au NBPs and SERS detection. (**b**) SERS spectra measured with the ZIF-8 nanocrystals, Au NBPs, NBP/end-ZIF, NBP/waist-ZIF, and NBP@ZIF nanostructures for the detection of aniline and 4-ATP, respectively [71]. Copyright 2021, American Chemical Society.

3.2. Other Effects of MOFs for Improving Selectivity and Sensitivity

Owing to the large specific surface area [20], uniform porosity [54], structural adaptability and flexibility [73,114], and ease of functionalization [63], MOFs as shells can enhance the selectivity and specificity of SERS substrates through physical adsorption and chemical recognition [72,110]. The introduction of additional functional layers, such as aptamers and antibodies, can ensure selective adsorption through multiple molecular interactions [23]. In addition, other factors, such as the thickness of MOFs, the nature of the metal species, and the functional groups afforded by the organic linkers [23,126], can also affect the selectivity.

The MOF shell over the plasmonic SERS substrate can serve as a sieve to allow only the target of interest to diffuse to hotspots, as well as to facilitate an efficient reaction with the Raman label by prolonging the contact time. VOCs are important biomarkers for early diagnoses of diseases [127], but the low concentration and high mobility of gaseous molecules result in insufficient collisions between gas molecules and SERS substrates, significantly compromising the detection sensitivity [128]. Wang and colleagues [64] used a GSPs@ZIF-8 SERS substrate to selectively detect aldehydes, a lung cancer biomarker in patients' exhalation. The detection of gaseous aldehydes is currently limited by its small Raman cross section and poor adsorptivity on SERS substrates. The ZIF-8 shell allows the small vapor-phase aromatic compounds such as benzaldehyde, glutaricdiadehyde, and 4-ethylbenzaldehyde, but not 2-naphthaldehyde, to diffuse into the channel due to the sieving effect (Figure 7c–e). The prolonged contact time between gaseous molecules and the GSPs hotspots can facilitate the reaction of the analyte with the Raman-active probe molecule p-aminothiophenol (4-ATP) through the Schiff base reaction to generate a distinguished Raman signal. Apart from the physical adsorption, the MOFs can also selectively adsorb analytes through chemical recognition. Yusuke and colleagues [72] prepared hierarchical mesoporous Au films coated with homochiral MOF, which were able to realize ultrasensitive and enantioselective sensing of pseudoephedrine (PE) in complex bio-samples. The chiral ligands (a kind of alanine derivative) were used to endow the MOF with the homochiral property to distinguish (+)-PE. For analysis of PE in blood serum, the matrix effect was reduced by taking the advantage of the pore size limit to prevent large molecules from entering the MOF shell. This smart SERS substrate enabled an extremely low detection limit (10^{-12} M) in the complex biomatrix without preliminary separation.

In reality, the size of most of the biomarkers does not match that of the MOF channel. Efforts have been made to detect the metabolites as an indicator of the amount of biomarkers. By using an in situ reduction strategy, Zhou and colleagues [69] constructed a AuNPs@MIL-101@GOx (or AuNPs@MIL-101@LOx) nanoplatform for the detection of glucose/lactate, which are important neurochemicals associated with many physiological and pathological brain functions, such as ischemia, learning, and memory. The Raman-inactive reporter leucomalachite green (LMG) was oxidized into the active malachite green (MG) through a cascade of catalytic processes, and the signal intensity was used to indicate the amount of glucose/lactate. This nanoplatform has also been used to evaluate the therapeutic effects of astaxanthin for the purpose of alleviating cerebral ischemic injuries.

The metal-MOF-based SERS substrates have also been used in drug delivery and bioimaging due to the high surface area and porosity, excellent biocompatibility, and stability of MOFs [5]. The exposed active sites on the surface of the MOFs can be used for functionalization with recognition units, and the large specific surface area provides sufficient accessible binding sites for target molecules [80]. For example, the carboxyl group on the surface of the Au@Cu$_3$(BTC)$_2$ nanoparticles was used for functionalization with the aptamers [5]. In another example, the ZIF-8 shell on the Au@Ag surface was used to conjugate with IgG antibodies and recombinant nanobodies [70], based on the high affinity of polyhistidine toward transition metal ions [129]. The Au@Ag@ZIF-8 substrate was used to detect CD44 and EGFR biomarkers in mixed cell cultures, indicating the potential of the nanoprobes for SERS imaging and multiplexed bio-detection.

Farha and colleagues [113] reported the controlled encapsulation of gold nanorods (AuNRs) with a scu-topology Zr-MOF (NU-901) via the room-temperature assembly of MOF on AuNRs seeds. After incubating AuNR@NU-901 with a mixture of thiolated polystyrene (PST-SH; Mw = 5000 g/mol) and 4′-mercaptobiphenylcarbonitrile (BPTCN) molecules (roughly 15×7 Å along the thiol$-$CN axis and phenyl axis), the resulting spectrum closely matched with that of BPTCN alone. This result demonstrated that the prepared AuNR@MOFs were able to take-up molecules with suitable sizes and block large molecules from the pores, thus facilitating highly selective SERS detection at the AuNR ends.

3.3. Enhancement of the Stability, Homogeneity, and Reproducibility of SERS Substrates

The instability and reproducibility of the plasmonic nanoparticles under harsh environment represent inherent challenges in SERS detection [4,63,65,72]; therefore, a protective layer is necessary. Due to their excellent chemical and thermal stability [130], as well as their mechanical robustness [131], MOFs are an ideal candidate to act as a stabilizing layer. For example, Liang and colleagues [65] prepared a dense MIL-101(Cr) film on the rough titanium oxide foil via a secondary growth method, and then the Ag$^+$ was reduced to Ag on the surface of the film to form the Ag@MIL-101(Cr) film SERS substrate. In such a way, the excellent SERS effect and the high reproducibility of the SERS substrate were achieved to realize the detection of nitrofurantoin (down to 10^{-7} M) without any complex subsequent procedures. Li and colleagues reported a highly sensitive and continuously stable 3D substrate (Cu$_2$O@SiO$_2$@ZIF-8@Ag) for SERS detection of phenol red with a LOD of 5.76×10^{-12} mol·L^{-1} and LOQ of 1.92×10^{-12} mol·L^{-1}, and a high enhancement factor of 1.7×10^{7} was achieved even after 35 days [73].

The sensitivity and quantification performance of the SERS technique often contradict one another due to the modest reproducibility of the SERS substrate. Yang and colleagues constructed an integrated SERS platform with analyte enrichment and analyte filtration functions (referred to as AEF-SERS) to simultaneously achieve a good quantification performance and ultra-high sensitivity (Figure 9) [75]. In their work, single Au NRs were separated from each other through the coating of a thick ZIF-8 shell to form a AuNR@ZIF-8 submicroscale truncated rhombic dodecahedron (TRD); thus, a homogeneous SERS substrate was produced to improve the reproducibility of the detection. The separation of Au NRs may reduce the number of hotspots, thus compromising the sensitivity. However, the authors were still able to successfully realize a highly sensitive detection by constructing a polydimethylsiloxane (PDMS) brush surface that was capable of shrinking the analyte dispersion area by a million-fold in order to enrich the analyte.

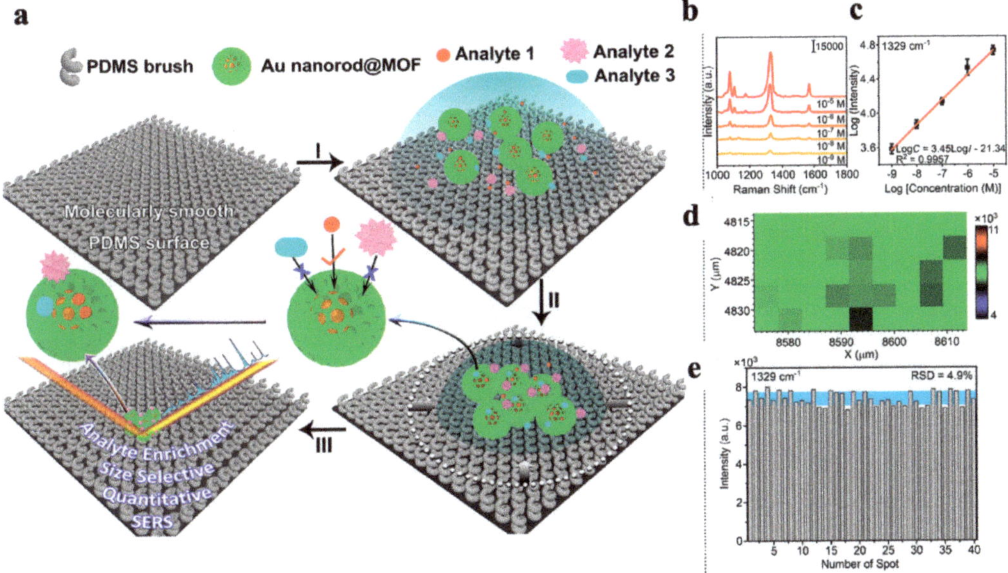

Figure 9. (**a**) Schematic illustration of the working principle of the AEF-SERS platform, showing application of the test solution (I), the enrichment and sieving effect (II), and the formation of analyte-SERS substrate aggregates (III) for selective and sensitive quantification of the analyte. (**b**) SERS spectra of 4-NBT at different concentrations using AuNR@ZIF-8 TRDs on the PDMS brush surface. (**c**) The calibration curve of 4-NBT at 1329 cm^{-1}. (**d**) SERS mapping of 10 nM 4-NBT at 1329 cm^{-1} using AuNR@ZIF-8 TRDs. (**e**) Intensity variations at 1329 cm^{-1} from randomly chosen 40 SERS spectra using AuNR@ZIF-8 TRDs [75]. Copyright 2020, American Chemical Society.

The integration of noble metals and MOFs can speed up the development of the SERS technique. The high sensitivity can be partially explained by pre-concentration of the analyte through physical adsorption and chemical recognition. Aptamers, antibodies, and other recognition units can be easily used to modify MOFs in order to increase the molecular recognition specificity. In addition, when used in complex environment such as the biomatrix, the MOF shells provide a physical defense to improve the stability and reproducibility of the substrates, as well as to reduce the nonspecific adsorption and, thus, improving the detection sensitivity.

4. Conclusions and Future Perspectives

Applications of MOFs in the analytical and bioanalytical fields have experienced rapid growth due to their unique structural features. This mini-review summarized the advances of MOF-based optical detection methods, including luminescence and SERS, from the following aspects: the development of MOF-based luminophores, including the single luminophore signal, ratiometric signal and multi-modality signals; and the SERS effect of including MOFs as enhancement substrates and as auxiliary moieties for target molecule concentration, selective separation, and SERS substrate homogeneity for the purpose of improving the method's robustness.

Compared to detection based on single parameter, the multiplexed detection with which multiple target or parameter detection is achieved in one sample volume can be more informational and can help to draw solid conclusions in the analysis of biological samples. Optical-based analytical methods have been used in the field of multiplexed detection due to their non-invasiveness, excellent spatiotemporal resolution, and, most importantly, their multiple coding elements, including intensity, wavelength, lifetime, location, and combinations of the above. Luminescent MOFs should be developed as an excellent type of multiplexing probe, because the broad choice of guest molecules and the structural diversity of MOFs provide diversified coding elements. However, the multiplexing capability of photoluminescent MOFs has been less frequently studied, if at all. Therefore, more efforts should be contributed to designing luminescent MOFs with multiple signal sources to facilitate the necessary analytical and bioanalytical applications.

Lanthanide ion-based MOFs, including mixed Ln ions, exhibit tunable luminescence peaks and lifetimes, making them suitable for ratiometric, multiplexed, and multi-modal measurements. However, as luminescent nanoprobes, the modest luminescence quantum yield in aqueous media impedes the application of these MOF-based luminescence sensors in aqueous solution as well as in biological samples. Future efforts should also include the engineering of the building elements of MOF structures to create more MOF-based optical nanosensors with improved performance in terms of factors such as physical and chemical stability, photostability, easiness in functionalization, quantum yield, red/near infrared emission wavelength, and tuned luminescence lifetime.

For the applications of MOFs in SERS measurements, the SERS mechanism of MOFs needs to be explored in more depth and breadth to rationally achieve the maximum SERS sensitivity. Further novel and facile approaches are expected to produce distinctive Raman signals via chemistry of the MOF with the analyte of interest. Efforts towards reproducible MOF substrates with high enhancement factors will be crucial to applications of MOF-based SERS in practical samples.

In both the photoluminescence and SERS fields, the breadth of applications need to be further explored in order to most effectively utilize the excellent physical and chemical properties of MOFs. The aforementioned needs represent challenges, but they also represent opportunities for MOF-based optical nanosensors to play a more significant role in bioanalytical applications.

Author Contributions: Conceptualization, writing and editing, N.L.; discussion, writing and editing, C.W.; discussion and editing, R.L.; discussion and editing, X.C. All authors have read and agreed to the published version of the manuscript.

Funding: This research was funded by National Natural Science Foundation of China grant number [21974006] and National Natural Science Foundation of China grant number [22134005].

Institutional Review Board Statement: Not applicable.

Informed Consent Statement: Not applicable.

Data Availability Statement: Not applicable.

Conflicts of Interest: The authors declare no conflict of interest.

References

1. Stich, M.I.J.; Fischer, L.H.; Wolfbeis, O.S. Multiple fluorescent chemical sensing and imaging. *Chem. Soc. Rev.* **2010**, *39*, 3102–3114.
2. Huang, X.; Song, J.; Yung, B.C.; Huang, X.; Xiong, Y.; Chen, X. Ratiometric optical nanoprobes enable accurate molecular detection and imaging. *Chem. Soc. Rev.* **2018**, *47*, 2873–2920.
3. Deng, X.; Liang, S.; Cai, X.; Huang, S.; Cheng, Z.; Shi, Y.; Pang, M.; Ma, P.; Lin, J. Yolk-shell structured Au nanostar@metal-organic framework for synergistic chemo-photothermal therapy in the second near-infrared window. *Nano Lett.* **2019**, *19*, 6772–6780. [CrossRef]
4. Sun, H.; Yu, B.; Pan, X.; Zhu, X.; Liu, Z. Recent progress in metal–organic frameworks-based materials toward surface-enhanced Raman spectroscopy. *Appl. Spectrosc. Rev.* **2022**, *57*, 513–528. [CrossRef]
5. He, J.; Dong, J.; Hu, Y.; Li, G.; Hu, Y. Design of Raman tag-bridged core-shell Au@Cu$_3$(BTC)$_2$ nanoparticles for Raman imaging and synergistic chemo-photothermal therapy. *Nanoscale* **2019**, *11*, 6089–6100. [PubMed]
6. Zhu, Z.H.; Ni, Z.; Zou, H.H.; Feng, G.; Tang, B.Z. Smart metal–organic frameworks with reversible luminescence/magnetic switch behavior for HCl vapor detection. *Adv. Funct. Mater.* **2021**, *31*, 2106925. [CrossRef]
7. Eddaoudi, M.; Moler, D.B.; Li, H.; Chen, B.; Reineke, T.M.; O'keeffe, M.; Yaghi, O.M. Modular chemistry: Secondary building units as a basis for the design of highly porous androbust metal-organic carboxylate frameworks. *Acc. Chem. Res.* **2001**, *341*, 319–330. [CrossRef] [PubMed]
8. Liu, J.; Fan, Y.Z.; Zhang, K.; Zhang, L.; Su, C.Y. Engineering porphyrin metal-organic framework composites as multifunctional platforms for CO$_2$ adsorption and activation. *J. Am. Chem. Soc.* **2020**, *142*, 14548–14556.
9. Liu, C.; Xing, J.; Akakuru, O.U.; Luo, L.; Sun, S.; Zou, R.; Yu, Z.; Fang, Q.; Wu, A. Nanozymes-engineered metal-organic frameworks for catalytic cascades-enhanced synergistic cancer therapy. *Nano Lett.* **2019**, *19*, 5674–5682. [CrossRef]
10. Liu, X.; Pan, Y.; Yang, J.; Gao, Y.; Huang, T.; Luan, X.; Wang, Y.; Song, Y. Gold nanoparticles doped metal-organic frameworks as near-infrared light-enhanced cascade nanozyme against hypoxic tumors. *Nano Res.* **2020**, *13*, 653–660. [CrossRef]
11. Shao, Y.; Liu, B.; Di, Z.; Zhang, G.; Sun, L.D.; Li, L.; Yan, C.H. Engineering of upconverted metal-organic frameworks for near-infrared light-triggered combinational photodynamic/chemo-/immunotherapy against hypoxic tumors. *J. Am. Chem. Soc.* **2020**, *142*, 3939–3946. [PubMed]
12. Zhou, H.; Fu, C.; Chen, X.; Tan, L.; Yu, J.; Wu, Q.; Su, L.; Huang, Z.; Cao, F.; Ren, X.; et al. Mitochondria-targeted zirconium metal-organic frameworks for enhancing the efficacy of microwave thermal therapy against tumors. *Biomater. Sci.* **2018**, *6*, 1535–1545. [CrossRef] [PubMed]
13. Liu, B.; Hu, F.; Zhang, J.; Wang, C.; Li, L. A biomimetic coordination nanoplatform for controlled encapsulation and delivery of drug-gene combinations. *Angew. Chem. Int. Ed.* **2019**, *58*, 8804–8808. [CrossRef]
14. Alsaiari, S.K.; Patil, S.; Alyami, M.; Alamoudi, K.O.; Aleisa, F.A.; Merzaban, J.S.; Li, M.; Khashab, N.M. Endosomal escape and delivery of CRISPR/Cas9 genome editing machinery enabled by nanoscale zeolitic imidazolate framework. *J. Am. Chem. Soc.* **2017**, *140*, 143–146. [CrossRef]
15. Wang, S.; Chen, Y.; Wang, S.; Li, P.; Mirkin, C.A.; Farha, O.K. DNA-functionalized metal-organic framework nanoparticles for intracellular delivery of proteins. *J. Am. Chem. Soc.* **2019**, *141*, 2215–2219. [CrossRef] [PubMed]
16. Wang, C.; Zhao, P.; Yang, G.; Chen, X.; Jiang, Y.; Jiang, X.; Wu, Y.; Liu, Y.; Zhang, W.; Bu, W. Reconstructing the intracellular pH microenvironment for enhancing photodynamic therapy. *Mater. Horiz.* **2020**, *7*, 1180–1185. [CrossRef]
17. Chen, Q.W.; Liu, X.H.; Fan, J.X.; Peng, S.Y.; Wang, J.W.; Wang, X.N.; Zhang, C.; Liu, C.J.; Zhang, X.Z. Self-mineralized photothermal bacteria hybridizing with mitochondria-targeted metal–organic frameworks for augmenting photothermal tumor therapy. *Adv. Funct. Mater.* **2020**, *30*, 1909806. [CrossRef]
18. Yu, H.; Cheng, Y.; Wen, C.; Sun, Y.Q.; Yin, X.B. Triple cascade nanocatalyst with laser-activatable O$_2$ supply and photothermal enhancement for effective catalytic therapy against hypoxic tumor. *Biomaterials* **2022**, *280*, 121308. [CrossRef]
19. Liu, Y.; Zhang, C.; Xu, C.; Lin, C.; Sun, K.; Wang, J.; Chen, X.; Li, L.; Whittaker, A.K.; Xu, H.B. Controlled synthesis of up-conversion luminescent Gd/Tm-MOFs for pH-responsive drug delivery and UCL/MRI dual-modal imaging. *Dalton Trans.* **2018**, *47*, 11253–11263. [CrossRef]
20. Deng, J.; Wang, K.; Wang, M.; Yu, P.; Mao, L. Mitochondria targeted nanoscale zeolitic imidazole framework-90 for ATP imaging in live cells. *J. Am. Chem. Soc.* **2017**, *139*, 5877–5882.
21. Wang, Y.-M.; Liu, W.; Yin, X.-B. Self-limiting growth nanoscale coordination polymers for fluorescence and magnetic resonance dual-modality imaging. *Adv. Funct. Mater.* **2016**, *26*, 8463–8470. [CrossRef]
22. Zhang, G.; Shan, D.; Dong, H.; Cosnier, S.; Al-Ghanim, K.A.; Ahmad, Z.; Mahboob, S.; Zhang, X. DNA-Mediated Nanoscale Metal—Organic frameworks for ultrasensitive photoelectrochemical enzyme-free immunoassay. *Anal. Chem.* **2018**, *90*, 12284–12291. [CrossRef] [PubMed]
23. Sen Bishwas, M.; Malik, M.; Poddar, P. Raman spectroscopy-based sensitive, fast and reversible vapour phase detection of explosives adsorbed on metal–organic frameworks UiO-67. *New J. Chem.* **2021**, *45*, 7145–7153. [CrossRef]
24. Li, L.; Zou, J.; Han, Y.; Liao, Z.; Lu, P.; Nezamzadeh-Ejhieh, A.; Liu, J.; Peng, Y. Recent advances in Al(iii)/In(iii)-based MOFs for the detection of pollutants. *New J. Chem.* **2022**, *46*, 19577–19592. [CrossRef]

25. Dong, X.; Li, Y.; Li, D.; Liao, D.; Qin, T.; Prakash, O.; Kumar, A.; Liu, J. A new 3D 8-connected Cd(ii) MOF as a potent photocatalyst for oxytetracycline antibiotic degradation. *CrystEngComm* **2022**, *24*, 6933–6943. [CrossRef]

26. Qin, L.; Li, Y.; Liang, F.; Li, L.; Lan, Y.; Li, Z.; Lu, X.; Yang, M.; Ma, D. A microporous 2D cobalt-based MOF with pyridyl sites and open metal sites for selective adsorption of CO_2. *Micropor. Mesopor. Mater.* **2022**, *341*, 112098. [CrossRef]

27. Qin, L.; Liang, F.; Li, Y.; Wu, J.; Guan, S.; Wu, M.; Xie, S.; Luo, M.; Ma, D. A 2D porous zinc-organic framework platform for loading of 5-fluorouracil. *Inorganics* **2022**, *10*, 202. [CrossRef]

28. Li, M.; Yin, S.; Lin, M.; Chen, X.; Pan, Y.; Peng, Y.; Sun, J.; Kumar, A.; Liu, J. Current status and prospects of metal-organic frameworks for bone therapy and bone repair. *J. Mater. Chem. B* **2022**, *10*, 5105–5128. [CrossRef]

29. Cui, Y.; Zhang, J.; He, H.; Qian, G. Photonic functional metal–organic frameworks. *Chem. Soc. Rev.* **2018**, *47*, 5740–5785. [CrossRef]

30. Dong, M.-J.; Zhao, M.; Ou, S.; Zou, C.; Wu, C.-D. A luminescent Dye@MOF platform: Emission fingerprint relationships of volatile organic molecules. *Angew. Chem. Int. Ed.* **2014**, *53*, 1575–1579. [CrossRef]

31. Lustig, W.P.; Mukherjee, S.; Rudd, N.D.; Desai, A.V.; Li, J.; Ghosh, S.K. Metal–organic frameworks: Functional luminescent and photonic materials for sensing applications. *Chem. Soc. Rev.* **2017**, *46*, 3242–3285. [CrossRef] [PubMed]

32. Hao, J.N.; Niu, D.; Gu, J.; Lin, S.; Li, Y.; Shi, J. Structure engineering of a lanthanide-based metal-organic framework for the regulation of dynamic ranges and sensitivities for pheochromocytoma diagnosis. *Adv. Mater.* **2020**, *32*, e2000791. [CrossRef]

33. Hu, M.-L.; Razavi, S.A.A.; Piroozzadeh, M.; Morsali, A. Sensing organic analytes by metal–organic frameworks: A new way of considering the topic. *Inorg. Chem. Front.* **2020**, *7*, 1598–1632. [CrossRef]

34. Guo, B.B.; Yin, J.C.; Li, N.; Fu, Z.X.; Han, X.; Xu, J.; Bu, X.H. Recent progress in luminous particle-encapsulated host–guest metal-organic frameworks for optical applications. *Adv. Opt. Mater.* **2021**, *9*, 2100283. [CrossRef]

35. Wu, S.; Min, H.; Shi, W.; Cheng, P. Multicenter metal–organic framework-based ratiometric fluorescent sensors. *Adv. Mater.* **2019**, *32*, 1805871. [CrossRef]

36. Xu, J.; Zhou, J.; Chen, Y.; Yang, P.; Lin, J. Lanthanide-activated nanoconstructs for optical multiplexing. *Coordin. Chem. Rev.* **2020**, *415*, 213328. [CrossRef]

37. Liu, X.Y.; Yin, X.M.; Yang, S.L.; Zhang, L.; Bu, R.; Gao, E.Q. Chromic and fluorescence-responsive metal-organic frameworks afforded by N-amination modification. *Acs Appl. Mater. Inter.* **2021**, *13*, 20380–20387. [CrossRef] [PubMed]

38. Fan, C.; Lv, X.; Liu, F.; Feng, L.; Liu, M.; Cai, Y.; Liu, H.; Wang, J.; Yang, Y.; Wang, H. Silver nanoclusters encapsulated into metal–organic frameworks with enhanced fluorescence and specific ion accumulation toward the microdot array-based fluorimetric analysis of copper in blood. *ACS Sensors* **2018**, *3*, 441–450. [CrossRef]

39. Xia, T.; Song, T.; Zhang, G.; Cui, Y.; Yang, Y.; Wang, Z.; Qian, G. A terbium metal-organic framework for highly selective and sensitive luminescence sensing of Hg^{2+} ions in aqueous solution. *Chem. Eur. J.* **2016**, *22*, 18429–18434. [CrossRef]

40. Xu, L.; Fang, G.; Liu, J.; Pan, M.; Wang, R.; Wang, S. One-pot synthesis of nanoscale carbon dots-embedded metal–organic frameworks at room temperature for enhanced chemical sensing. *J. Mater. Chem. A* **2016**, *4*, 15880–15887. [CrossRef]

41. Zhao, Y.; Zhang, N.; Wang, Y.; Bai, F.Y.; Xing, Y.H.; Sun, L.X. Ln-MOFs with window-shaped channels based on triazine tricarboxylic acid as a linker for the highly efficient capture of cationic dyes and iodine. *Inorg. Chem. Front.* **2021**, *8*, 1736–1746. [CrossRef]

42. Dong, X.-Y.; Si, Y.; Yang, J.-S.; Zhang, C.; Han, Z.; Luo, P.; Wang, Z.-Y.; Zang, S.-Q.; Mak, T.C.W. Ligand engineering to achieve enhanced ratiometric oxygen sensing in a silver cluster-based metal-organic framework. *Nat. Commun.* **2020**, *11*, 3678. [CrossRef]

43. Xiao, Y.; Chen, C.; Wu, Y.; Yin, Y.; Wu, H.; Li, H.; Fan, Y.; Wu, J.; Li, S.; Huang, X.; et al. Fabrication of two-dimensional metal-organic framework nanosheets through crystal dissolution-growth kinetics. *Acs Appl. Mater. Inter.* **2022**, *14*, 7192–7199. [CrossRef] [PubMed]

44. Cui, Y.; Yue, Y.; Qian, G.; Chen, B. Luminescent functional metal–organic frameworks. *Chem. Rev.* **2011**, *112*, 1126–1162. [CrossRef]

45. Wu, S.; Lin, Y.; Liu, J.; Shi, W.; Yang, G.; Cheng, P. Rapid detection of the biomarkers for carcinoid tumors by a water stable luminescent lanthanide metal-organic framework sensor. *Adv. Funct. Mater.* **2018**, *28*, 1707169. [CrossRef]

46. Yan, B. Luminescence response mode and chemical sensing mechanism for lanthanide-functionalized metal–organic framework hybrids. *Inorg. Chem. Front.* **2021**, *8*, 201–233. [CrossRef]

47. Qu, X.-L.; Yan, B. Cd-based metal–organic framework containing uncoordinated carbonyl groups as lanthanide postsynthetic modification sites and chemical sensing of diphenyl phosphate as a flame-retardant biomarker. *Inorg. Chem.* **2020**, *59*, 15088–15100. [CrossRef]

48. Thorarinsdottir, A.E.; Harris, T.D. Metal–organic framework magnets. *Chem. Rev.* **2020**, *120*, 8716–8789. [CrossRef]

49. Zhu, L.; Zhu, B.; Luo, J.; Liu, B. Design and property modulation of metal–organic frameworks with aggregation-induced emission. *ACS Mater. Lett.* **2020**, *3*, 77–89. [CrossRef]

50. Zhou, C.; Xu, F.; Wang, W.; Nie, W.; You, W.; Ye, X. Simple synthesis of dual-emission CsPbBr$_3$@EuBTC composite for latent fingerprints and optical anti-counterfeiting applications. *Mater. Today Commun.* **2022**, *33*, 104493. [CrossRef]

51. Jiang, W.; Zhang, H.; Wu, J.; Zhai, G.; Li, Z.; Luan, Y.; Garg, S. CuS@MOF-based well-designed quercetin delivery system for chemo-photothermal therapy. *Acs Appl. Mater. Inter.* **2018**, *10*, 34513–34523. [CrossRef] [PubMed]

52. Yang, J.C.; Shang, Y.; Li, Y.H.; Cui, Y.; Yin, X.B. An "all-in-one" antitumor and anti-recurrence/metastasis nanomedicine with multi-drug co-loading and burst drug release for multi-modality therapy. *Chem. Sci.* **2018**, *9*, 7210–7217. [CrossRef] [PubMed]
53. Hao, C.; Wu, X.; Sun, M.; Zhang, H.; Yuan, A.; Xu, L.; Xu, C.; Kuang, H. Chiral core-shell upconversion nanoparticle@MOF nanoassemblies for quantification and bioimaging of reactive oxygen species in vivo. *J. Am. Chem. Soc.* **2019**, *141*, 19373–19378. [CrossRef]
54. He, L.; Brasino, M.; Mao, C.; Cho, S.; Park, W.; Goodwin, A.P.; Cha, J.N. DNA-assembled core-satellite upconverting-metal-organic framework nanoparticle superstructures for efficient photodynamic therapy. *Small* **2017**, *13*, 1700504. [CrossRef] [PubMed]
55. Li, Y.; Tang, J.; He, L.; Liu, Y.; Liu, Y.; Chen, C.; Tang, Z. Core-shell upconversion nanoparticle@metal-organic framework nanoprobes for luminescent/magnetic dual-mode targeted imaging. *Adv. Mater.* **2015**, *27*, 4075–4080. [CrossRef]
56. Zhou, J.; Wang, P.; Wang, C.; Goh, Y.T.; Fang, Z.; Messersmith, P.B.; Duan, H. Versatile core-shell nanoparticle@metal-organic framework nanohybrids: Exploiting mussel-inspired polydopamine for tailored structural integration. *ACS Nano* **2015**, *9*, 6951–6960. [CrossRef]
57. Xie, Z.; Cai, X.; Sun, C.; Liang, S.; Shao, S.; Huang, S.; Cheng, Z.; Pang, M.; Xing, B.; Kheraif, A.A.A.; et al. O$_2$-loaded pH-responsive multifunctional nanodrug carrier for overcoming hypoxia and highly efficient chemo-photodynamic cancer therapy. *Chem. Mater.* **2018**, *31*, 483–490. [CrossRef]
58. Lu, G.; Li, S.; Guo, Z.; Farha, O.K.; Hauser, B.G.; Qi, X.; Wang, Y.; Wang, X.; Han, S.; Liu, X.; et al. Imparting functionality to a metal-organic framework material by controlled nanoparticle encapsulation. *Nat. Chem.* **2012**, *4*, 310–316. [CrossRef]
59. Liang, S.; Sun, C.; Yang, P.; Ma, P.; Huang, S.; Cheng, Z.; Yu, X.; Lin, J. Core-shell structured upconversion nanocrystal-dendrimer composite as a carrier for mitochondria targeting and catalase enhanced anti-cancer photodynamic therapy. *Biomaterials* **2020**, *240*, 119850. [CrossRef]
60. Liu, S.; Huo, Y.; Deng, S.; Li, G.; Li, S.; Huang, L.; Ren, S.; Gao, Z. A facile dual-mode aptasensor based on AuNPs@MIL-101 nanohybrids for ultrasensitive fluorescence and surface-enhanced Raman spectroscopy detection of tetrodotoxin. *Biosens. Bioelectron.* **2022**, *201*, 113891. [CrossRef]
61. Shang, W.; Zeng, C.; Du, Y.; Hui, H.; Liang, X.; Chi, C.; Wang, K.; Wang, Z.; Tian, J. Core-shell gold nanorod@metal-organic framework nanoprobes for multimodality diagnosis of glioma. *Adv. Mater.* **2017**, *29*, 1604381. [CrossRef] [PubMed]
62. An, L.; Cao, M.; Zhang, X.; Lin, J.; Tian, Q.; Yang, S. pH and glutathione synergistically triggered release and self-assembly of Au nanospheres for tumor theranostics. *Acs Appl. Mater. Inter.* **2020**, *12*, 8050–8061. [CrossRef] [PubMed]
63. Huang, C.; Li, A.; Chen, X.; Wang, T. Understanding the role of metal-organic frameworks in surface-enhanced raman scattering application. *Small* **2020**, *16*, e2004802. [CrossRef] [PubMed]
64. Qiao, X.; Su, B.; Liu, C.; Song, Q.; Luo, D.; Mo, G.; Wang, T. Selective surface enhanced Raman scattering for quantitative detection of lung cancer biomarkers in superparticle@MOF structure. *Adv. Mater.* **2018**, *30*, 1702275. [CrossRef] [PubMed]
65. Shao, Q.; Zhang, D.; Wang, C.-e.; Tang, Z.; Zou, M.; Yang, X.; Gong, H.; Yu, Z.; Jin, S.; Liang, P. Ag@MIL-101(Cr) film substrate with high SERS enhancement effect and uniformity. *J. Phys. Chem. C* **2021**, *125*, 7297–7304. [CrossRef]
66. Biswas, S.; Chen, Y.; Xie, Y.; Sun, X.; Wang, Y. Ultrasmall Au(0) inserted hollow PCN-222 MOF for the high-sensitive detection of estradiol. *Anal. Chem.* **2020**, *92*, 4566–4572. [CrossRef]
67. Cao, J.; Yang, Z.; Xiong, W.; Zhou, Y.; Wu, Y.; Jia, M.; Zhou, C.; Xu, Z. Ultrafine metal species confined in metal–organic frameworks: Fabrication, characterization and photocatalytic applications. *Coordin. Chem. Rev.* **2021**, *439*, 213924. [CrossRef]
68. Zhang, X.; Zhi, H.; Wang, F.; Zhu, M.; Meng, H.; Wan, P.; Feng, L. Target-responsive smart nanomaterials via a Au–S binding encapsulation strategy for electrochemical/colorimetric dual-mode paper-based analytical devices. *Anal. Chem.* **2022**, *94*, 2569–2577. [CrossRef]
69. Hu, Y.; Cheng, H.; Zhao, X.; Wu, J.; Muhammad, F.; Lin, S.; He, J.; Zhou, L.; Zhang, C.; Deng, Y.; et al. Surface-enhanced raman scattering active gold nanoparticles with enzyme-mimicking activities for measuring glucose and lactate in living tissues. *ACS Nano* **2017**, *11*, 5558–5566. [CrossRef]
70. De Marchi, S.; Vázquez-Iglesias, L.; Bodelón, G.; Pérez-Juste, I.; Fernández, L.Á.; Pérez-Juste, J.; Pastoriza-Santos, I. Programmable modular assembly of functional proteins on Raman-encoded zeolitic imidazolate framework-8 (ZIF-8) nanoparticles as SERS tags. *Chem. Mater.* **2020**, *32*, 5739–5749. [CrossRef]
71. Yang, X.; Liu, Y.; Lam, S.H.; Wang, J.; Wen, S.; Yam, C.; Shao, L.; Wang, J. Site-selective deposition of metal–organic frameworks on gold nanobipyramids for surface-enhanced Raman scattering. *Nano Lett.* **2021**, *21*, 8205–8212. [CrossRef] [PubMed]
72. Guselnikova, O.; Lim, H.; Na, J.; Eguchi, M.; Kim, H.-J.; Elashnikov, R.; Postnikov, P.; Svorcik, V.; Semyonov, O.; Miliutina, E.; et al. Enantioselective SERS sensing of pseudoephedrine in blood plasma biomatrix by hierarchical mesoporous Au films coated with a homochiral MOF. *Biosens. Bioelectron.* **2021**, *180*, 113109. [CrossRef] [PubMed]
73. Chen, X.; Qin, L.; Kang, S.-Z.; Li, X. A special zinc metal-organic frameworks-controlled composite nanosensor for highly sensitive and stable SERS detection. *Appl. Surf. Sci.* **2021**, *550*, 149302. [CrossRef]
74. Zhang, Y.; Hu, Y.; Li, G.; Zhang, R. A composite prepared from gold nanoparticles and a metal organic framework (type MOF-74) for determination of 4-nitrothiophenol by surface-enhanced Raman spectroscopy. *Microchim. Acta* **2019**, *186*, 477. [CrossRef] [PubMed]

75. Ding, Q.; Wang, J.; Chen, X.; Liu, H.; Li, Q.; Wang, Y.; Yang, S. Quantitative and sensitive SERS platform with analyte enrichment and filtration function. *Nano Lett.* **2020**, *20*, 7304–7312. [CrossRef]

76. Jiang, Z.; Gao, P.; Yang, L.; Huang, C.; Li, Y. Facile in situ synthesis of silver nanoparticles on the surface of metal–organic framework for ultrasensitive surface-Enhanced Raman scattering detection of dopamine. *Anal. Chem.* **2015**, *87*, 12177–12182. [CrossRef]

77. Cao, W.; Cui, Y.; Yang, Y.; Qian, G. Dyes encapsulated nanoscale metal–organic frameworks for multimode temperature sensing with high spatial resolution. *ACS Mater. Lett.* **2021**, *3*, 1426–1432. [CrossRef]

78. Ding, Z.; Tan, J.; Feng, G.; Yuan, Z.; Wu, C.; Zhang, X. Nanoscale metal–organic frameworks coated with poly(vinyl alcohol) for ratiometric peroxynitrite sensing through FRET. *Chem. Sci.* **2017**, *8*, 5101–5106. [CrossRef]

79. Xu, R.; Wang, Y.; Duan, X.; Lu, K.; Micheroni, D.; Hu, A.; Lin, W. Nanoscale metal–organic frameworks for ratiometric oxygen sensing in live cells. *J. Am. Chem. Soc.* **2016**, *138*, 2158–2161. [CrossRef]

80. Wang, J.; Li, D.; Ye, Y.; Qiu, Y.; Liu, J.; Huang, L.; Liang, B.; Chen, B. A fluorescent metal–organic framework for food real-time visual monitoring. *Adv. Mater.* **2021**, *33*, 2008020. [CrossRef]

81. Guan, Q.; Zhou, L.L.; Li, Y.A.; Dong, Y.B. Diiodo-bodipy-encapsulated nanoscale metal-organic framework for pH-driven selective and mitochondria targeted photodynamic therapy. *Inorg. Chem.* **2018**, *57*, 10137–10145. [CrossRef]

82. Wang, D.; Wu, H.; Lim, W.Q.; Phua, S.Z.F.; Xu, P.; Chen, Q.; Guo, Z.; Zhao, Y. A mesoporous nanoenzyme derived from metal-organic frameworks with endogenous oxygen generation to alleviate tumor hhypoxia for significantly enhanced photodynamic therapy. *Adv. Mater.* **2019**, *31*, e1901893.

83. Su, Y.; Yu, J.; Li, Y.; Phua, S.F.Z.; Liu, G.; Lim, W.Q.; Yang, X.; Ganguly, R.; Dang, C.; Yang, C.; et al. Versatile bimetallic lanthanide metal-organic frameworks for tunable emission and efficient fluorescence sensing. *Commun. Chem.* **2018**, *1*, 12. [CrossRef]

84. Cui, Y.; Xu, H.; Yue, Y.; Guo, Z.; Yu, J.; Chen, Z.; Gao, J.; Yang, Y.; Qian, G.; Chen, B. A luminescent mixed-lanthanide metal-organic framework thermometer. *J. Am. Chem. Soc.* **2012**, *134*, 3979–3982. [CrossRef]

85. Feng, T.; Ye, Y.; Liu, X.; Cui, H.; Li, Z.; Zhang, Y.; Liang, B.; Li, H.; Chen, B. A robust mixed-lanthanide polyMOF membrane for ratiometric temperature sensing. *Angew. Chem. Int. Ed.* **2020**, *59*, 21752–21757. [CrossRef] [PubMed]

86. Yin, X.-B.; Sun, Y.-Q.; Yu, H.; Cheng, Y.; Wen, C. Design and multiple applications of mixed-ligand metal–organic frameworks with dual emission. *Anal. Chem.* **2022**, *94*, 4938–4947. [CrossRef]

87. Yin, H.Q.; Yin, X.B. Metal-organic frameworks with multiple luminescence emissions: Designs and applications. *Acc. Chem. Res.* **2020**, *53*, 485–495. [CrossRef] [PubMed]

88. Cui, Y.; Chen, F.; Yin, X.-B. A ratiometric fluorescence platform based on boric-acid-functional Eu-MOF for sensitive detection of H_2O_2 and glucose. *Biosens. Bioelectron.* **2019**, *135*, 208–215. [CrossRef]

89. Yin, H.-Q.; Yang, J.-C.; Yin, X.-B. Ratiometric fluorescence sensing and real-time detection of water in organic solvents with one-pot synthesis of Ru@MIL-101(Al)–NH_2. *Anal. Chem.* **2017**, *89*, 13434–13440. [CrossRef]

90. Wen, C.; Yin, F.; Cheng, Y.; Yu, H.; Sun, Y.Q.; Yin, X.B. Construction of $NaYF_4$ Library for Morphology-Controlled Multimodality Applications. *Small* **2021**, *17*, e2103206. [CrossRef]

91. Zhao, D.; Yu, K.; Han, X.; He, Y.; Chen, B. Recent progress on porous MOFs for process-efficient hydrocarbon separation, luminescent sensing, and information encryption. *Chem. Commun.* **2022**, *58*, 747–770. [CrossRef] [PubMed]

92. Wang, Z.; Zhu, C.Y.; Mo, J.T.; Fu, P.Y.; Zhao, Y.W.; Yin, S.Y.; Jiang, J.J.; Pan, M.; Su, C.Y. White-light emission from dual-way photon energy conversion in a dye-encapsulated metal–organic framework. *Angew. Chem. Int. Ed.* **2019**, *58*, 9752–9757. [CrossRef]

93. Zhou, J.; del Rosal, B.; Jaque, D.; Uchiyama, S.; Jin, D. Advances and challenges for fluorescence nanothermometry. *Nat. Methods* **2020**, *17*, 967–980. [CrossRef] [PubMed]

94. Liu, Y.; Jia, Q.; Zhai, X.; Mao, F.; Jiang, A.; Zhou, J. Rationally designed pure-inorganic upconversion nanoprobes for ultra-highly selective hydrogen sulfide imaging and elimination in vivo. *Chem. Sci.* **2019**, *10*, 1193–1200. [CrossRef]

95. Zhang, H.; Yin, X.-B. Mixed-ligand metal–organic frameworks for all-in-one theranostics with controlled drug delivery and enhanced photodynamic therapy. *ACS Appl. Mater. Interfaces* **2022**, *14*, 26528–26535. [CrossRef]

96. Wang, Y.M.; Liu, W.; Yin, X.B. Multifunctional mixed-metal nanoscale coordination polymers for triple-modality imaging-guided photodynamic therapy. *Chem. Sci.* **2017**, *8*, 3891–3897. [CrossRef]

97. Wei, X.; Li, N.; Wang, Y.; Xie, Z.; Huang, H.; Yang, G.; Li, T.; Qin, X.; Li, S.; Yang, H.; et al. Zeolitic imidazolate frameworks-based nanomaterials for biosensing, cancer imaging and phototheranostics. *Appl. Mater. Today* **2021**, *23*, 100995. [CrossRef]

98. Cheng, Y.; Wen, C.; Sun, Y.Q.; Yu, H.; Yin, X.B. Mixed-metal MOF-derived hollow porous nanocomposite for trimodality imaging guided reactive oxygen species-augmented synergistic therapy. *Adv. Funct. Mater.* **2021**, *31*, 2104378. [CrossRef]

99. Li, S.; Wang, K.; Shi, Y.; Cui, Y.; Chen, B.; He, B.; Dai, W.; Zhang, H.; Wang, X.; Zhong, C.; et al. Novel biological functions of ZIF-NP as a delivery vehicle: High pulmonary accumulation, favorable biocompatibility, and improved therapeutic outcome. *Adv. Funct. Mater.* **2016**, *26*, 2715–2727. [CrossRef]

100. Yang, P.; Men, Y.; Tian, Y.; Cao, Y.; Zhang, L.; Yao, X.; Yang, W. Metal–organic framework nanoparticles with near-infrared dye for multimodal imaging and guided phototherapy. *Acs Appl. Mater. Inter.* **2019**, *11*, 11209–11219. [CrossRef]

101. Wang, F.; Zhang, F.; Zhao, Z.; Sun, Z.; Pu, Y.; Wang, Y.; Wang, X. Multifunctional MOF-based probes for efficient detection and discrimination of Pb^{2+}, Fe^{3+} and $Cr_2O_7^{2-}/CrO_4^{2-}$. *Dalton Trans.* **2021**, *50*, 12197–12207. [CrossRef]

102. Xia, T.; Wan, Y.; Li, Y.; Zhang, J. Highly stable lanthanide metal–organic framework as an internal calibrated luminescent sensor for glutamic acid, a neuropathy biomarker. *Inorg. Chem.* **2020**, *59*, 8809–8817. [CrossRef]

103. He, C.; Lu, K.; Lin, W. Nanoscale metal–organic frameworks for real-time intracellular pH sensing in live cells. *J. Am. Chem. Soc.* **2014**, *136*, 12253–12256. [CrossRef]

104. Zhang, J.; He, M.; Nie, C.; He, M.; Pan, Q.; Liu, C.; Hu, Y.; Chen, T.; Chu, X. Biomineralized metal–organic framework nanoparticles enable a primer exchange reaction-based DNA machine to work in living cells for imaging and gene therapy. *Chem. Sci.* **2020**, *11*, 7092–7101. [CrossRef]

105. Zong, C.; Xu, M.; Xu, L.-J.; Wei, T.; Ma, X.; Zheng, X.-S.; Hu, R.; Ren, B. Surface-enhanced Raman spectroscopy for bioanalysis: Reliability and challenges. *Chem. Rev.* **2018**, *118*, 4946–4980. [CrossRef]

106. Su, H.-S.; Chang, X.; Xu, B. Surface-enhanced vibrational spectroscopies in electrocatalysis: Fundamentals, challenges, and perspectives. *Chin. J. Catal.* **2022**, *43*, 2757–2771. [CrossRef]

107. Chang, X.; Vijay, S.; Zhao, Y.; Oliveira, N.J.; Chan, K.; Xu, B. Understanding the complementarities of surface-enhanced infrared and Raman spectroscopies in CO adsorption and electrochemical reduction. *Nat. Commun.* **2022**, *13*, 2656. [CrossRef]

108. Chen, S.-H.; He, W.-J.; Zhu, Y.-J.; Song, H.-T. A luminescent turn-off sensor for Cr(VI) anions recognition derived from a Zn(II)-based metal–organic framework. *Inorg. Chim. Acta* **2021**, *525*, 120498. [CrossRef]

109. Liu, S.; Li, H.; Hassan, M.M.; Ali, S.; Chen, Q. SERS based artificial peroxidase enzyme regulated multiple signal amplified system for quantitative detection of foodborne pathogens. *Food Control* **2021**, *123*, 107733. [CrossRef]

110. Hu, Y.; Liao, J.; Wang, D.; Li, G. Fabrication of gold nanoparticle-embedded metal-organic framework for highly sensitive surface-enhanced Raman scattering detection. *Anal. Chem.* **2014**, *86*, 3955–3963. [CrossRef] [PubMed]

111. Chen, Z.; Su, L.; Ma, X.; Duan, Z.; Xiong, Y. A mixed valence state Mo-based metal–organic framework from photoactivation as a surface-enhanced Raman scattering substrate. *New J. Chem.* **2021**, *45*, 5121–5126. [CrossRef]

112. Fu, J.H.; Zhong, Z.; Xie, D.; Guo, Y.J.; Kong, D.X.; Zhao, Z.X.; Zhao, Z.X.; Li, M. SERS-active MIL-100(Fe) sensory array for ultrasensitive and multiplex detection of VOCs. *Angew. Chem. Int. Ed.* **2020**, *59*, 20489–20498. [CrossRef]

113. Osterrieth, J.W.M.; Wright, D.; Noh, H.; Kung, C.W.; Vulpe, D.; Li, A.; Park, J.E.; Van Duyne, R.P.; Moghadam, P.Z.; Baumberg, J.J.; et al. Core-shell gold nanorod@zirconium-based metal-organic framework composites as in situ size-selective Raman probes. *J. Am. Chem. Soc.* **2019**, *141*, 3893–3900. [CrossRef] [PubMed]

114. Sun, H.; Cong, S.; Zheng, Z.; Wang, Z.; Chen, Z.; Zhao, Z. Metal–organic frameworks as surface enhanced Raman scattering substrates with high tailorability. *J. Am. Chem. Soc.* **2018**, *141*, 870–878. [CrossRef] [PubMed]

115. Jiang, L.; Hu, Y.; Zhang, H.; Luo, X.; Yuan, R.; Yang, X. Charge-transfer resonance and surface defect-dominated WO_3 hollow microspheres as SERS substrates for the miRNA 155 assay. *Anal. Chem.* **2022**, *94*, 6967–6975. [CrossRef] [PubMed]

116. Dong, Y.; Gong, M.; Shah, F.U.; Laaksonen, A.; An, R.; Ji, X. Phosphonium-based ionic liquid significantly enhances SERS of cytochrome c on TiO_2 nanotube arrays. *Acs Appl. Mater. Inter.* **2022**, *14*, 27456–27465. [CrossRef]

117. Cong, S.; Yuan, Y.; Chen, Z.; Hou, J.; Yang, M.; Su, Y.; Zhang, Y.; Li, L.; Li, Q.; Geng, F.; et al. Noble metal-comparable SERS enhancement from semiconducting metal oxides by making oxygen vacancies. *Nat. Commun.* **2015**, *6*, 7800. [CrossRef]

118. Li, W.; Zamani, R.; Rivera Gil, P.; Pelaz, B.; Ibanez, M.; Cadavid, D.; Shavel, A.; Alvarez-Puebla, R.A.; Parak, W.J.; Arbiol, J.; et al. CuTe nanocrystals: Shape and size control, plasmonic properties, and use as SERS probes and photothermal agents. *J. Am. Chem. Soc.* **2013**, *135*, 7098–7101. [CrossRef]

119. Qiu, B.; Xing, M.; Yi, Q.; Zhang, J. Chiral carbonaceous nanotubes modified with titania nanocrystals: Plasmon-free and recyclable SERS sensitivity. *Angew. Chem. Int. Ed.* **2015**, *54*, 10643–10647. [CrossRef]

120. Kim, J.; Jang, Y.; Kim, N.J.; Kim, H.; Yi, G.C.; Shin, Y.; Kim, M.H.; Yoon, S. Study of chemical enhancement mechanism in non-plasmonic surface enhanced Raman spectroscopy (SERS). *Front. Chem.* **2019**, *7*, 582. [CrossRef]

121. Yu, T.-H.; Ho, C.-H.; Wu, C.-Y.; Chien, C.-H.; Lin, C.-H.; Lee, S. Metal-organic frameworks: A novel SERS substrate. *J. Raman Spectrosc.* **2013**, *44*, 1506–1511. [CrossRef]

122. Yu, X.; Cai, H.; Zhang, W.; Li, X.; Pan, N.; Luo, Y.; Hou, J.G. Tuning chemical enhancement of SERS by controlling the chemical reduction of graphene oxide nanosheets. *ACS Nano* **2011**, *5*, 952–958. [CrossRef] [PubMed]

123. Doering, W.E.; Nie, S. Single-molecule and single-nanoparticle SERS: Examining the roles of surface active sites and chemical enhancement. *J. Phys. Chem. B* **2002**, *106*, 311–317. [CrossRef]

124. ÇELİK, Y.; Kurt, A. Three dimensional porous expanded graphite/silver nanoparticles nanocomposite platform as a SERS substrate. *Appl. Surf. Sci.* **2021**, *568*, 150946. [CrossRef]

125. Xu, F.; Shang, W.; Xuan, M.; Ma, G.; Ben, Z. Layered filter paper-silver nanoparticle-ZIF-8 composite for efficient multi-mode enrichment and sensitive SERS detection of thiram. *Chemosphere* **2022**, *288*, 132635. [CrossRef]

126. Chen, X.; Zhang, Y.; Kong, X.; Yao, K.; Liu, L.; Zhang, J.; Guo, Z.; Xu, W.; Fang, Z.; Liu, Y. Photocatalytic performance of the MOF-coating layer on SPR-excited Ag nanowires. *ACS Omega* **2021**, *6*, 2882–2889. [CrossRef]

127. Wang, X.; Wang, Y.; Qi, H.; Chen, Y.; Guo, W.; Yu, H.; Chen, H.; Ying, Y. Humidity-independent artificial olfactory array enabled by hydrophobic core-shell dye/MOFs@COFs composites for plant disease diagnosis. *ACS Nano* **2022**, *16*, 14297–14307. [CrossRef]

128. Yang, F.P.; He, Q.T.; Jiang, H.; Li, Z.; Chen, W.; Chen, R.L.; Tang, X.Y.; Cai, Y.P.; Hong, X.J. Rapid and specific enhanced luminescent switch of aniline gas by MOFs assembled from a planar binuclear cadmium(II) metalloligand. *Inorg. Chem.* **2022**, *61*, 10844–10851. [CrossRef]

129. Cheung, R.C.; Wong, J.H.; Ng, T.B. Immobilized metal ion affinity chromatography: A review on its applications. *Appl. Microbiol. Biotechnol.* **2012**, *96*, 1411–1420. [CrossRef] [PubMed]

130. Ahmadijokani, F.; Mohammadkhani, R.; Ahmadipouya, S.; Shokrgozar, A.; Rezakazemi, M.; Molavi, H.; Aminabhavi, T.M.; Arjmand, M. Superior chemical stability of UiO-66 metal-organic frameworks (MOFs) for selective dye adsorption. *Chem. Eng.J.* **2020**, *399*, 125346. [CrossRef]
131. Du, L.; Zhang, B.; Deng, W.; Cheng, Y.; Xu, L.; Mai, L. Hierarchically self-assembled MOF network enables continuous ion transport and high mechanical strength. *Adv. Energy Mater.* **2022**, *12*, 2200501. [CrossRef]

Review

Recent Trends in Metal Nanoparticles Decorated 2D Materials for Electrochemical Biomarker Detection

Aneesh Koyappayil [†], Ajay Kumar Yagati [†] and Min-Ho Lee *

School of Integrative Engineering, Chung-Ang University, 84 Heuseok-ro, Dongjak-Gu, Seoul 06974, Republic of Korea
* Correspondence: mhlee7@cau.ac.kr; Tel.: +82-2-820-5503; Fax: +82-2-814-2651
† These authors contributed equally to this work.

Abstract: Technological advancements in the healthcare sector have pushed for improved sensors and devices for disease diagnosis and treatment. Recently, with the discovery of numerous biomarkers for various specific physiological conditions, early disease screening has become a possibility. Biomarkers are the body's early warning systems, which are indicators of a biological state that provides a standardized and precise way of evaluating the progression of disease or infection. Owing to the extremely low concentrations of various biomarkers in bodily fluids, signal amplification strategies have become crucial for the detection of biomarkers. Metal nanoparticles are commonly applied on 2D platforms to anchor antibodies and enhance the signals for electrochemical biomarker detection. In this context, this review will discuss the recent trends and advances in metal nanoparticle decorated 2D materials for electrochemical biomarker detection. The prospects, advantages, and limitations of this strategy also will be discussed in the concluding section of this review.

Keywords: metal nanoparticles; immunosensor; MXene; MoS_2; graphene; MOF; biomarkers; graphitic carbon nitride; black phosphorous; 2D-LDHs; boron nitrides; graphdiyne

1. Introduction

The definition of biomarkers has evolved over time, and a broader definition was suggested by the World Health Organization as "a biomarker is any substance, structure, or process that can be measured in the body or its products and influence or predict the incidence of outcome or disease" [1,2]. More specific definitions such as "a biological molecule found in blood, other body fluids, or tissues that is a sign of a normal or abnormal process, or of a condition or disease and can be tested to see how well the body responds to treatment for a disease or condition" [3], and "a characteristic that can be objectively measured and quantitatively evaluated as an indicator of a normal biological and pathological process, or pharmacological responses to a therapeutic intervention" [4] were coined by the US National Cancer Institute, and the US National Institutes of Health, respectively. Biomarkers can be biological, chemical, or physical, and are measurable parameters indicative of a specific biological state. The detection of biomarkers is crucial for the diagnosis and treatment of numerous diseases [5]. Biomarkers are classified broadly into imaging biomarkers and molecular biomarkers based on their characteristics. Imaging biomarkers are often used in combination with various imaging tools, whereas molecular biomarkers comprise RNA, DNA, and proteins [6]. Molecular biomarkers are easily quantifiable from biological samples and can complement clinical characteristics [7,8]. Another category, known as pharmacodynamic biomarkers, is applied in drug development during dose optimization studies [9]. Based on the application, biomarkers are classified into prognostic biomarkers, diagnostic biomarkers, predictive biomarkers, and monitoring biomarkers [10]. Prognostic biomarkers help to identify the risk of disease progression in the future [11]. Diagnostic biomarkers help physicians to identify a specific disease condition [12], and predictive biomarkers predict the responses related to therapeutic interventions [11], whereas

Citation: Koyappayil, A.; Yagati, A.K.; Lee, M.-H. Recent Trends in Metal Nanoparticles Decorated 2D Materials for Electrochemical Biomarker Detection. *Biosensors* **2023**, *13*, 91. https://doi.org/10.3390/bios13010091

Received: 7 November 2022
Revised: 27 December 2022
Accepted: 1 January 2023
Published: 5 January 2023

a monitoring biomarker is usually measured for assessing the status of a medical condition or disease [13].

An ideal biomarker sensor must capture the biomarker selectively from the complex biological matrix of interfering molecules. Although nonspecific binding is still a concern, electrochemical detection methods, specifically electrochemical impedance spectroscopy (EIS), allow the selective analysis of biomarker detections by the resistive and/or capacitive changes due to physical and/or biomolecular interactions of the electrode surfaces coated with nanomaterials, DNA, proteins, etc. [14–16]. It is one of the basic and widely used approaches to determine the fundamental redox events at the electrode-electrolyte interface. However, evaluations are made by comparing the results of the EIS with cyclic voltammetry (CV) measurements. Also, differential pulse voltammetry (DPV) and square wave voltammetry (SWV) techniques are used in biomarker detection systems for both label and label-free approaches [17,18]. Among these techniques, CV-based detection sensing is widely reported due to its ability to explain the electrochemical events, such as oxidation-reduction reactions and electron-transfer kinetics occurring at the electrode-electrolyte interface, and the mass transport towards the electrode surface [19–21]. The search for advanced functional materials for electrochemical biomarker detection has sparked a research interest in layered 2D materials over the past few years and several novel approaches were reported for the synthesis of various 2D materials and their nanocomposites with exciting immunosensor applications. The interest and demand for 2D materials have increased significantly, and the global market for 2D materials is expected to grow rapidly with a CAGR of 3.9% between 2020 and 2027 and a corresponding increase in valuation from 2.27 billion to 2.86 billion USD [22]. In this context, this review discusses the recent advances and challenges of metal nanoparticle decorated 2D materials for biomarker detection.

2. Metal Nanoparticles on 2D Materials for Biomarker Detection

Nanoparticles used separately or in conjugation with other nanomaterials on 2D materials fulfill various roles in the design and development of electrochemical immunosensors. Also, they improve the analytical characteristics of the developed sensors such as linear range, LOD, and sensitivity [23]. For instance, nanoparticles deposited on the surface of the working electrode result in an enhancement of the surface area, thereby leading to an increased molecule loading capacity [24,25]. Additionally, the unique properties of nanoparticles could enhance the signal for the sensitive determination of biomarkers [23]. Also, the high electrical conductivity of metal nanoparticles at the electrode surface accelerates the redox electron transfer process. In some cases, nanoparticles could act as platforms for anchoring antibodies [26]. Metal nanoparticles were also used as a transport medium to capture the analyte from the sample, thereby concentrating the analyte molecules towards the electrode surface to improve the analytical signal [27]. Among various metal nanoparticles, AuNPs were extensively used to immobilize antibodies on the electrode surface to effectively amplify the immunosensor signal, anchor antibodies, and improve electrocatalytic activity [28,29].

2.1. Graphene Oxide Conjugated with Nanoparticles for Electrochemical Biomarker Detection

Graphene, a single layer (monolayer) of SP^2 carbon atoms with a molecular bond length of 0.142 nm, is tightly bound in a hexagonal honeycomb lattice. It is basically extracted from graphite and is merely a sheet of graphite. Graphene possesses excellent electrical conductivity (200,000 cm^2/Vs) due to its bonding and antibonding of pi orbitals, with the strongest compound around 100–130 times stronger than steel with a tensile strength of 130 GPa and a Young's Modulus of 1 TPa-150,000,000 psi. It is also one of the best conductors of heat at room temperature (at (4.84 $\times$ 10^3–5.30 $\times$ 10^3 W/mK). As graphene is a subunit of graphite it can be synthesized by direct extraction from bulk graphite. From the high-quality sample of graphite, graphene can be extracted by micromechanical cleavage or the scotch tape method of production. It is a straightforward method that doesn't need any specialized equipment. A piece of adhesive tape is placed onto and then peeled off the surface of a sample of graphite, resulting in a single to few layers of graphene. Other methods include the dispersion of graphite, exfoliation of graphite oxide, epitaxial growth, and chemical vapor deposition (CVD) as shown in Figure 1.

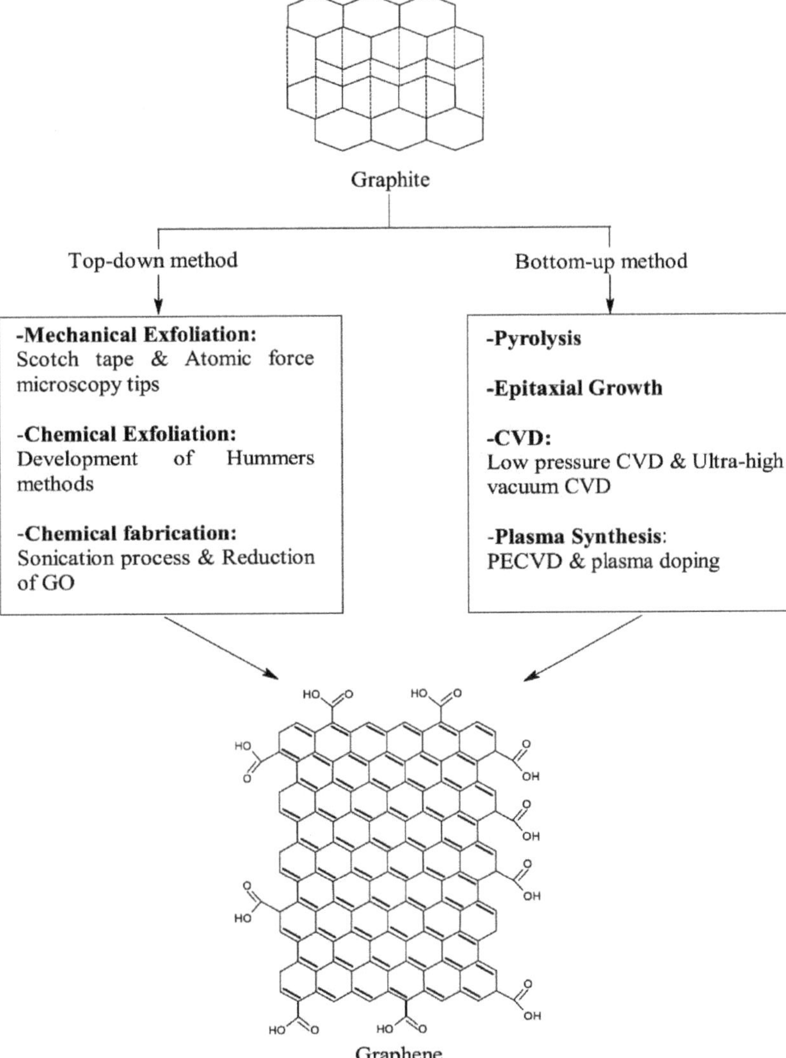

Figure 1. The schematic diagram for the synthesis of graphene. Reprinted with permission from Ref. [30]. Copyright 2018, Elsevier.

Graphene oxide is a form of graphene that includes oxygen functional groups and possesses interesting properties that are different from graphene. By reducing graphene oxide, these functional groups can be removed resulting in reduced graphene oxide. The production of reduced graphene oxide can be done in (i) chemical reduction, (ii) Thermal reduction; (iii) microwave and photoreduction; (iv) photocatalyst reduction; (v) solvothermal/hydrothermal reduction. The detailed information for various synthesis routes can be found elsewhere [31–33] and is beyond the scope of this review.

In this section, we discuss the development of various types of electrochemical sensors based on graphene oxide conjugated with nanoparticles that have been reported recently for various types of biomarkers. The development of biosensors that accurately measure the desired biomarker at high sensitivity and selectivity is crucial. However, sensitivity and selectivity are the two main factors that limit accuracy when performing the detections at the point of care with meager volumes of biological test solutions. For cancer cell analysis, the sensors should be able to detect tumors within the range of 100–1000 cell counts. To overcome these difficulties, innovative biosensor approaches with the optical, electrochemical, and piezoelectric transducer occupy the place of benchtop protocols adopted by the classical detection methods. Among these biosensors, electrochemical-based approaches competed with optical sensors which are widely used for the analysis of cancer biomarkers due to the characteristics of high sensitivity, selectivity, fast response, ease of use, low cost, and minimal fabrication procedures. In electrochemical biosensors, the right choice of transducer material is crucial, since it is the transducer that mainly influences the overall sensitivity [34] with minimal contributions from labeling methods.

Recently, Ranjan et al. [35] reported on the detection of breast cancer CD44 biomarkers using a gold-graphene oxide nanocomposite with ionic liquid with differential pulse voltammetry and electrochemical impedance spectroscopy. In this work, the authors reported the synthesis of RGO, ionic liquid (IL), and Au nanoparticles (Au NPs) by the citrate reduction method and other chemical procedures to form a nanocomposite on a glassy carbon electrode (GCE), as shown in Figure 2. In this work, the addition of 1-butyl-3-methylimidazolium tetrafluoroborate, an ionic liquid in conjugation with Au nanoparticles enabled the enhancement in the overall sensitivity of the developed sensor. Once the nanocomposite is deposited on GCE, the surface is activated with EDC/NHS to covalently bind the anti-CD44 antibodies. After the surface is blocked with BSA for nonspecific binding, then different concentrations of CD44 antigen were allowed for electrochemical investigation with CV, DPV, and EIS. The sensor possessed a linear range of 5 fg/mL to 50 µg/mL with a LOD of 2.7 fg/mL and 2.0 fg/mL in serum and PBS samples, respectively. This sensor is a promising candidate for the onsite detection of CD44 in breast cancer patients.

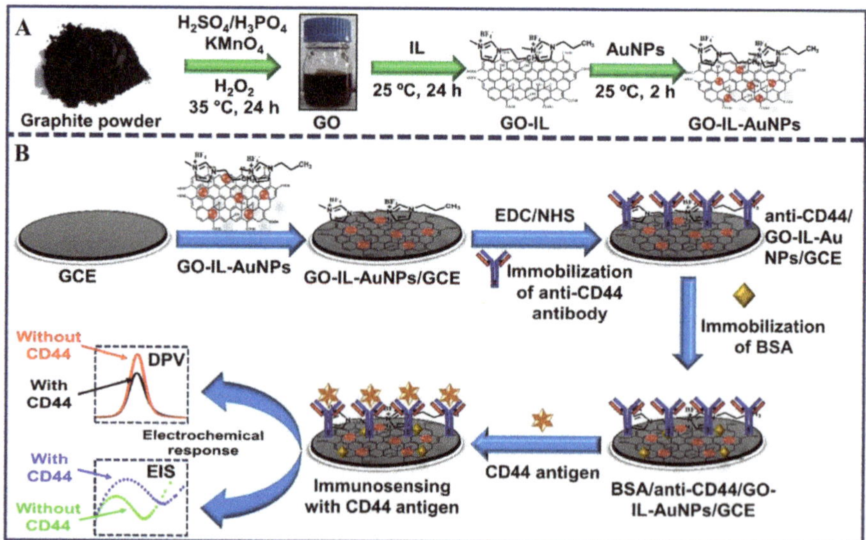

Figure 2. (**A**) Schematic diagram shows the synthesis of GO-IL-AuNPs hybrid nanocomposite and (**B**) Stepwise fabrication shows the surface modification procedures for the fabrication of BSA/anti-CD44/GO-IL-AuNPs/GCE Immunosensor. Reprinted with permission from Ref. [35] Copyright 2022, ACS.

In another study, Yagati et al. [36] proposed indium tin oxide (ITO)-based electrodes modified with reduced graphene oxide-gold nanoparticles that were used for the electrochemical impedance sensing of the C-reactive protein in serum samples. This biomarker detection is crucial in analyzing the inflammation due to an infection, and the risk of heart disease. In this study, graphene oxide-Au nanoparticles were electrodeposited on ITO microdisk electrodes fabricated using standard photolithography techniques. Subsequently, the modified electrodes were coated with a self-assembled monolayer of 3-MPA and activated with EDC/NHS. After the surface-blocking protocol was performed, then the selective antibodies were immobilized on the rGO-NP surface. Once the transducer surface is ready, a different concentration of CRP in human serum (1: 200) was detected with the help of impedance spectroscopy (Figure 3). The key feature of this sensor is that by forming the nanohybrid materials (RGO-NP hybrid) on the electrode, it results in an enhanced sensitivity toward CRP detection. The linear range of the sensor is 1–1000 ng/mL with an LOD of 0.08 ng/mL in serum samples. Based on the findings, it has the feasibility to employ multiplexed assay detection of biomarkers for point-of-care applications.

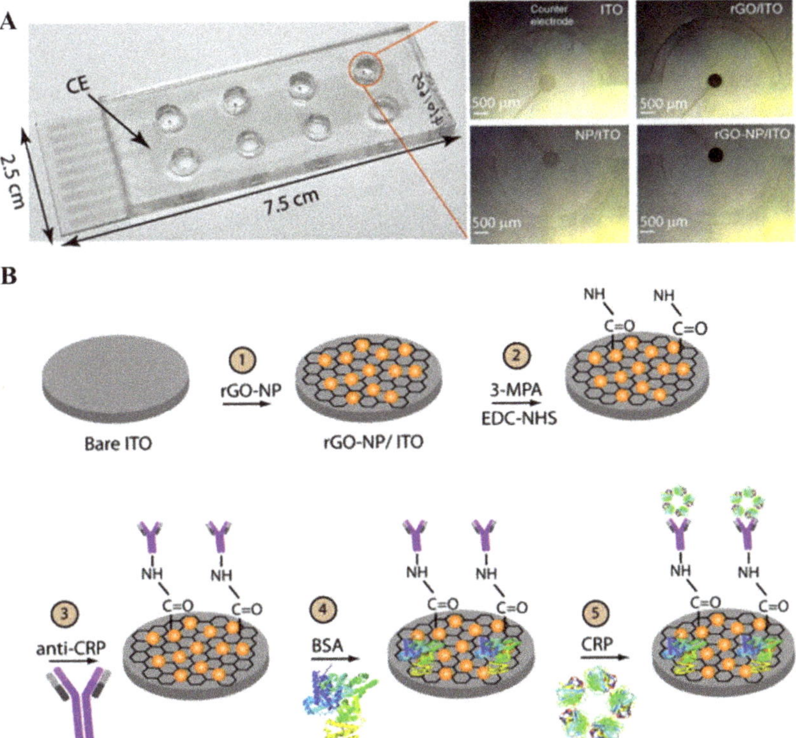

Figure 3. (**A**) Fabrication of 8-channel Indium-tin oxide electrodeposited with reduced graphene oxide-nanoparticle microdisk electrode array as working electrodes with a shared counter electrode. (**B**) Chemical functionalization of modified ITO electrode with EDC/NHS to couple antibodies for CRP detection in real samples. Reprinted with permission from Ref. [36]. Copyright 2016, Elsevier.

Jonous et al. [37] reported on the detection of prostate-specific antigen (PSA) by using a sandwich-type transducer composed of graphene oxide (GO) and gold nanoparticles (AuNPs). In this work the authors utilized an 11-mercaptoundecanoic acid for self-assembled monolayer formation on the GO-coated glassy carbon electrode (GCE) and a subsequent modification with EDC/NHS to convert -COOH to -NH for antibody bindings (Figure 4). After blocking with 1% BSA, different concentrations of PSA were allowed to bind to the electrode and with square wave voltammetry, and the quantification was made. The sensor possessed a limit of detection estimated to be around 0.2 and 0.07 ng/mL for total and free PSA antigens, respectively. The incorporation of AuNPs on GO/GCE enabled double functionality, i.e., specific recognition and signal amplification, for sensitive determination of PSA.

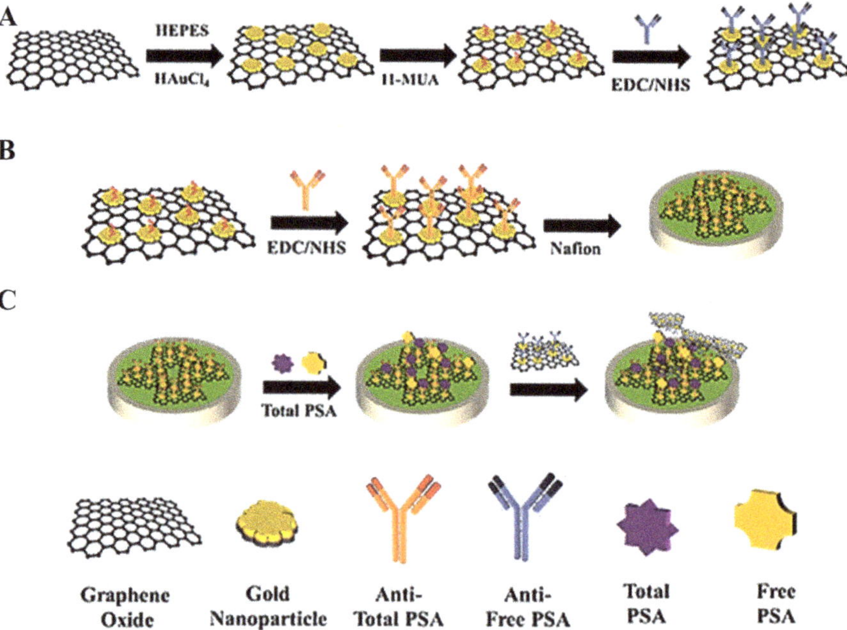

Figure 4. (**A**) Procedures for the fabrication of Go/GNP/Ab. (**B**) Procedure for preparing the electrochemical sensor. (**C**) Schematic illustration of the novel electrochemical sensor for PSA marker detection. Reprinted with permission from Ref. [37]. Copyright 2019, Wiley.

Also, Kasturi et al. [38] reported on the development of a biosensor for the detection of microRNA-122 (miRNA-122) with AuNPs-decorated reduced graphene oxide (rGO) on the Au electrode surface (Figure 5). The thiol-labeled DNA probes were attached to the Au-rGO transducer surface by forming a SAM layer, with subsequent blocking with 1% BSA. Then, the target miRNA was allowed to bind to the transducer surface to quantify the biomarker for liver diseases.

The sensor possessed a linear range from 10 μM to 10 pM and had a detection limit of 1.73 pM. The sensor possessed good biocompatibility, superior electron transfer characteristics, large surface area, and selective conjugation with biomarkers. Also, the sensor design can be applied to construct other types of biomarker detection. Furthermore, it can be integrated with a lab on a chip platform. It is also applicable to the large-scale production of sensors with a focus on the early detection of diseases.

In another interesting work, Rauf et al. [39] reported on the use of laser-induced graphene oxide [34] as a new-generation electrode in cancer research for the detection of human epidermal growth factor receptor 2 (HER-2). In this study, with laser printing technology, the structures of working, counter, and reference electrodes were formed on a polyimide sheet, then the gold nanostructures (Christmas-tree-like structures) were formed by electrodeposition on the working electrode (Figure 6). Subsequently, the sensor surface is modified with thiol labeled HER-2 aptamer and blocked with BSA for any nonspecific bindings. Then, the HER-2 protein was allowed, in different concentrations, to interact with the aptamer immobilized surface. The electrochemical signals were then recorded for the aptamer surface after bindings with different concentrations with $[Fe(CN)_6]^{3-/4-}$

redox probe. The CV analysis showed a decrease in current upon bindings of various concentrations of HER-2, and from the calibration, the limit of detection was found to be 0.008 ng/mL. It is claimed that with the incorporation of 3D Au nanostructures the sensor possessed a high electron transfer rate, which resulted in achieving a lower LOD and possessing high sensitivity and accuracy in detecting HER-2 in human serum samples. Furthermore, special software was developed to make it a POC device, in which the laboratory aptasensor could be converted into a hand-held aptasensor.

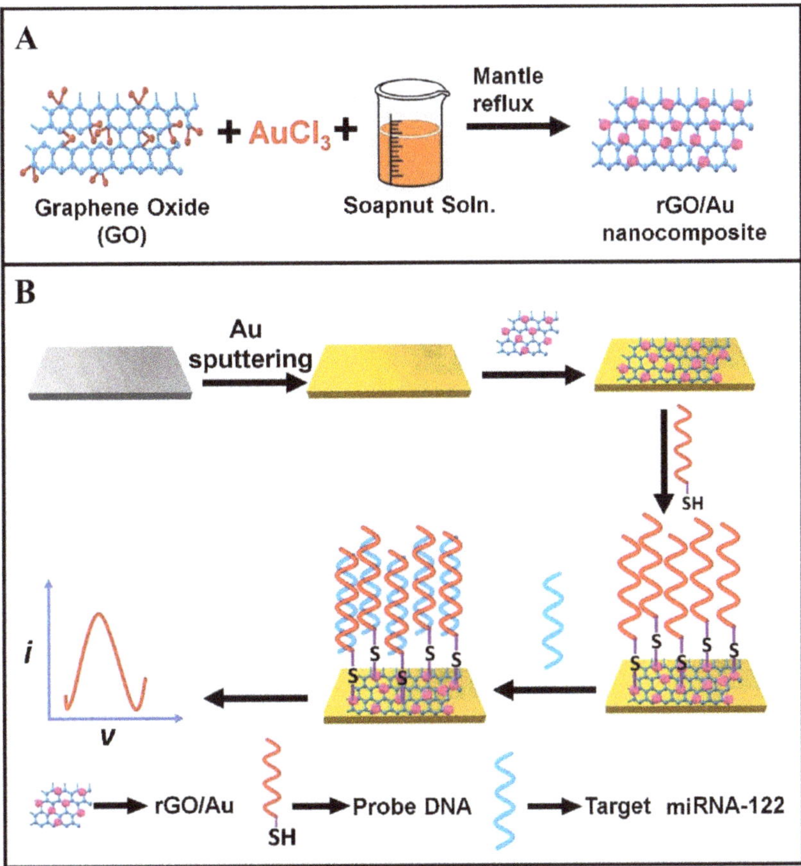

Figure 5. Schematic representation of the (**A**) Synthesis of rGO/Au nanocomposite, (**B**) Fabrication of rGO/Au nanocomposite-based miRNA-122 electrochemical detection platform. Reprinted with permission from Ref. [38]. Copyright 2021, Elsevier.

Also, Hasanjani et al. [40] reported on the development of Zidovudine (ZDV). A modified pencil graphite electrode (PGE) was made using deoxyribonucleic acid/Au-Pt bimetallic nanoparticles/graphene oxide-chitosan (DNA/Au-Pt BNPs/GO-chit/PGE) (Figure 7). The PGE was immersed in the GO-chit solution to create the graphene oxide-chitosan/pencil graphite electrode (GO-chit/PGE). Later, the electrodeposition of Au-Pt bimetallic nanoparticles (Au-Pt BNPs) was accomplished on the surface of the GO-chit/PGE-modified electrode. Subsequently, DNA was immobilized on the Au-Pt BNPs/GO-chit/PGE, applying a constant potential of 0.5 V.

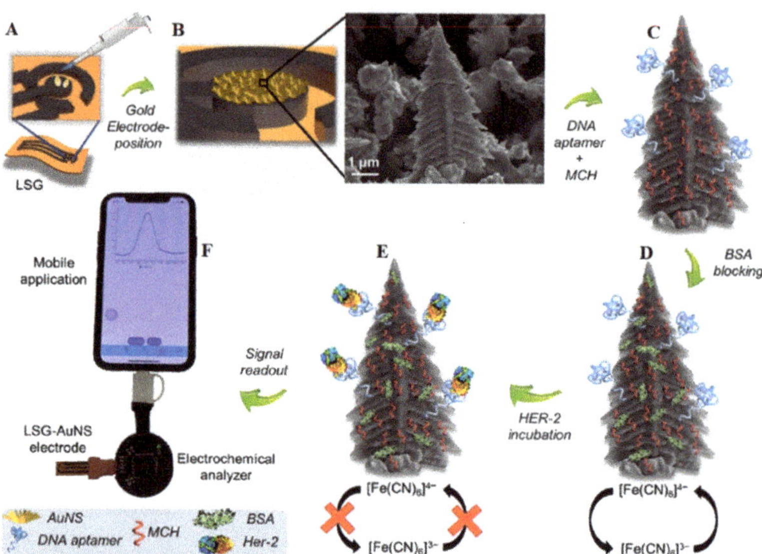

Figure 6. The schematic diagram for the formation of laser-induced graphene (LIG) electrode sensor. (**A**) LIG electrode on polyimide sheet, (**B**) Formation of Au nanostructures on working electrode area with electrodeposition, inset shows the SEM images of the tree-like structure of Au. (**C**) Bindings of DNA aptamer on the electrode through self-assembly of mecaptohexanol (MCH), (**D**) Surface blocking procedures with BSA and measurement of electrochemical signal with $[Fe(CN)_6]^{3-/4-}$ redox probe, (**E**) Incubation with the HER-2 antigen and measurement of EC signal, and (**F**) Quantification of HER-2 by evaluating the electrochemical signal. Reprinted with permission from Ref. [39]. Copyright 2021, Elsevier.

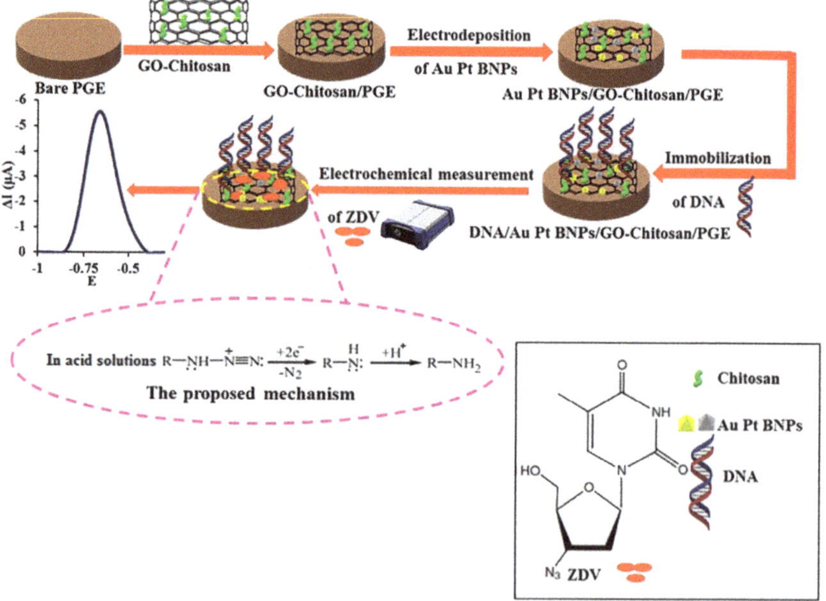

Figure 7. Schematic route for the fabrication of DNA/Au−Pt BNPs/GO−chit/PGE transducer surface for the development of an electrochemical biosensor for the detection of ZDV. Reprinted with permission from Ref. [40]. Copyright 2021, Elsevier.

Using differential pulse voltammetry, the $I-V$ response was recorded for different concentrations of ZDV. The sensor showed a linear dynamic range from 0.01 pM to 10.0 nM, with a detection limit of 0.003 pM in human serum samples.

Recently, Kangavalli and Veerapandian reported on the development of a dengue biomarker using ruthenium bipyridine complex on the surface of graphene oxide [41]. They also reported on various EC-based techniques for the electrodeposition and electroless deposition procedures of graphene oxide as a nanoarchitecture for a label-free biosensor platform [42]. Some more information on electrochemical biosensors developed for biomarker detection that contain graphene oxide and metal nanoparticles can be found in some valuable studies recently reported, and are available in the literature [43–47]. Graphene oxide-based nanomaterials offer a wide range of possibilities for developing sensitive electrochemical biosensors for biomarker detection. In recent years, significant advances in graphene-nanoparticle-based electrochemical sensors are made for the detection of cancer biomarkers, and here we analyze the analytical parameters of those sensors, as shown in Table 1.

Table 1. Literature reports on the analytical parameters of graphene oxide conjugated nanoparticles for various biomarker detection.

Sensing Platform	Biomarker	Technique	Linear Range	LOD	Real Sample	Ref.
RGO-NP/ITO	CRP	EIS	1–10,000 ng/mL	0.08 ng/mL	Human serum	[36]
GO-CoPP	CPEB4	DPV	0.1 pg/mL–10 ng/mL	0.074 pg/mL	Human serum	[48]
AuNP-RGO/ITO	TNF-α	EIS	1–1000 pg/mL	0.43 pg/mL	Human serum	[49]
rGO@AgNPs	LA	CV	10–250 μM	0.726 μM	Human serum	[50]
AgPdNPs/rGO	RAC	LSA	0.01–100 ng/mL	1.52 pg/mL	—	[51]
	SAL			1.44 pg/mL		
	CLB			1.38 pg/mL		
MWCNTs-AuNPs/CS-AuNPs/rGO-AuNPs	OTC	DPV	1.00–540 nM	30 pM	—	[52]
GO-Fe$_3$O$_4$-β-CD	MGMT	DPV	0.001–1000 nM	0.0825 pM	Human plasma	[53]
AuNPs/GQDs/GO/SPCE	miRNA-21	SWV	0.001–1000 pM	0.04 fM	Human serum	[54]
	miRNA-155			0.33 fM		
	miRNA210			0.28 fM		
rGO/RhNPs/GE	HER-2-ECD	DPV	10–500 ng/mL	0.667 ng/mL	Human serum	[55]
AuNPs-rGO/ITO	IL8	DPV	500 fg/mL–4 ng/mL	72.73 pg/mL	—	[56]
Pd@Au@Pt/rGO	CEA	DPV	12 pg/mL–85 ng/mL	8 pg/mL	Human serum	[57]
	PSA		3 pg/mL–60 ng/mL	2 pg/mL		
AgNPs/GO/SPCE	PSA	DPV	0.75–100 ng/mL	0.27 ng/mL	Human serum	[58]
rGO-GNPs-Cr.6/GCE	L-Trp	SWV	0.1–2.5 μM	0.48 μM	Human serum	[59]
GO/AgNPs/Au	PSA	LSV	5–20,000 pg/mL	0.33 pg/mL	Human serum	[60]
AuNP/RGO/GCE	CA125	SWV	0.0001–300 U/mL	0.000042 U/mL	Human serum	[61]
ErGO-SWCNT/AuNPs	HER2	EIS	0.1 pg/mL–1 ng/mL	50 fg/mL	Human serum	[62]
Au-PtBNPs/CGO/FTO	MUC1	DPV	1 fM–100 nM	0.79 fM	Human serum	[63]
BNPAu-Fe-rGO/GCE	Acetaminophen	DPV	50–800 nM	0.14 nM	Human urine	[64]

2.2. MoS$_2$ Conjugated Nanoparticles for Electrochemical Biomarker Detection

Recently, transition metal dichalcogenides (TMDCs) found their applications in various biosensors due to their large surface-to-volume ratio, tunable electronic and optical properties, low toxicity, and unique van der Waals layered structure [65]. In TMDCs, one layer of transition metal atoms (M) lies between two layers of chalcogen atoms (X) resulting in a formula MX$_2$. Various kinds of TMDCs can be realized by altering the chalcogen atoms such as Sulphur (S), Selenium (Se), and Tellurium (Te), and metal atoms like Molybdenum (Mo) and Tungsten (W). Among these, MoS$_2$ is commonly used because its fundamental constituents are surplus and innoxious [66]. MoS$_2$ molybdenum (Mo) atoms lie between the two sulfide atoms layers (S-Mo-S) and atoms in the crystal are associated by strong covalent bonding and adjacent layers of MoS_2 are held by weak van der Waals forces. MoS_2 possesses a mobility of 200 cm^2/Vs at room temperature, high on/off current ratio of 10^8, and a direct band gap of 1.8 eV. Based on these properties, MoS_2 becomes a promising alternative to graphene and is applied in various electrochemical and optical sensors [67–69]. MoS_2 can be synthesized in both top-down and bottom-up approaches (Figure 8). The top-down approach includes the exfoliation of MoS_2 [70], while the bottom-up approaches include (i) chemical vapor deposition [71]; (ii) physical vapor deposition [72]; (iii) solution-based processing [73]. For a more detailed synthesis of MoS_2, readers are encouraged to go through the literature survey of the desired synthesis approach. Thus, like graphene, MoS_2 offers a large surface area that enhances its biosensing performance.

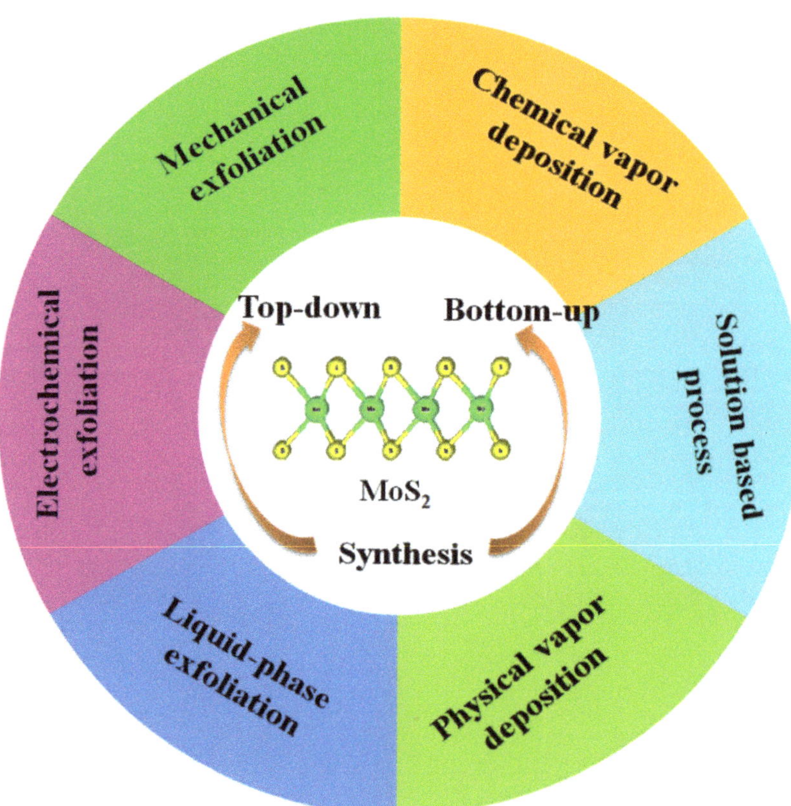

Figure 8. Various synthetic methods for MoS_2 preparation. Reprinted with permission for Ref. [74]. Copyright 2022 MDPI.

MoS_2 possesses a direct band gap of 1.8 eV in the monolayer, lattice defects of zero dimensionality, grain boundary defects, and an enhanced surface-to-volume ratio. Also, the feasibility of surface modification and chemical functionalization makes these characteristics of MoS_2 to adopt and study in scientific and industrial fields [75] (Figure 9). Furthermore, to increase the electroactivity/conductivity of graphene and/or other 2D materials, mostly nanoparticles were incorporated to achieve the synergistic effects from both nanomaterials, which ultimately resulted in an improvement in the overall analytical performance of the biosensor. In this section, we review various types of biosensors that incorporate metal nanoparticles on MoS_2 for the detection of various biomarkers.

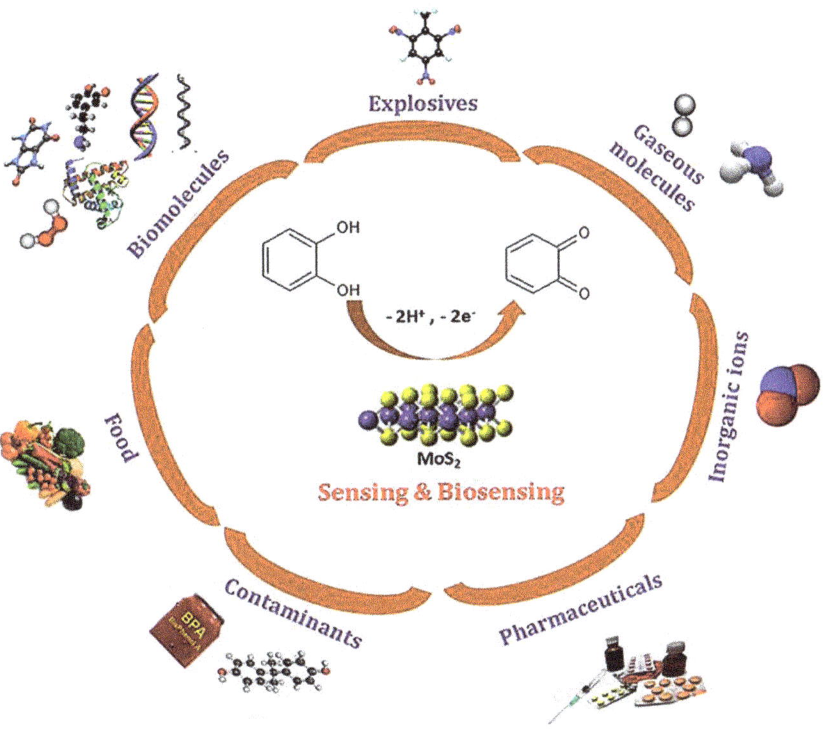

Figure 9. MoS_2 nanostructures-based electrochemical sensing application in various fields. Reprinted with permission from Ref. [76]. Copyright 2018, Elsevier.

In a recent report that mentions the usage of MoS_2-Au nanoparticles, Yagati et al. [77] reported on the applications of MoS_2 conjugated Au nanoparticles on indium tin oxide (ITO) electrodes for the detection of the thyroid-stimulating hormone biomarker, triiodothyronine (T_3), as shown in Figure 10. Electrodeposition procedures allowed the formation of MoS_2 and Au nanostructures on the ITO electrode. Subsequently, T_3 antibodies were immobilized on the MoS_2-Au/ITO surface by forming a self-assembled monolayer of dithiobis (succinimidyl propionate) (DSP). For any nonspecific bindings, the surface is coated with casein and then subjected to different concentrations of the T_3 biomarker diluted in both PBS and serum samples. Electrochemical impedance spectroscopy was used to analyze the bindings of T_3 to its antibodies and a linear correlation was observed for different concentrations. Based on the quantifications made by this sensor for the detection of T_3, a linear range of 0.01–100 ng/mL with a detection limit of 2.5 pg/mL was observed. The sensor also showed a good correlation with data observed by the conventional method (Roche Cobas) and possessed high sensitivity and selectivity in discriminating the healthy and cancer samples. Based on the findings, the developed sensor could apply to cancer-related biomolecule analysis.

Su et al. [78] developed dual target sensing (adenosine triphosphate (ATP) and thrombin) detection electrochemical biosensors based on gold nanoparticles-decorated MoS_2 ($AuNPs$–MoS_2) nanocomposites which feature both "signal-on" and "signal-off" elements in the detection system, and thrombin and ATP could act as inputs to activate an AND logic gate (Figure 11). In this approach, two different aptamer probes labeled with redox tags (ferrocene (Fc) and methylene blue (MB)) were simultaneously immobilized on an $AuNPs$-MoS_2 modified glassy carbon electrode (GCE) through Au-S bond formations. Subsequently, the electrode was immersed in 6-mercaptohexanol to block the uncovered spots of $AuNPs$–MoS_2/GCE. Square wave voltammetry (SWV) was used to determine the various concentrations of ATP and thrombin applied to the GCE. From concentration vs. change in the current results, it was evaluated that the sensor had a linear range for the determination of ATP, which was 1 nM to 10 mM with a detection limit of 0.32 nM, while for the thrombin determination, the linear range was 0.01 nM to 10 µM with a detection limit of 0.0014 nM.

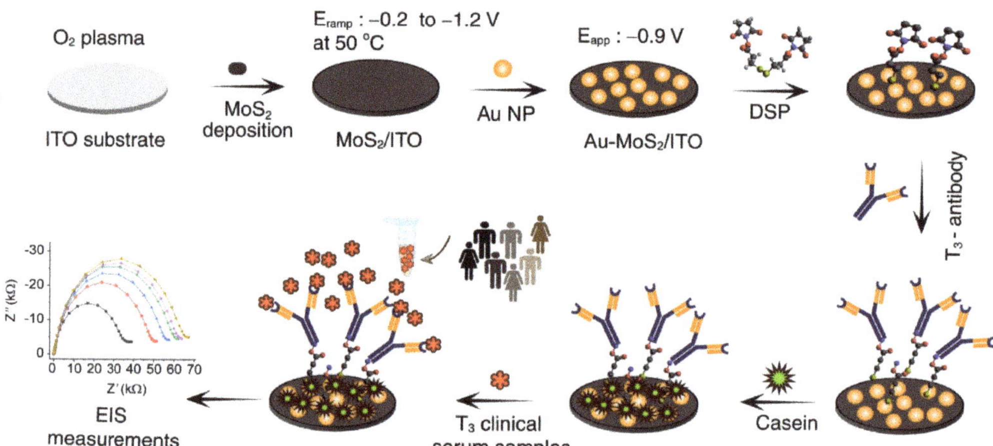

Figure 10. Schematic illustration of the total triiodothyronine (T$_3$) receptive interface fabrication through the immobilization of the antibody on a step-by-step modification process of MoS$_2$–Au formation and subsequent functionalization with a dithiobis (succinimidyl propionate) monolayer on an indium tin oxide electrode surface. With increasing concentration of the T3 analyte in serum, the EIS (Nyquist plot) shows increased semi-circle (Rct) for quantification. Reprinted with permission from Ref. [77]. Copyright 2020, Elsevier.

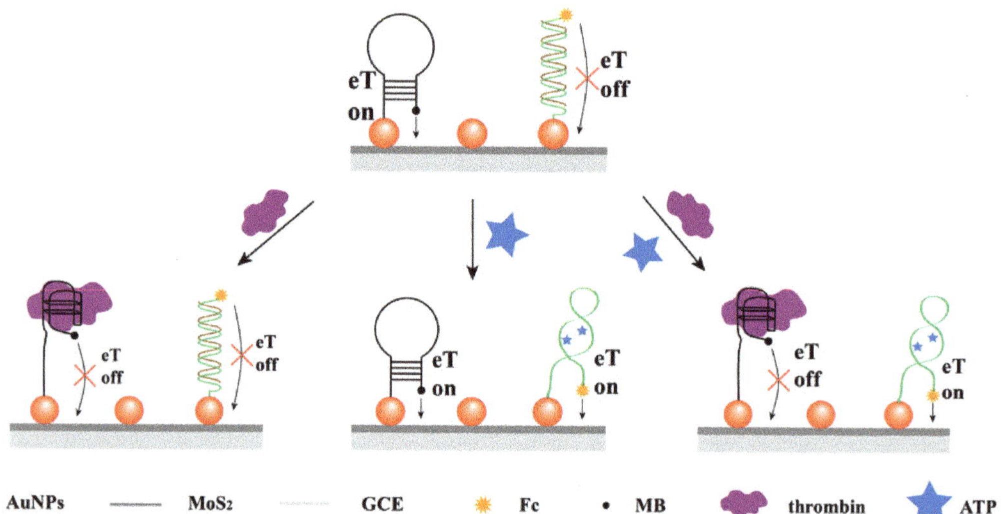

Figure 11. Schematic representation for the development of the aptasensor for the determination of ATP and thrombin. Reprinted with permission from Ref. [78]. Copyright 2016, ACS.

The authors also suggested that this mechanism can be acted as an AND logic gate by using ATP and thrombin as inputs and the electrochemical signals of Fc and MB as outputs (Figure 12). The logic gate works on the structural conversion of the aptamer probe triggered by ATP and thrombin. The working mechanism was the individual peak current enhancement of Fc or the suppression of MB as electron transfer OFF (eTOFF) or "zero" output, and the simultaneous peak current enhancement of Fc and suppression of MB as electron transfer ON (ON) or "one" output. From the inset table, a "one" output was achieved only when both inputs were "one". When there were no inputs (0, 0) or only one input (0, 1 or 1, 0), the result was "zero" output. Thus, the MoS$_2$-based multiplexed aptasensor could also serve as an "AND" gate.

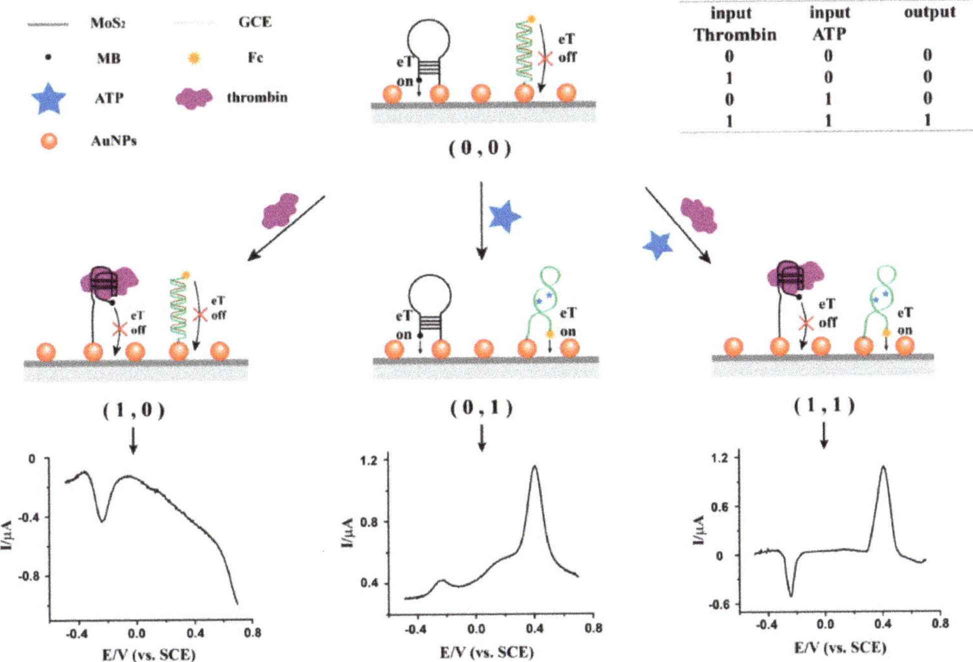

Figure 12. Schematic description of the MoS$_2$-Based AND logic gate for determination of ATP and thrombin. Reprinted with permission from Ref. [78]. Copyright 2016, ACS.

In another work, Chen et al. [79] reported on the development of a growth differentiation factor-15 (GDF-15) expression sensor which is a potential biomarker for the diagnosis, risk stratification, and prognosis of various cardiovascular diseases (Figure 13). Here, a sandwich-type immunosensor was constructed using amine-modified graphene-supported gold nanorods (NG/AuNPs) as a substrate platform, and the durian-shaped MoS$_2$/AuPtPd nanodendrite (NDs) as a label for secondary antibodies (Ab$_2$) for the quantification of growth differentiation factor-15 (GDF-15). NG/AuNPs are used to enhance the surface area and for the immobilization of primary antibodies through the binding of amino or sulfhydryl groups. Subsequently, the electrodes were blocked with 1wt% BSA. Finally, the signal probe MoS$_2$/AuPtPd-Ab$_2$ was added to the sample.

The developed sensor was also applied to evaluate the efficacy towards the clinical sample analysis and compared with traditional sensing methods, such as ELISA, to evaluate the accuracy of the results. The sensor showed a linear range of 1.5 pg/mL to 1.5 μg/mL with a detection limit of 0.9 pg/mL. Due to its high sensitivity, rapid response, and feasibility to miniaturization, the proposed sensor could be applied to a point-of-care diagnostic tool for cardiovascular diseases and paves the path toward "liquid biopsies".

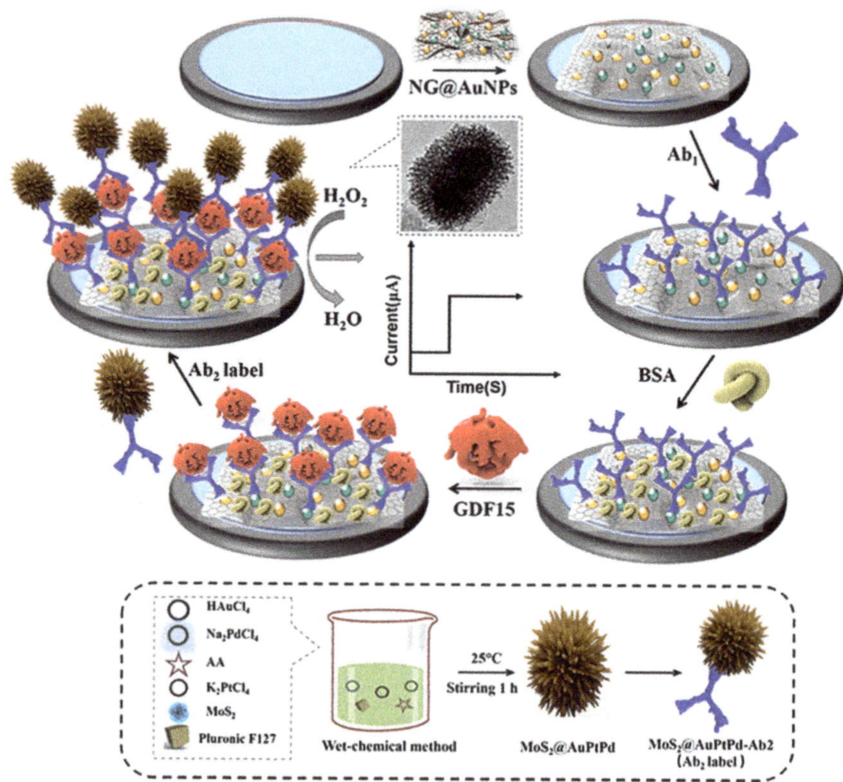

Figure 13. Schematic illustration for the development of a sandwich-type electrochemical sensor for GDF-15 detection sensor. Reprinted with permission from Ref. [79]. Copyright 2022, Elsevier.

Nong et al. [80] reported on the detection of cortisol which is a glucocorticoid hormone that adrenal glands produce and release, and this hormone regulates stress, inflammation, blood pressure, sugar, and overall metabolism. In this work, copper tungstate-molybdenum sulfide ($CuWO_4@MoS_2$) and chitosan-gold (Chit-Au) nanocomposite were synthesized and applied to GCE (Figure 14). Subsequently, the cortisol antibody (C-Mab) was immobilized using the EDC/NHS reaction and subsequent blocking with BSA. Once the transducer surface was fabricated, SWV was performed to analyze the bindings of various concentrations of cortisol and a linear relationship was observed concerning different concentrations. The sensor showed a linear range of 0.1 fg/mL to 1 µg/mL with a detection limit of 0.014 fg/mL (S/N = 3). The sensor showed excellent storage stability and reproducibility and it can detect the content of cortisol in saliva.

Su et al. [81] reported on the use of a MoS_2-Au nanocomposite for the detection of a carcinoembryonic antigen (CEA). In this work, CEA antibodies labeled with horseradish peroxidase resulted in an amplified electrochemical signal by catalyzing o-phenylenediamine (o-PD) in the presence of hydrogen peroxide (H_2O_2). As can be seen in Figure 15, the MoS_2-Au conjugated HRP labeled antibodies enhance the overall sensitivity when the different concentrations of CEA were measured using cyclic voltammetry. From the analytical performance, the sensor displayed a linear range of 10 fg/mL to 1 ng/mL with a detection limit of 1.2 fg/mL. The sensor also exhibited good stability, and high selectivity suggesting that the proposed immunosensor could detect CEA in real samples.

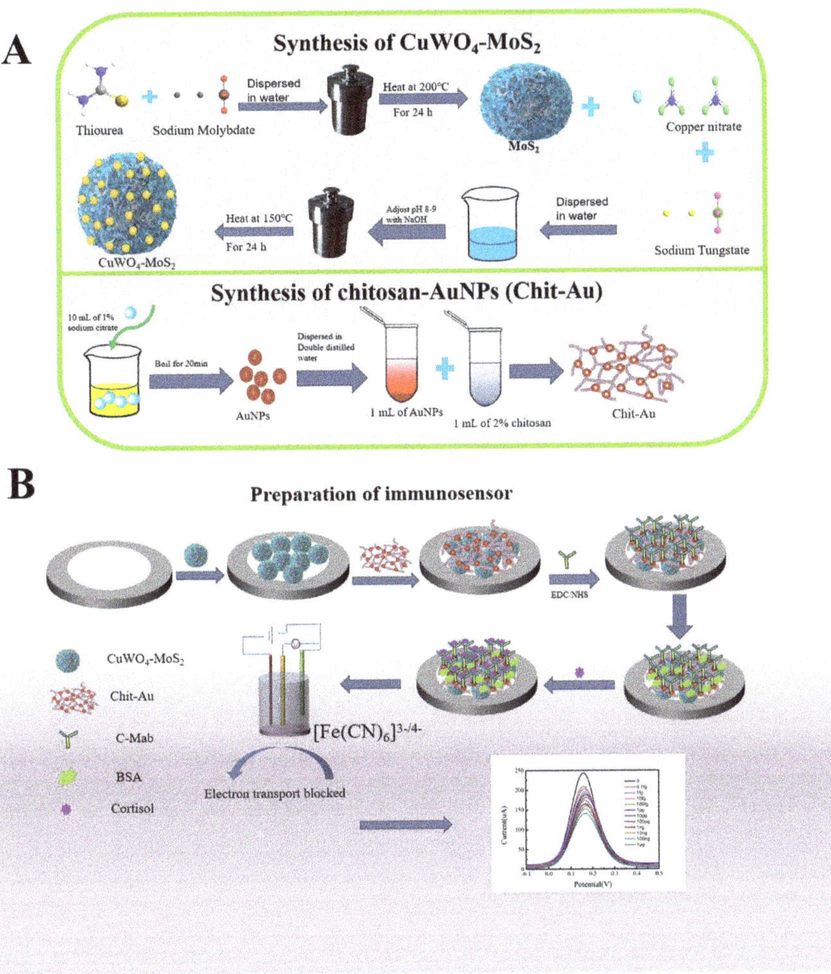

Figure 14. Schematic representation for (**A**) Synthesis of MoS_2, $CuWO_4@MoS_2$, AuNPs, and Chit-Au nanocomposites; (**B**) Preparation process of the immune electrode. Reprinted with permission for Ref. [80]. Copyright 2022, Elsevier.

Also, Ma et al. [82] reported similar works using $MoS_2@Cu_2O$-Au nanoparticles for the detection of alpha-fetoprotein (AFP), a tumor marker to identify adult primary liver cancer (Figure 16). In this work, AuNPs were electrodeposited on GCE which acted as antibody carriers and sensing platforms. Further, $MoS_2@Cu_2O$ was combined with the AuNPs as a strategy to obtain the signal amplification resulting in a composite MoS_2-Cu_2O-Au as a triamplification electrochemical signal. A sandwich immunosensor was developed by immobilizing primary antibodies on Au-deposited GCE and blocked with a surface with BSA for nonspecific bindings. Then, the electrodes were dipped with different concentrations of AFP. Subsequently, the HRP-labeled secondary antibodies coupled with $MoS_2@Cu_2O$ were then allowed to conjugate with the electrode. Amperometric response, under suitable experimental conditions, exhibited that the sensor possessed a linear range of 0.1 pg/mL to 50 ng/mL and a detection limit of 0.037 pg/mL (S/N = 3). The sensor showed satisfactory recoveries when tested in human serum samples, and the proposed approach could extend the potential application of electrochemical immunosensors to medical applications.

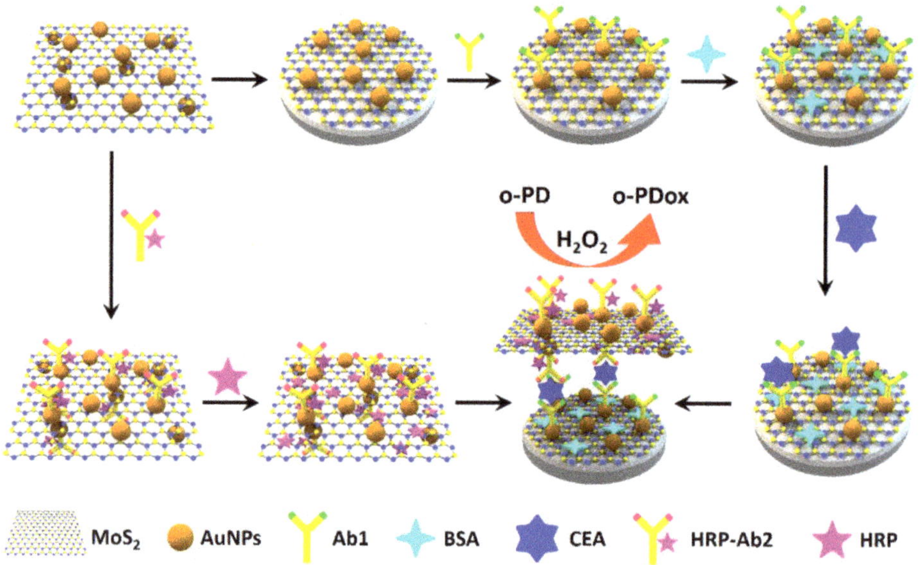

Figure 15. The schematic diagram for the stepwise modification of the GCE with MoS$_2$ and Au nanoparticles for anti-CEA antibody immobilization for developing a CEA detection sensor. Reprinted with permission from Ref. [81]. Copyright 2019, Elsevier.

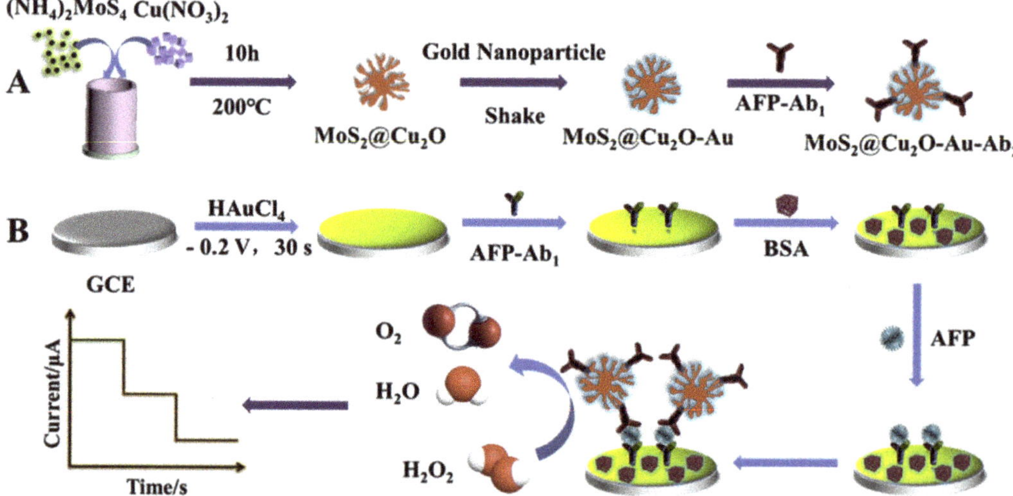

Figure 16. The schematic diagram for the preparation procedure for the sandwich-type electrochemical immunosensor. Reprinted with permission from Ref. [82]. Copyright 2019, Elsevier.

Likewise, several reports demonstrated the usage of a MoS$_2$-Au nanocomposite for the detection of electrochemical biosensors for various types of biomarker detection in clinical applications. However, very few reports show the possibility of point-of-care applications. Here, we analyzed the analytical parameters of the reports that adopt the MoS$_2$-Au nanocomposite used for electrochemical sensors and presented them in the following Table 2.

Table 2. Literature reports on the analytical parameters of MoS_2 conjugated nanoparticles for various biomarker detections.

Sensing Platform	Biomarker	Technique	Linear Range	LOD	Real Sample	Ref.
Au-NPs/MoS$_2$	CRP	EIS	1 fg/mL–1 µg/mL	0.01 fg/mL	—	[83]
Fe$_3$O$_4$@MoS$_2$-AuNPs	H$_2$O$_2$	SWV	1–120 µM	80 nM	Human serum	[84]
Au/MoS$_2$/Au/PET	GP120	SWV	0.1 pg/mL–10 ng/mL	0.066 pg/mL	Human serum	[85]
MoS$_2$/Pt@Au-nanoprism/PDA	free-PSA; total-PSA	DPV	0.0001–100 ng/mL	0.1 pg/mL; 0.0011 fg/mL	Human serum	[86]
MoS$_2$ NFs/Au@AgPt YNCs	CEA	i-t curve	10 fg/mL–100 ng/mL	3.09 fg/mL	Human serum	[87]
Au/Co-BDC[f]/MoS$_2$	CTnI[g]	i-t curve	10 fg/mL–100 ng/mL	3.02 fg/mL	Human serum	[88]
Au/MoS$_2$/rGO	CA 27-29 BCA	i-t curve	0.1–100 U/mL	0.08 U/mL	Human serum	[89]
MoS$_2$-AnNPs/GCE	CEA	DPV	1 pg/mL–50 ng/mL	0.27 pg/mL	Human serum	[90]
Ce-MoS$_2$/AgNRs	PSA	CV	0.1–1000 ng/mL	0.051 ng/mL	Human serum	[91]
MoS$_2$@Au	Siglec-5	ECL	10 pM–500 pM	8.9 pM	Human serum	[92]
MoS$_2$/PPY/AuNPs	Glucose	DPV	0.1–80 nM	0.08 nM	Human serum	[93]
AgPt/MoS$_2$	H$_2$O$_2$	i-t curve	20 µM–4 mM	1.0 µM	—	[94]

2.3. Biomarker Detection on MXenes Conjugated with Metal Nanoparticles

MXenes are transition-metal carbides/nitrides/carbonitrides with a 2D structure and general formula $M_{n+1}X_nT_x$ (n = 1–3), where M is an early transition metal, X can be carbon or nitrogen, and T_x corresponds to the surface terminations (Figure 17A,B). The ideal electronic structure [95], structural stability [96], high surface-to-volume ratios [97], outstanding mechanical [98] and optical properties [99], versatile surface chemistries [100], tunable bandgap [101], and high thermal and chemical stability [102,103] make them promising materials for biomarker detection (Table 3). The initial synthesis approach for MXenes was realized based on the etching of Ti_3AlC_2 with 50% HF for 2 h at room temperature [104]. Later many environmentally friendly approaches were formulated [105] (Figure 17C). However, similar to any other pristine 2D materials, MXenes suffer from poor selectivity, low sensitivity, and slow response [106]. These disadvantages were usually overcome by synthesizing MXene-metal nanoparticle nanocomposites. MXene-metal nanoparticle nanocomposites possess a large specific surface area, superior electron conductivity, and enhanced electron transfer properties for biosensing applications [107]. To expand beyond the limitations of MXenes, Liu et al. [108] reported the covalent grafting of PAMAM onto MXene (MXene@PAMAM) (Figure 18A). Here, the PAMAM acted as an efficient stabilizer and spacer, thereby preventing the restacking and oxidation of the MXene. Moreover, the aminoterminals of PAMAM acted as adsorption sites for AuNPs. The AuNPs@MXene@PAMAM nanobiosensing platform was applied for the detection of the cardiovascular disease biomarker cTnT. The sensor performance was remarkable with a wide detection range (0.1–1000 ng/mL) and a very low detection limit (0.069 ng/mL). Medetalibeyoglu et al. [109] fabricated a d-Ti$_3$C$_2$T$_X$ MXene@AuNPs/Ab2 bioconjugate-based sandwich-type electrochemical immunosensor for the detection of PSA. Here, AuNPs at the bioconjugate were used to label PSA secondary antibody-2 for signal amplification (Figure 18B). In one study, Laochai et al. [110] fabricated thread-based L-Cys/AuNPs/MXene working electrodes for the noninvasive electrochemical detection of sweat cortisol, which is an important biomarker for identifying adrenal gland disorders (Figure 18C). Here, MXene served as a 2D platform to anchor the monoclonal anticortisol antibodies, whereas AuNPs increased the specific surface area, and thereby the sensitivity of the detection system. Mesoporous nanoparticles (MNPs), comprising metallic and nonmetallic counterparts, show better catalytic performance compared to their bulk nanoparticles [111]. Liu et al. [112] reported sandwich-type PdPtBP MNPs/MXene-based immunosensor for the ultrasensitive detection of urine kidney injury molecule-1(KIM-1) (Figure 18D). Yang et al. [113] reported an interesting cascaded signal amplification strategy on in situ reduced gold nanoparticle deposited Ti$_3$C$_2$ MXene (Figure 18E), where MXene acted as a stabilizer and reductant. Here, AuNPs with the predominant (111) facet on MXene provided high electrocatalytic activity and were also used as a carrier of the C-DNA and to make DNA hybridization. Mohsen et al. [114] reported Au nanoparticles on Ti_3C_2 MXene for synergistic signal amplification (Figure 18F). Here, the perfectly distributed Au nanoparticles on the flaky architecture of MXene contributed to the enhanced electrochemical performance and the attomolar detection of multiple micro-RNAs (miRNAs) achieved on an AuNP@MXene/Au electrode. Wang et al. [115] proposed a competitive electrochemical aptasensor for the breast cancer biomarker Mucin1 based on Au nanoparticles decorated Ti_3C_2 MXene. Here, aptamer binding to the electrode surface was achieved through Au-S bonds by the electrodeposited gold nanoparticles. The electrochemical aptasensor reported a wide linear range (1.0 pM–10 µM) and a low detection limit (0.33 pM) with promising clinical applications. Cheng et al. [116] demonstrated a gold nanoparticle-modified MXene-based sandwich-type immunosensor platform for squamous cell lung cancer cytokeratin fragment antigen 21-1 (CYFRA 21-1).

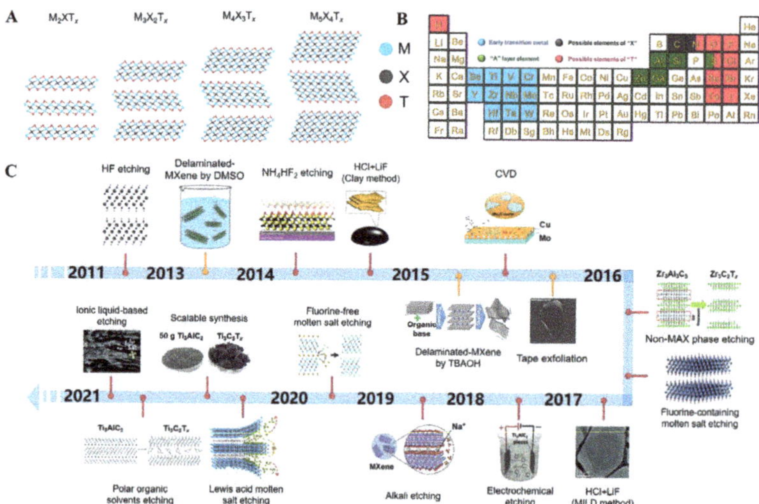

Figure 17. (**A**) Structure of various MXenes with surface terminations. (**B**) Periodic table elements experimentally used for the synthesis of MXenes, and (**C**) Timeline of the various synthesis routes to MXenes. Reproduced with permission from Ref. [117]. Copyright 2021, Wiley.

Table 3. Recent literature reports on metal nanoparticles incorporated MXenes for electrochemical biomarker detection.

Sensing Platform	Biomarker	Technique	Linear Range	LOD	Real Sample	Ref.
AuNPs/Ti_3C_2@PAMAM	cTnT	DPV	0.1–1000 ng/mL	0.069 ng/mL	Human serum	[108]
Ti_3C_2@AuNPs	PSA	DPV	pg/mL	3.0 fg/mL	Plasma	[109]
L-cys/AuNP/Ti_3C_2	Cortisol	CA	5–40 ng/mL	0.54 ng/mL	Artificial sweat	[110]
PdPtBP MNPs/Ti_3C_2	KIM-1	DPV	0.5–100 ng/mL	86 pg/mL	Human urine	[112]
AuNPs-Ti_3C_2/AuE	miRNA-21	DPV	100 aM–1 nM	50 aM	—	[113]
AuNP@MXene/Au	miRNA-21 miRNA-141	DPV	500 aM–50 nM	204 aM 138 aM	Total plasma	[114]
cDNA-Fc/MXene/Apt/Au/GCE	MUC1	SWV	$0.001–1.0 \times 10^4$ nM	0.33×10^{-3} nM	Human serum	[115]
AuNP-Ti_3C_2	CYFRA21-1	SWV	$0.5–1.0 \times 10^4$ pg/mL	0.1 pg/mL	Human serum	[116]
MCH/CP/MXene-Au/GCE	miRNA-377	SWV	10 aM–100 pM	1.35 aM	Human serum	[118]
Ti_3C_2-AuNPs/GCE	PSA	DPV	1–50,000 pg/mL	0.31 pg/mL	—	[119]
AuNPs/d-S-Ti_3C_2	PCT	DPV	0.01–1.0	2.0 fg/mL	—	[120]
MB/DNA/HT/HP1/AuNPs/Ti_3C_2/$BiVO_4$/GCE	$VEGF_{165}$	PEC	10 fM–100 nM	3.3 fM	—	[121]

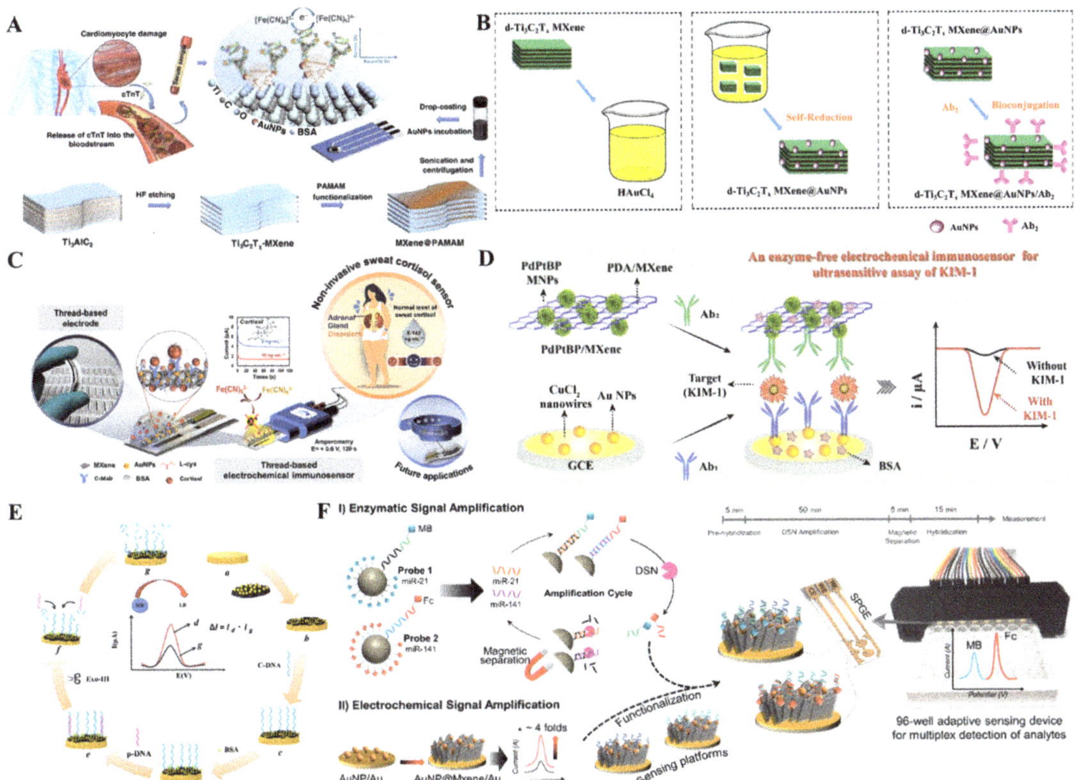

Figure 18. (**A**) Schematic illustration of the fabrication of AuNPs/MXene@PAMAM for the electrochemical detection of cTnT. Reproduced with permission from Ref. [108]. Copyright 2022, Nature. (**B**) Preparation of d-Ti$_3$C$_2$ MXene@AuNPs/Ab2 for the detection of PSA. Reproduced with permission from Ref. [109]. Copyright 2020, Elsevier. (**C**) Fabrication of L-cys/AuNPs/MXene on a thread-based electrochemical biosensor for noninvasive sweat cortisol detection. Reproduced with permission from Ref. [110]. Copyright 2022, Elsevier. (**D**) Fabrication of PdPtBP nanoparticles/MXene-based enzyme-free electrochemical biosensor for the detection of kidney injury molecule-1 (KIM-1). Reproduced with permission from Ref. [111]. Copyright 2021, Elsevier. (**E**) Schematics of the AuNPs-based cascaded signal amplification process for the detection of miRNA-21. Reproduced with permission from Ref. [113]. Copyright 2022, ECS, and (**F**) Schematic diagram based on AuNPs decorated MXene for the multiplex and concurrent detection of miR-21 and miR-141. Reproduced with permission from Ref. [114]. Copyright 2020, Elsevier.

2.4. MOFs Conjugated Metal Nanoparticles for Electrochemical Biomarker Detection

As an emerging material with exceptional properties, metal-organic frameworks (MOFs) have been studied exceptionally during the past decades. MOFs are porous materials comprising a framework of metal ions or metal-containing clusters and organic ligands [122]. MOFs have been reported to have excellent properties such as a tunable structure [123], large surface area [124], abundant functional groups [125], high porosity [126], good conductivity [127], and thermal stability [128]. MOFs have been traditionally synthesized by hydrothermal/solvothermal methods [129]. The solvothermal method is a general concept where a solvent other than water is used, and the synthesis is usually performed at a temperature above the boiling temperature of the solvent in closed chemical reactors at higher pressures. Moreover, the greater pressure inside the closed reactor results in enhanced salt solubility. The benefits of the solvothermal process allowed researchers to develop reproducible protocols with total control of the long-term synthesis processes. The solvothermal method has the advantage of higher product yield with improved crystallinity [130]. The hydrothermal/solvothermal method has been optimized for the synthesis of MOFs such as Ni-MOF [131], Co-MOF [131], Fe-MOF [132], Cu-MOF [133], Zn-MOF [134], and mixed-ligand metal-organic frameworks [135]. In recent years, electrochemical synthesis gained attention, and several MOFs such as $Cu_3(HHTP)_2$ [136], Mn-DABDC(ES) [137], 2D/3D Zn(II)-MOF hybrid [138], Fe-MIL-101 and Fe-MIL-101-NH$_2$ [139], etc. have been reported for various MOFs' electrocatalytic applications. Electrochemical synthesis has the advantages of mild synthesis conditions, shorter synthesis times, and controllability of morphology and thickness by the applied current/voltage [140]. During electrochemical synthesis, the metal ions enter the solution through the dissolution of the anode and the process is usually continuous with the availability of dissolved linker molecules [141]. Researchers have also developed a variety of other synthesis approaches such as ultrasound and microwave-assisted [142], mechanochemical [143], and sonochemical [144] methods for the synthesis of MOFs with different morphology and applications (Figure 19). As shown in Table 4, modified MOF nanocomposites often outperform unmodified MOF and are often exploited for diverse biosensor applications [145]. MOFs are often decorated with metal nanoparticles in immunosensor applications for anchoring antibodies and enhancing the electrochemical signal. Nanoparticles decorated MOFs with versatile ligands and metal clusters, low cost, and simple operation provide researchers with an adequate 2D platform for biosensing applications. Li et al. [146] fabricated such an interesting immunosensor platform with core-shell $Cu_2O@Cu$-MOF@AuNPs nanostructures for the sensitive detection of CEA (Figure 20A). Here, the sandwich-type electrochemical immunosensor achieved a tripled electrical signal amplification due to the synergistic effect of Cu-MOF, Cu_2O, and AuNPs. Nanowires had more surface area to accommodate proteins and were used to fabricate label-free sensors with exceptional performance [147,148]. Li et al. [149] constructed such an ultrasensitive label-free platform for the detection of NMP-22 based on CuAu nanowires decorated Co-MOFs (Figure 20B). The outstanding catalytic capabilities of Co-MOFs/CuAu NWs achieved a highly sensitive immunosensor with a good linear response (0.1 pg/mL–1 ng/mL), with a lower detection limit (33 fg/mL) suitable for the detection of NMP-22 from human urine samples. An immunoprobe based on AuNPs decorated Fe-MOF for the detection of PSA was reported by Feng et al. [150]. In this study, the labeling antibody was immobilized on AuNPs/Fe-MOF, and methylene blue (MB) covered by a thin layer of AuNPs-rGO served to covalently attach the coating antibodies. An amperometric signal at 0.18 V was measured to quantitatively measure PSA from urine samples (Figure 20C). Zhang et al. [27] reported a similar MB-based strategy for the detection of PSA (Figure 20D). Here, the MOF-325 adsorbed and stabilized MB, thereby solving the problem of MB leakage. A similar nanocomposite comprising MOF, rGO, and AuNPs was reported by Mehmandoust et al. [151] for the detection of a GFAP biomarker (Figure 20F). Here, AuNPs were anchored onto zeolitic imidazolate MOFs and were deployed as a recognition element for the detection of GFAP in urine samples. The intrinsic properties of unique nanomaterials are advantageous for specific immunosensor applications. Zhao et al. [152] fabricated an immunosensor for the detection of NMP-22 based on AuNPs and PtNPs decorated MOFs. The nanoparticles decorated MOF sowed an increased surface area to anchor antibodies through Pt-S and Au-N bonding (Figure 20E), and the immunosensor reported a sensitive response towards NMP-22.

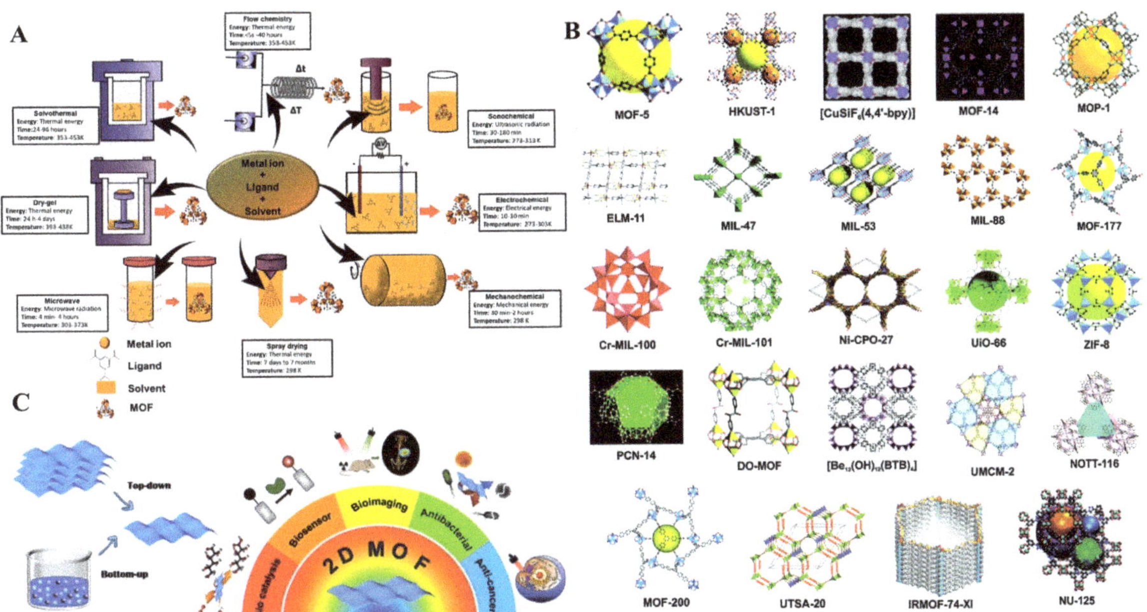

Figure 19. (**A**) Various literature reported conditions and approaches for the synthesis of MOFs. Reprinted with permission from Ref. [153]. Copyright 2021, Elsevier. (**B**) Structures of porous MOFs reported by several research groups. Reprinted with permission from Ref. [154]. Copyright 2015, Royal Society of Chemistry. (**C**) Various biomedical applications of 2D MOFs. Reprinted with permission from Ref. [155]. Copyright 2022, BMC (Springer).

Table 4. Recent literature reports on metal nanoparticles incorporated MOFs for electrochemical biomarker detection.

Sensing Platform	Biomarker	Technique	Linear Range	LOD	Real Sample	Ref.
Au/MOF-235/MB	PSA	DPV	0.01–1.2 ng/mL	3 pg/mL	Human serum	[28]
Co-MOFs/CuAu NWs	NMP-22	CA	10^{-4}–1 ng/mL	33 fg/mL	Human urine	[149]
AuNPs/Fe-MOF	PSA	SWV	0.001–100 ng/mL	0.13 pg/mL	Human serum	[150]
Au@ZIF-8@rGO/SPE	GFAP	EIS	50–10,000 fg/mL	50 fg/mL	Human urine	[151]
rGO-TEPA/AuNPs-PtNPs-MOFs	NMP-22	DPV	0.005–20 ng/mL	1.7 pg/mL	Human urine	[152]
PtNPs/Fe-MOF	Thrombin	DPV	1 fM–10 nM	0.33 fM	Human serum	[156]
Fe$_3$O$_4$@UiO-66/Cu@Au	cTnI	DPV	0.05–100 ng/mL	16 pg/mL	Human serum	[157]
SiO$_2$-Fc-COOH-Au/UiO-66-TB	PCT	DPV	1 pg/mL–100 ng/mL	0.3 pg/mL	Human serum	[158]
Au-MoS$_2$/MOF	NSE	CA	1 pg/mL–100 ng/mL	0.37 pg/mL	Human serum	[159]
AgNPs@Co/Ni-MOF	AFP	ECL	1 pg/mL–100 ng/mL	0.417 pg/mL	Human serum	[160]
BSA/Ab-AgNPs/CdS@MOF-5/PDDA/FTO	cTnI	ECL	0.01–1000 pg/mL	5.01 fg/mL	Human serum	[161]
Pd/NH$_2$-ZIF-67	PSA	CA	100 fg/mL–50 ng/mL	0.03 pg/mL	Human serum	[162]

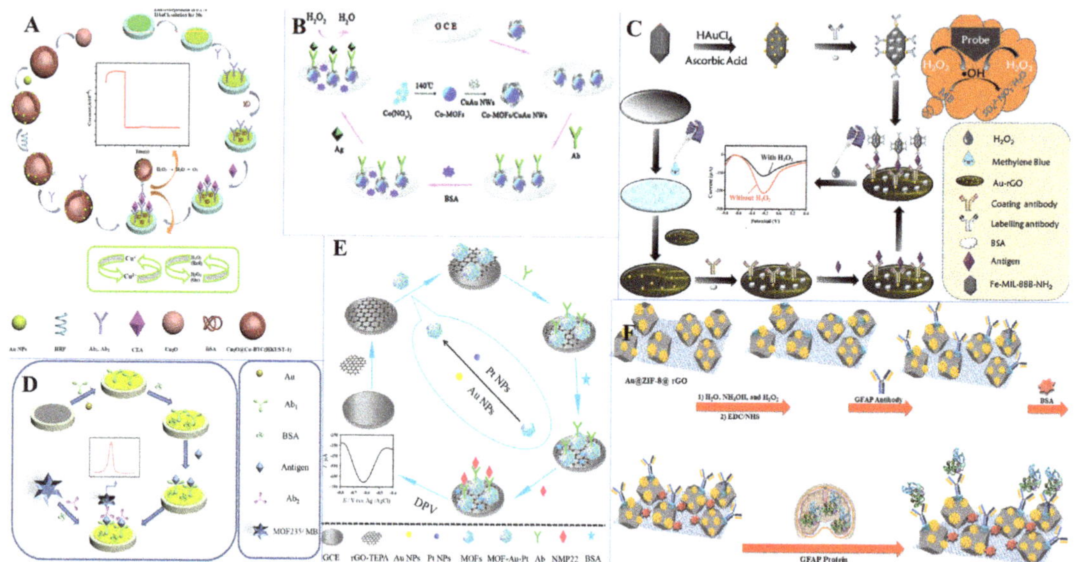

Figure 20. Schematic illustrations of (**A**) Fabrication of core-shell Cu$_2$O@Cu-MOF@AuNPs-based electrochemical immunosensor for CEA detection. Reproduced with permission from Ref. [146]. Copyright 2020 Springer, (**B**) Preparation of Co-MOFs/CuAu NWs based label-free immunosensor for the detection of NMP-22. Reproduced with permission from Ref. [149]. Copyright 2019 Royal society of chemistry, (**C**) Fabrication of Au-MOF-based amperometric immunosensor for the detection of PSA. Reproduced with permission from Ref. [150]. Copyright 2020 Springer, (**D**) Preparation steps of AuNPs decorated MOF235/MB based electrochemical immunosensor for PSA detection. Reproduced with permission from Ref. [28]. Copyright 2021 Elsevier, (**E**) Stepwise assembly of AuNPs-PtNPs-MOFs based electrochemical immunosensor for the detection of NMP-22 in urine samples. Reproduced with permission from Ref. [152]. Copyright 2019 Elsevier, and (**F**) Preparation of GFAP-BSA-Anti-GFAP-Au@ZIF-8@rGO/SPE based electrochemical immunosensor for the detection of GFAP. Reproduced with permission from Ref. [151]. Copyright 2022 ACS.

2.5. Biomarker Detection on Other 2D Materials Conjugated with Metal Nanoparticles

2D materials such as graphitic carbon nitride, black phosphorous, 2D layered double hydroxides (LDHs), boron nitrides, graphdiyne, etc. have also been explored in conjunction with metal nanoparticles for immunosensor applications with interesting biomarker targets (Figure 21, Table 5). Graphdiyne, the new 2D carbon allotrope with its unique sp-sp2 carbon network and highly π-conjugated structure has been receiving increased attention [163]. A graphdiyne-based self-powered biosensor platform was constructed by Hou et al. [164] for the determination of miRNA-21. Here, both the cathode and bioanode were fabricated by different modifications of AuNPs/GDY (Figure 21A). The 2D hexagonal boron nitride nanosheets, due to their electronic conductivity and large surface area were explored for immunosensor applications [165]. A label-free aptasensor for the detection of cardiac biomarker myoglobin on AuNPs decorated 2D-Boron nitride nanosheets was reported by Adeel et al. [166]. Here, the boron nitride nanosheets modified electrode AuNPs/BNNSs/FTO acted as a transducer for the immobilization of thiol-functionalized DNA aptamer for the specific binding of myoglobin (Figure 21B). Carbon nitrides are polymeric materials mainly consisting of carbon and nitrogen [167,168]. At ambient temperature, graphitic carbon nitride (g-C$_3$N$_4$) is the most stable allotrope of carbon nitrides. Due to the presence of basic surface groups and rich surface properties, g-C$_3$N$_4$ is attractive for many applications including catalysis [169]. Neto et al. [170] fabricated a miniaturized PEC system based on AuNPs decorated g-C$_3$N$_4$ for the detection of the breast cancer biomarker CA15-3 (Figure 21C). In this work, AuNPs on the g-C$_3$N$_4$ platform acted as a linker to

11-mercaptoundecanoic acid for the effective adsorption of antibodies. The performance of the PEC sensor was remarkable with a long linear range (0.1 fg/mL–10 ng/mL) and a very low detection limit (0.04 fg/mL). One of the promising candidates for immunosensor applications is 2D-Black phosphorus (BP) with high carrier mobility and controllable bandgap [171]. The unique properties of BP at atomic thickness are valuable for diverse applications [172–174]. Li et al. [175] reported a 2D-black phosphorous-supported Pt-Pd nanoelectrocatalyst for the determination of 4-AP, a potent biomarker for aniline exposure. Layered double hydroxides (LDHs) received attention because of their tunable chemistry and high charge density [176]. In one study, an electrochemical immunosensor based on AuNPs decorated ferrocene carboxylic acid conjugated MgAl layered double hydroxides for the label-free detection of CA-125 was reported by Wu et al. [177]. In this work, an LBL approach was used to increase the number of ferrocenes and antibodies, thereby amplifying the signal. The sensor reportedly displayed a wide linear range (0.01–1000 U/mL) and LOD (0.004 U/mL) and was tested for clinical cancer diagnostics (Figure 21D).

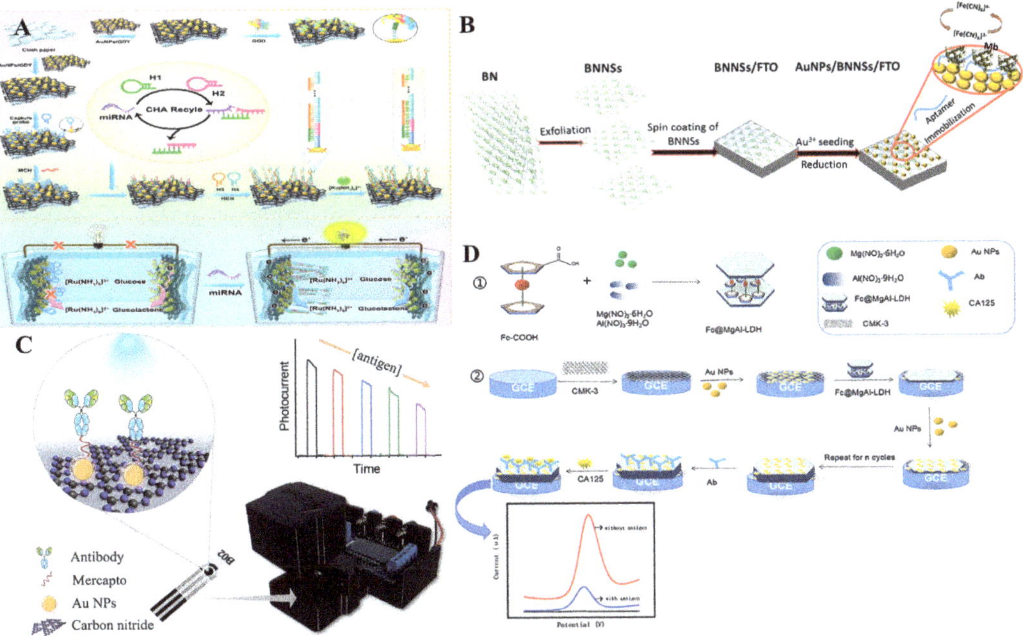

Figure 21. Schematic illustration of (**A**) Fabrication of a GDY-based self-powered device for miRNA-21 detection. Reprinted with permission from Ref. [164]. Copyright 2021 ACS, (**B**) Fabrication of AuNPs decorated boron nitride nanosheets based label-free aptasensor for the detection of the cardiac biomarker myoglobin. Reprinted with permission from Ref. [165]. Copyright 2019 Elsevier, (**C**) Graphitic carbon nitride sensitized with AuNPs for the PEC detection of CA15-3. Reprinted with permission from Ref. [170]. Copyright 2022 Elsevier, and (**D**) Fabrication of label-free electrochemical immunosensor based on LBL assembly of mesoporous carbon, AuNPs, and MgAl LDHs containing ferrocenecarboxylic acid. Reprinted with permission from Ref. [177]. Copyright 2022 Elsevier.

Table 5. Recent literature reports on biomarker detection based on various metal nanoparticles decorated 2D materials.

Sensing Platform	Biomarker	Technique	Linear Range	LOD	Real Sample	Ref.
AuNPs/GDY	miRNA-21	OCV	0.1–100,000 fM	0.034 fM	Human serum	[164]
Au-NPs/2D-hBN/FTO	Mb	DPV	0.1–100 μg/mL	34.6 ng/mL	Human serum	[166]
AuNPs-g-C_3N_4	CA15-3	PEC	10^{-7}–10^1 ng/mL	0.04 fg/mL	Human serum	[170]
Pt-Pd/BP	4-AP	DPV	0.02–5 μM	14.1 nM	—	[175]
Au/Fc@MgAl-LDH	CA-125	DPV	0.01 U/mL–1000 U/mL	0.004 U/mL	Human serum	[177]
AuNRs-g-C_3N_4	NS1	EIS	0.6–216 ng/mL	0.09 ng/mL	Human serum	[178]

3. Conclusions

In this review, we have discussed various electrochemical sensors that have been reported in recent years which incorporate various 2D nanomaterials conjugated with metal nanoparticles towards biomarker detection that have potential suitability for clinical use and some for point-of-care applications for cancer diagnosis. Although much research has been done in the synthesis of graphene, MoS_2, MXenes, MOFs, and other 2D materials incorporated with metal nanoparticles for an in vitro analysis of biomarkers. However, significant progress needs to be done in performing an in vivo analysis. Moreover, due to their inherent conductivity, these 2D nanomaterials are significantly used in electrochemical or even optical sensing. However, they are often doped with other nanomaterials to improve their electroactivity/conductivity. Further, new approaches such as nanofabrication and clinical applicability are most crucial for developing an open-use-dispose type of sensor at low cost. Furthermore, electrode-to-electrode variations upon modifications with nanomaterials largely depend on the type of functionalization method adopted, which also needed to be studied for developing electrochemical transducers with greater stability and reproducibility. Finally, the paper-based electrochemical and wearable electrochemical sensing approaches for biomarker detections are also promising due to their improved sensitivity, selectivity, and portability, such as a simple paper-based sensor that can measure with an application able to get the electrochemical signal downloaded into a smartphone is best suitable for clinical/point-of-care applications [179,180]. Though the integration of microfluidic devices with electrochemical systems possesses numerous advantages, including rapid manipulation of sample fluid, reduced reagent consumption, and low cost, commercialization of these electrochemical sensors is still in its infancy due to the challenges that these techniques are facing, such as miniaturization (multiple electrodes and channels) and integration of microfluidic systems (miniaturized flow controllers). Therefore, it is necessary to develop manufacturable biosensors that can provide accurate quantification of a biomarker of interest with a meager quantity of solutions at point-of-care with simple fabrication steps by avoiding multiple modifications on the electrode surface.

Author Contributions: A.K.: conceptualization, methodology, validation, writing—original draft preparation; A.K.Y.: conceptualization, methodology, validation, writing—original draft preparation; M.-H.L.: writing—review and editing, project administration, funding acquisition. All authors have read and agreed to the published version of the manuscript.

Funding: This work was supported by the Ministry of Trade, Industry, and Energy (Grant no. 20008763 and 20009860).

Institutional Review Board Statement: Not applicable.

Informed Consent Statement: Not applicable.

Data Availability Statement: Not applicable.

Conflicts of Interest: The authors declare no conflict of interest.

Abbreviations

2D	Two dimensional
2D-Hbn	2D-hexagonal boron nitride
AgNPs	Silver nanoparticles
AgNRs	Silver nanorods
Apt	Aptamer
Au	Gold
AuE	Gold electrode
AuNP-RGO	Au nanoparticle-reduced graphene oxide
AuNPs	Gold nanoparticles
AuPtBNPs	Gold platinum bimetallic nanoparticles
BDC	1,4-benzenedicarboxylate
$BiVO_4$	Bismuth vanadate
BNNSs	Boron nitride nanosheets
BP	Black phosphorous
BSA	Bovine serum albumin
CA 27-29 BCA	Cancer antigen 27-29 breast cancer antigen
CA	Chronoamperometry
CA125	Cancer antigen 125
CA15-3	Cancer antigen 15-3
C-DNA	Capture DNA
CEA	Carcinoembryonic antigen
ce-MoS_2	Chemical exfoliated MoS_2
CGO	Carboxylic groups
CLB	Clenbuterol
CoPP	Cobalt protoporphyrin
CP	Capture probe
CPEB4	Cytoplasmic polyadenylate element-binding protein 4
Cr.6	18-crown-6
CRP	C-reactive protein
CS	Chitosan
CTnI	Cardiac troponin I
CTnT	Cardiac troponin T
CV	Cyclic voltammetry
CYFRA21-1	Cytokeratin 19 fragment
DNA	Deoxyribonucleic acid
DPV	Differential pulse voltammetry
ECD	Extracellular domain
ECL	Electrochemiluminescence
EIS	Electrochemical impedance spectroscopy
ELISA	Enzyme-linked immunosorbent assay
eT	Electron transfer
Fc	Ferrocene
FTO	Fluorine doped tin oxide
g-C_3N_4	Graphitic carbon nitride
GCE	Glassy carbon electrode
GDY	Graphdiyne
GE	Graphite electrode
GFAP	Glial fibrillary acidic protein
GP120	Glycoprotein GP120
4-AP	p-Aminophenol
HER-2	Human epidermal growth factor receptor-2
HP1	Hairpin DNA
HT	Hexane thiol
IL8	Interleukin-8
i-t curve	Amperometric current-time response

ITO	Indium tin oxide
LA	Lactic acid
LBL	Layer by layer
L-cys	L-Cysteine
LOD	Limit of detection
LSV	Linear sweep voltammetry
L-Trp	L-tryptophan
Mb	Myoglobin
MCH	6-mercaptohexanol
MgAl-LDH	Mg-Al-Layered double hydroxide
MGMT	O6-methylguanine-DNA methyltransferase
miRNA-141	micro-RNA-141
miRNA-21	micro-RNA-21
miRNA-377	micro-RNA-377
miRNAs	micro-RNAs
MNPs	Mesoporous nanoparticles
MOFs	Metal organic frameworks
MUC1	Mucin1
MWCNT	Multiwalled carbon nanotubes
NMP-22	Nuclear matrix protein 22
NS1	Non-structural 1
NSE	Neuron-specific enolase
OCV	Open circuit voltage
OTC	Oxytetracycline
PAMAM	Polyamidoamine
PCT	Procalcitonin
PDA	Polydopamine
PdPtBP MNPs	Pd-Pt-Black phosphorous-mesoporous nanoparticles
PEC	Photoelectrochemical
PET	Polyethylene terephthalate
PPY	Polypyrrole
PSA	Prostate specific antigen
PtNPs	Platinum nanoparticles
RAC	Ractopamine
rGO	Reduced graphene oxide
RhNPs	Rhodium nanoparticles
RNA	Ribonucleic acid
S/N	Signal-to-noise ratio
SAL	Salbutamol
SPCE	Screen-printed carbon electrode
SWV	Square wave voltammetry
TEPA	Tetraethylenepentamine
VEGF165	Vascular endothelial growth factor 165
YNCs	Yolk-shell nanocubes
β-CD	β-cyclodextrin

References

1. Filice, M.; Ruiz-Cabello, J. *Nucleic Acid Nanotheranostics: Biomedical Applications*; Elsevier: Amsterdam, The Netherlands, 2019.
2. WHO. *Biomarkers in Risk Assessment: Validity and Validation-Environmental Health Criteria 222*; WHO: Geneva, Switzerland, 2001.
3. National Cancer Institute. Biomarker. Available online: https://www.cancer.gov/publications/dictionaries/cancer-terms/def/biomarker (accessed on 5 October 2022).
4. Biomarkers Definitions Working Group. Biomarkers and surrogate endpoints: Preferred definitions and conceptual framework. *Clin. Pharmacol. Ther.* **2001**, *69*, 89–95. [CrossRef]
5. Kikkeri, K.; Wu, D.; Voldman, J. A sample-to-answer electrochemical biosensor system for biomarker detection. *Lab Chip* **2022**, *22*, 100–107. [CrossRef] [PubMed]
6. Ziemssen, T.; Akgün, K.; Brück, W. Molecular biomarkers in multiple sclerosis. *J. Neuroinflamm.* **2019**, *16*, 272. [CrossRef] [PubMed]

7. Pachner, A.R.; DiSano, K.; Royce, D.B.; Gilli, F. Clinical utility of a molecular signature in inflammatory demyelinating disease. *Neurol. Neuroimmunol. Neuroinflamm.* **2019**, *6*, e520. [CrossRef] [PubMed]

8. Sant, G.R.; Knopf, K.B.; Albala, D.M. Live-single-cell phenotypic cancer biomarkers-future role in precision oncology? *NPJ Precis. Oncol.* **2017**, *1*, 21. [CrossRef] [PubMed]

9. Huss, R. Chapter 19—Biomarkers. In *Translational Regenerative Medicine*; Atala, A., Allickson, J.G., Eds.; Academic Press: Boston, MA, USA, 2015; pp. 235–241.

10. Manzanares, J.; Sala, F.; Gutiérrez, M.S.G.; Rueda, F.N. 2.30—Biomarkers. In *Comprehensive Pharmacology*; Kenakin, T., Ed.; Elsevier: Oxford, UK, 2022; pp. 693–724.

11. Goossens, N.; Nakagawa, S.; Sun, X.; Hoshida, Y. Cancer biomarker discovery and validation. *Transl. Cancer Res.* **2015**, *4*, 256.

12. Califf, R.M. Biomarker definitions and their applications. *Exp. Biol. Med.* **2018**, *243*, 213–221. [CrossRef]

13. FDA-NIH Biomarker Working Group. BEST (Biomarkers, Endpoints, and Other Tools). Available online: https://www.ncbi.nlm.nih.gov/books/NBK326791/ (accessed on 7 November 2022).

14. Magar, H.S.; Hassan, R.Y.A.; Mulchandani, A. Electrochemical Impedance Spectroscopy (EIS): Principles, Construction, and Biosensing Applications. *Sensors* **2021**, *21*, 6578. [CrossRef]

15. Bertok, T.; Lorencova, L.; Chocholova, E.; Jane, E.; Vikartovska, A.; Kasak, P.; Tkac, J. Electrochemical Impedance Spectroscopy Based Biosensors: Mechanistic Principles, Analytical Examples and Challenges towards Commercialization for Assays of Protein Cancer Biomarkers. *ChemElectroChem* **2019**, *6*, 989–1003. [CrossRef]

16. Lisdat, F.; Schafer, D. The use of electrochemical impedance spectroscopy for biosensing. *Anal. Bioanal. Chem.* **2008**, *391*, 1555–1567. [CrossRef]

17. Grieshaber, D.; MacKenzie, R.; Voros, J.; Reimhult, E. Electrochemical biosensors—Sensor principles and architectures. *Sensors* **2008**, *8*, 1400–1458. [CrossRef]

18. Rezaei, B.; Irannejad, N. Chapter 2—Electrochemical detection techniques in biosensor applications. In *Electrochemical Biosensors*; Ensafi, A.A., Ed.; Elsevier: Amsterdam, The Netherlands, 2019; pp. 11–43.

19. Speiser, B. Electroanalytical Methods. 2. Cyclic Voltametry. *Chem. Unserer. Zeit.* **1981**, *15*, 62–67. [CrossRef]

20. Huan, T.N.; Ha, V.T.T.; Hung, L.Q.; Yoon, M.Y.; Han, S.H.; Chung, H. Square wave voltammetric detection of Anthrax utilizing a peptide for selective recognition of a protein biomarker. *Biosens. Bioelectron.* **2009**, *25*, 469–474. [CrossRef]

21. Kumar, S.; Kalkal, A. 3—Electrochemical detection: Cyclic voltammetry/differential pulse voltammetry/impedance spectroscopy. In *Nanotechnology in Cancer Management*; Khondakar, K.R., Kaushik, A.K., Eds.; Elsevier: Amsterdam, The Netherlands, 2021; pp. 43–71.

22. Moore, S. Current Global Market of 2D Materials. Available online: https://www.azonano.com/article.aspx?ArticleID=6294 (accessed on 26 December 2022).

23. Popov, A.; Brasiunas, B.; Kausaite-Minkstimiene, A.; Ramanaviciene, A. Metal Nanoparticle and Quantum Dot Tags for Signal Amplification in Electrochemical Immunosensors for Biomarker Detection. *Chemosensors* **2021**, *9*, 85. [CrossRef]

24. Enoch, I.V.M.V.; Ramasamy, S.; Mohiyuddin, S.; Gopinath, P.; Manoharan, R. Cyclodextrin–PEG conjugate-wrapped magnetic ferrite nanoparticles for enhanced drug loading and release. *Appl. Nanosci.* **2018**, *8*, 273–284. [CrossRef]

25. Liu, Y.; Yang, G.; Jin, S.; Xu, L.; Zhao, C.X. Development of high-drug-loading nanoparticles. *ChemPlusChem* **2020**, *85*, 2143–2157. [CrossRef]

26. Aido, A.; Wajant, H.; Buzgo, M.; Simaite, A. Development of anti-TNFR antibody-conjugated nanoparticles. *Multidiscip. Digit. Publ. Inst. Proc.* **2020**, *78*, 55.

27. Li, Y.; Wang, Y.; Zhang, N.; Fan, D.; Liu, L.; Yan, T.; Yang, X.; Ding, C.; Wei, Q.; Ju, H. Magnetic electrode-based electrochemical immunosensor using amorphous bimetallic sulfides of $CoSnS_x$ as signal amplifier for the NTpro BNP detection. *Biosens. Bioelectron.* **2019**, *131*, 250–256. [CrossRef]

28. Zhang, M.; Hu, X.; Mei, L.; Zhang, L.; Wang, X.; Liao, X.; Qiao, X.; Hong, C. PSA detection electrochemical immunosensor based on MOF-235 nanomaterial adsorption aggregation signal amplification strategy. *Microchem. J.* **2021**, *171*, 106870. [CrossRef]

29. Wang, H.; Zhang, S.; Li, S.; Qu, J. Electrochemical sensor based on palladium-reduced graphene oxide modified with gold nanoparticles for simultaneous determination of acetaminophen and 4-aminophenol. *Talanta* **2018**, *178*, 188–194. [CrossRef]

30. Lim, J.Y.; Mubarak, N.M.; Abdullah, E.C.; Nizamuddin, S.; Khalid, M.; Inamuddin. Recent trends in the synthesis of graphene and graphene oxide based nanomaterials for removal of heavy metals—A review. *J. Ind. Eng. Chem.* **2018**, *66*, 29–44. [CrossRef]

31. Mbayachi, V.B.; Ndayiragije, E.; Sammani, T.; Taj, S.; Mbuta, E.R.; Khan, A.U. Graphene synthesis, characterization and its applications: A review. *Results Chem.* **2021**, *3*, 100163. [CrossRef]

32. Bhuyan, M.S.A.; Uddin, M.N.; Islam, M.M.; Bipasha, F.A.; Hossain, S.S. Synthesis of graphene. *Int. Nano Lett.* **2016**, *6*, 65–83. [CrossRef]

33. Singh, R.K.; Kumar, R.; Singh, D.P. Graphene oxide: Strategies for synthesis, reduction and frontier applications. *RSC Adv.* **2016**, *6*, 64993–65011. [CrossRef]

34. Yagati, A.K.; Behrent, A.; Beck, S.; Rink, S.; Goepferich, A.M.; Min, J.; Lee, M.H.; Baeumner, A.J. Laser-induced graphene interdigitated electrodes for label-free or nanolabel-enhanced highly sensitive capacitive aptamer-based biosensors. *Biosens. Bioelectron.* **2020**, *164*, 112272. [CrossRef]

35. Ranjan, P.; Sadique, M.A.; Yadav, S.; Khan, R. An Electrochemical Immunosensor Based on Gold-Graphene Oxide Nanocomposites with Ionic Liquid for Detecting the Breast Cancer CD44 Biomarker. *ACS Appl. Mater. Inter.* **2022**, *14*, 20802–20812. [CrossRef]

36. Yagati, A.K.; Pyun, J.C.; Min, J.; Cho, S. Label-free and direct detection of C-reactive protein using reduced graphene oxide-nanoparticle hybrid impedimetric sensor. *Bioelectrochemistry* **2016**, *107*, 37–44. [CrossRef]

37. Jonous, Z.A.; Shayeh, J.S.; Yazdian, F.; Yadegari, A.; Hashemi, M.; Omidi, M. An electrochemical biosensor for prostate cancer biomarker detection using graphene oxide-gold nanostructures. *Eng. Life Sci.* **2019**, *19*, 206–216. [CrossRef]

38. Kasturi, S.; Eom, Y.; Torati, R.; Kim, C. Highly sensitive electrochemical biosensor based on naturally reduced rGO/Au nanocomposite for the detection of miRNA-122 biomarker. *J. Ind. Eng. Chem.* **2021**, *93*, 186–195. [CrossRef]

39. Rauf, S.; Lahcen, A.A.; Aljedaibi, A.; Beduk, T.; de Oliveira, J.I.; Salama, K.N. Gold nanostructured laser-scribed graphene: A new electrochemical biosensing platform for potential point-of-care testing of disease biomarkers. *Biosens. Bioelectron.* **2021**, *180*, 113116. [CrossRef]

40. Hasanjani, H.R.A.; Zarei, K. DNA/Au-Pt bimetallic nanoparticles/graphene oxide-chitosan composites modified pencil graphite electrode used as an electrochemical biosensor for sub-picomolar detection of anti-HIV drug zidovudine. *Microchem. J.* **2021**, *164*, 106005. [CrossRef]

41. Kanagavalli, P.; Veerapandian, M. Opto-electrochemical functionality of Ru(II)-reinforced graphene oxide nanosheets for immunosensing of dengue virus non-structural 1 protein. *Biosens. Bioelectron.* **2020**, *150*, 111878. [CrossRef] [PubMed]

42. Kanagavalli, P.; Andrew, C.; Veerapandian, M.; Jayakumar, M. In-situ redox-active hybrid graphene platform for label-free electrochemical biosensor: Insights from electrodeposition and electroless deposition. *TrAC-Trend Anal. Chem.* **2021**, *143*, 116413. [CrossRef]

43. Liang, Y.; Xu, Y.; Tong, Y.Y.; Chen, Y.; Chen, X.L.; Wu, S.M. Graphene-Based Electrochemical Sensor for Detection of Hepatocellular Carcinoma Markers. *Front. Chem.* **2022**, *10*, 883627. [CrossRef] [PubMed]

44. Ozkan-Ariksoysal, D. Current Perspectives in Graphene Oxide-Based Electrochemical Biosensors for Cancer Diagnostics. *Biosensors* **2022**, *12*, 607. [CrossRef]

45. Rashid, J.I.A.; Kannan, V.; Ahmad, M.H.; Mon, A.A.; Taufik, S.; Miskon, A.; Ong, K.K.; Yusof, N.A. An electrochemical sensor based on gold nanoparticles-functionalized reduced graphene oxide screen printed electrode for the detection of pyocyanin biomarker in Pseudomonas aeruginosa infection. *Mat. Sci. Eng. C Mater.* **2021**, *120*, 111625. [CrossRef]

46. Abolhasan, R.; Khalilzadeh, B.; Yousefi, H.; Samemaleki, S.; Chakari-Khiavi, F.; Ghorbani, F.; Pourakbari, R.; Kamrani, A.; Khataee, A.; Rad, T.S.; et al. Ultrasensitive and label free electrochemical immunosensor for detection of ROR1 as an oncofetal biomarker using gold nanoparticles assisted LDH/rGO nanocomposite. *Sci. Rep.* **2021**, *11*, 14921. [CrossRef]

47. Liu, X.K.; Lin, L.Y.; Tseng, F.Y.; Tan, Y.C.; Li, J.; Feng, L.; Song, L.J.; Lai, C.F.; Li, X.H.; He, J.H.; et al. Label-free electrochemical immunosensor based on gold nanoparticle/polyethyleneimine/reduced graphene oxide nanocomposites for the ultrasensitive detection of cancer biomarker matrix metalloproteinase-1. *Analyst* **2021**, *146*, 4066–4079. [CrossRef]

48. Ye, S.; Liu, Y.; Zeng, M.; Feng, W.; Yang, H.; Zheng, X. Electrochemical Immunoassay of Melanoma Biomarker CPEB4 Based on Cobalt Porphyrin Functionalized Graphene Oxide. *J. Electrochem. Soc.* **2022**, *169*, 027510. [CrossRef]

49. Yagati, A.K.; Lee, G.Y.; Ha, S.; Chang, K.A.; Pyun, J.C.; Cho, S. Impedimetric Tumor Necrosis Factor-alpha Sensor Based on a Reduced Graphene Oxide Nanoparticle-Modified Electrode Array. *J. Nanosci. Nanotechnol.* **2016**, *16*, 11921–11927. [CrossRef]

50. Ben Moussa, F.; Achi, F.; Meskher, H.; Henni, A.; Belkhalfa, H. Green one-step reduction approach to prepare rGO@AgNPs coupled with molecularly imprinted polymer for selective electrochemical detection of lactic acid as a cancer biomarker. *Mater. Chem. Phys.* **2022**, *289*, 126456. [CrossRef]

51. Wang, H.; Zhang, Y.; Li, H.; Du, B.; Ma, H.M.; Wu, D.; Wei, Q. A silver-palladium alloy nanoparticle-based electrochemical biosensor for simultaneous detection of ractopamine, clenbuterol and salbutamol. *Biosens. Bioelectron.* **2013**, *49*, 14–19. [CrossRef] [PubMed]

52. Akbarzadeh, S.; Khajehsharifi, H.; Hajihosseini, S. Detection of Oxytetracycline Using an Electrochemical Label-Free Aptamer-Based Biosensor. *Biosensors* **2022**, *12*, 468. [CrossRef] [PubMed]

53. Yang, H.; Ren, J.; Zhao, M.; Chen, C.; Wang, F.; Chen, Z. Novel electrochemical immunosensor for O^6-methylguanine-DNA methyltransferase gene methylation based on graphene oxide-magnetic nanoparticles-β-cyclodextrin nanocomposite. *Bioelectrochemistry* **2022**, *146*, 108111. [CrossRef]

54. Pothipor, C.; Jakmunee, J.; Bamrungsap, S.; Ounnunkad, K. An electrochemical biosensor for simultaneous detection of breast cancer clinically related microRNAs based on a gold nanoparticles/graphene quantum dots/graphene oxide film. *Analyst* **2021**, *146*, 4000–4009. [CrossRef]

55. Sadeghi, M.; Kashanian, S.; Naghib, S.M.; Arkan, E. A high-performance electrochemical aptasensor based on graphene-decorated rhodium nanoparticles to detect HER2-ECD oncomarker in liquid biopsy. *Sci. Rep.* **2022**, *12*, 3299. [CrossRef]

56. Verma, S.; Singh, A.; Shukla, A.; Kaswan, J.; Arora, K.; Ramirez-Vick, J.; Singh, P.; Singh, S.P. Anti-IL8/AuNPs-rGO/ITO as an Immunosensing Platform for Noninvasive Electrochemical Detection of Oral Cancer. *ACS Appl. Mater. Inter.* **2017**, *9*, 27462–27474. [CrossRef]

57. Barman, S.C.; Hossain, M.F.; Yoon, H.; Park, J.Y. Trimetallic Pd@Au@Pt nanocomposites platform on -COOH terminated reduced graphene oxide for highly sensitive CEA and PSA biomarkers detection. *Biosens. Bioelectron.* **2018**, *100*, 16–22. [CrossRef]

58. Thunkhamrak, C.; Chuntib, P.; Ounnunkad, K.; Banet, P.; Aubert, P.H.; Saianand, G.; Gopalan, A.I.; Jakmunee, J. Highly sensitive voltammetric immunosensor for the detection of prostate specific antigen based on silver nanoprobe assisted graphene oxide modified screen printed carbon electrode. *Talanta* **2020**, *208*, 120389. [CrossRef]

59. Khoshnevisan, K.; Torabi, F.; Baharifar, H.; Sajjadi-Jazi, S.M.; Afjeh, M.S.; Faridbod, F.; Larijani, B.; Khorramizadeh, M.R. Determination of the biomarker L-tryptophan level in diabetic and normal human serum based on an electrochemical sensing method using reduced graphene oxide/gold nanoparticles/18-crown-6. *Anal. Bioanal. Chem.* **2020**, *412*, 3615–3627. [CrossRef]

60. Meng, F.Y.; Sun, H.X.; Huang, Y.; Tang, Y.G.; Chen, Q.; Miao, P. Peptide cleavage-based electrochemical biosensor coupling graphene oxide and silver nanoparticles. *Anal. Chim. Acta* **2019**, *1047*, 45–51. [CrossRef] [PubMed]

61. Sangili, A.; Kalyani, T.; Chen, S.M.; Nanda, A.; Jana, S.K. Label-Free Electrochemical Immunosensor Based on One-Step Electrochemical Deposition of AuNP-RGO Nanocomposites for Detection of Endometriosis Marker CA 125. *ACS Appl. Bio Mater.* **2020**, *3*, 7620–7630. [CrossRef] [PubMed]

62. Rostamabadi, P.F.; Heydari-Bafrooei, E. Impedimetric aptasensing of the breast cancer biomarker HER2 using a glassy carbon electrode modified with gold nanoparticles in a composite consisting of electrochemically reduced graphene oxide and single-walled carbon nanotubes. *Microchim. Acta* **2019**, *186*, 495. [CrossRef] [PubMed]

63. Bharti, A.; Rana, S.; Dahiya, D.; Agnihotri, N.; Prabhakar, N. An electrochemical aptasensor for analysis of MUC1 using gold platinum bimetallic nanoparticles deposited carboxylated graphene oxide. *Anal. Chim. Acta* **2020**, *1097*, 186–195. [CrossRef] [PubMed]

64. Kumar, A.; Purohit, B.; Mahato, K.; Mandal, R.; Srivastava, A.; Chandra, P. Gold-Iron Bimetallic Nanoparticles Impregnated Reduced Graphene Oxide Based Nanosensor for Label-Free Detection of Biomarker Related to Non-Alcoholic Fatty Liver Disease. *Electroanalysis* **2019**, *31*, 2417–2428. [CrossRef]

65. Hu, H.W.; Zavabeti, A.; Quan, H.Y.; Zhu, W.Q.; Wei, H.Y.; Chen, D.C.; Ou, J.Z. Recent advances in two-dimensional transition metal dichalcogenides for biological sensing. *Biosens. Bioelectron.* **2019**, *142*, 111573. [CrossRef]

66. Subbaiah, Y.P.V.; Saji, K.J.; Tiwari, A. Atomically Thin MoS_2: A Versatile Nongraphene 2D Material. *Adv. Funct. Mater.* **2016**, *26*, 2046–2069. [CrossRef]

67. Kalantar-zadeh, K.; Ou, J.Z. Biosensors Based on Two-Dimensional MoS_2. *ACS Sens.* **2016**, *1*, 5–16. [CrossRef]

68. Van, T.D.; Thuy, N.D.T.; Phuong, T.D.V.; Thi, N.N.; Thi, T.N.; Phuong, T.N.; Van, T.V.; Vuong-Pham, H.; Dinh, T.P. High-performance nonenzymatic electrochemical glucose biosensor based on AgNP-decorated MoS_2 microflowers. *Curr. Appl. Phys.* **2022**, *43*, 116–123. [CrossRef]

69. Cui, Z.L.; Li, D.J.; Yang, W.H.; Fan, K.; Liu, H.Y.; Wen, F.; Li, L.L.; Dong, L.X.; Wang, G.F.; Wu, W. An electrochemical biosensor based on few-layer MoS_2 nanosheets for highly sensitive detection of tumor marker ctDNA. *Anal. Methods* **2022**, *14*, 1956–1962. [CrossRef]

70. Li, L.; Zhang, D.; Gao, Y.H.; Deng, J.P.; Gou, Y.C.; Fang, J.F. Electric field driven exfoliation of MoS_2. *J. Alloys Compd.* **2021**, *862*, 158551. [CrossRef]

71. Shi, Z.T.; Zhao, H.B.; Chen, X.Q.; Wu, G.M.; Wei, F.; Tu, H.L. Chemical vapor deposition growth and transport properties of MoS_2-2H thin layers using molybdenum and sulfur as precursors. *Rare Met.* **2022**, *41*, 3574–3578. [CrossRef]

72. Jagannadham, K.; Das, K.; Reynolds, C.L.; El-Masry, N. Nature of electrical conduction in MoS_2 films deposited by laser physical vapor deposition. *J. Mater. Sci. Mater. Electron.* **2018**, *29*, 14180–14191. [CrossRef]

73. Gomes, F.O.V.; Pokle, A.; Marinkovic, M.; Balster, T.; Canavan, M.; Fleischer, K.; Anselmann, R.; Nicolosi, V.; Wagner, V. Influence of temperature on morphological and optical properties of MoS_2 layers as grown based on solution processed precursor. *Thin Solid Films* **2018**, *645*, 38–44. [CrossRef]

74. Shah, S.A.; Khan, I.; Yuan, A.H. MoS_2 as a Co-Catalyst for Photocatalytic Hydrogen Production: A Mini Review. *Molecules* **2022**, *27*, 3289. [CrossRef] [PubMed]

75. Li, X.; Zhu, H. Two-dimensional MoS_2: Properties, preparation, and applications. *J. Mater.* **2015**, *1*, 33–44. [CrossRef]

76. Sinha, A.; Dhanjai; Tan, B.; Huang, Y.J.; Zhao, H.M.; Dang, X.M.; Chen, J.P.; Jain, R. MoS2 nanostructures for electrochemical sensing of multidisciplinary targets: A review. *TrAC-Trend Anal. Chem.* **2018**, *102*, 75–90. [CrossRef]

77. Yagati, A.K.; Go, A.; Vu, N.H.; Lee, M.H. A MoS_2-Au nanoparticle-modified immunosensor for T-3 biomarker detection in clinical serum samples. *Electrochim. Acta* **2020**, *342*, 136065. [CrossRef]

78. Su, S.; Sun, H.F.; Cao, W.F.; Chao, J.; Peng, H.Z.; Zuo, X.L.; Yuwen, L.H.; Fan, C.H.; Wang, L.H. Dual-Target Electrochemical Biosensing Based on DNA Structural Switching on Gold Nanoparticle-Decorated MoS_2 Nanosheets. *ACS Appl. Mater. Int.* **2016**, *8*, 6826–6833. [CrossRef]

79. Chen, M.; Zhao, L.; Wu, D.; Tu, S.; Chen, C.; Guo, H.; Xu, Y. Highly sensitive sandwich-type immunosensor with enhanced electrocatalytic durian-shaped MoS_2/AuPtPd nanoparticles for human growth differentiation factor-15 detection. *Anal. Chim. Acta* **2022**, *1223*, 340194. [CrossRef]

80. Nong, C.J.; Yang, B.; Li, X.K.; Feng, S.X.; Cui, H.X. An ultrasensitive electrochemical immunosensor based on in-situ growth of $CuWO_4$ nanoparticles on MoS_2 and chitosan-gold nanoparticles for cortisol detection. *Microchem. J.* **2022**, *179*, 107434. [CrossRef]

81. Su, S.; Sun, Q.; Wan, L.; Gu, X.; Zhu, D.; Zhou, Y.; Chao, J.; Wang, L. Ultrasensitive analysis of carcinoembryonic antigen based on MoS_2-based electrochemical immunosensor with triple signal amplification. *Biosens. Bioelectron.* **2019**, *140*, 111353. [CrossRef]

82. Ma, N.; Zhang, T.; Fan, D.W.; Kuang, X.; Ali, A.; Wu, D.; Wei, Q. Triple amplified ultrasensitive electrochemical immunosensor for alpha fetoprotein detection based on MoS_2@Cu_2O-Au nanoparticles. *Sens. Actuators B Chem.* **2019**, *297*, 126821. [CrossRef]

83. Dalila, R.N.; Arshad, M.K.M.; Gopinath, S.C.B.; Ibau, C.; Nuzaihan, M.M.N.; Fathil, M.F.M.; Azmi, U.Z.M.; Anbu, P. Faradaic electrochemical impedimetric analysis on MoS_2/Au-NPs decorated surface for C-reactive protein detection. *J. Taiwan Inst. Chem. E* **2022**, *138*, 104450. [CrossRef]

84. Xu, W.; Fei, J.W.; Yang, W.; Zheng, Y.N.; Dai, Y.; Sakran, M.; Zhang, J.; Zhu, W.Y.; Hong, J.L.; Zhou, X.M. A colorimetric/electrochemical dual-mode sensor based on Fe_3O_4@MoS_2-Au NPs for high-sensitivity detection of hydrogen peroxide. *Microchem. J.* **2022**, *181*, 107825. [CrossRef]

85. Shin, M.; Yoon, J.; Yi, C.Y.; Lee, T.; Choi, J.W. Flexible HIV-1 Biosensor Based on the Au/MoS_2 Nanoparticles/Au Nanolayer on the PET Substrate. *Nanomaterials* **2019**, *9*, 76. [CrossRef]

86. Li, S.; Zhang, J.W.; Tan, C.S.; Chen, C.; Hu, C.; Bai, Y.C.; Ming, D. Electrochemical immunosensor based on hybrid MoS_2/Pt@Au-nanoprism/PDA for simultaneous detection of free and total prostate specific antigen in serum. *Sens. Actuators B Chem.* **2022**, *357*, 131413. [CrossRef]

87. Ma, E.H.; Wang, P.; Yang, Q.S.; Yu, H.X.; Pei, F.B.; Li, Y.Y.; Liu, Q.; Dong, Y.H. Electrochemical immunosensor based on MoS_2 NFs/Au@AgPt YNCs as signal amplification label for sensitive detection of CEA. *Biosens. Bioelectron.* **2019**, *142*, 111580. [CrossRef]

88. Zhao, H.; Du, X.; Dong, H.; Jin, D.L.; Tang, F.; Liu, Q.; Wang, P.; Chen, L.; Zhao, P.Q.; Li, Y.Y. Electrochemical immunosensor based on Au/Co-BDC/MoS_2 and DPCN/MoS2 for the detection of cardiac troponin I. *Biosens. Bioelectron.* **2021**, *175*, 112883. [CrossRef]

89. Alarfaj, N.A.; El-Tohamy, M.F.; Oraby, H. New label-free ultrasensitive electrochemical immunosensor-based Au/MoS_2/rGO nanocomposites for CA 27-29 breast cancer antigen detection. *New J. Chem.* **2018**, *42*, 11046–11053. [CrossRef]

90. Wang, X.; Chu, C.C.; Shen, L.; Deng, W.P.; Yan, M.; Ge, S.G.; Yu, J.H.; Song, X.R. An ultrasensitive electrochemical immunosensor based on the catalytical activity of MoS_2-Au composite using Ag nanospheres as labels. *Sens. Actuators B Chem.* **2015**, *206*, 30–36. [CrossRef]

91. Gui, J.C.; Han, L.; Du, C.X.; Yu, X.N.; Hu, K.; Li, L.H. An efficient label-free immunosensor based on ce-MoS_2/AgNR composites and screen-printed electrodes for PSA detection. *J. Solid State Electrochem.* **2021**, *25*, 973–982. [CrossRef]

92. Fan, Z.Q.; Yao, B.; Ding, Y.D.; Xie, M.H.; Zhao, J.F.; Zhang, K.; Huang, W. Electrochemiluminescence aptasensor for Siglec-5 detection based on MoS_2@Au nanocomposites emitter and exonuclease III-powered DNA walker. *Sens. Actuators B Chem.* **2021**, *334*, 129592. [CrossRef] [PubMed]

93. Ma, K.X.; Sinha, A.; Dang, X.M.; Zhao, H.M. Electrochemical Preparation of Gold Nanoparticles-Polypyrrole Co-Decorated 2D MoS_2 Nanocomposite Sensor for Sensitive Detection of Glucose. *J. Electrochem. Soc.* **2019**, *166*, B147–B154. [CrossRef]

94. Zhu, X.N.; Wang, Z.G.; Gao, M.Y.; Wang, Y.Q.; Hu, J.; Song, Z.X.; Wang, Z.B.; Dong, M.D. AgPt/MoS2 hybrid as electrochemical sensor for detecting H_2O_2 release from living cells. *New J. Chem.* **2022**, *46*, 15032–15041. [CrossRef]

95. Kim, H.; Wang, Z.; Alshareef, H.N. MXetronics: Electronic and photonic applications of MXenes. *Nano Energy* **2019**, *60*, 179–197. [CrossRef]

96. Palisaitis, J.; Persson, I.; Halim, J.; Rosen, J.; Persson, P.O. On the structural stability of MXene and the role of transition metal adatoms. *Nanoscale* **2018**, *10*, 10850–10855. [CrossRef]

97. Mehdi Aghaei, S.; Aasi, A.; Panchapakesan, B. Experimental and theoretical advances in MXene-based gas sensors. *ACS Omega* **2021**, *6*, 2450–2461. [CrossRef]

98. Ibrahim, Y.; Mohamed, A.; Abdelgawad, A.M.; Eid, K.; Abdullah, A.M.; Elzatahry, A. The recent advances in the mechanical properties of self-standing two-dimensional MXene-based nanostructures: Deep insights into the supercapacitor. *Nanomaterials* **2020**, *10*, 1916. [CrossRef]

99. Fu, B.; Sun, J.; Wang, C.; Shang, C.; Xu, L.; Li, J.; Zhang, H. MXenes: Synthesis, optical properties, and applications in ultrafast photonics. *Small* **2021**, *17*, 2006054. [CrossRef]

100. Yu, S.; Tang, H.; Zhang, D.; Wang, S.; Qiu, M.; Song, G.; Fu, D.; Hu, B.; Wang, X. MXenes as emerging nanomaterials in water purification and environmental remediation. *Sci. Total Environ.* **2021**, *811*, 152280. [CrossRef] [PubMed]

101. Zhang, Y.; Xia, W.; Wu, Y.; Zhang, P. Prediction of MXene based 2D tunable band gap semiconductors: GW quasiparticle calculations. *Nanoscale* **2019**, *11*, 3993–4000. [CrossRef] [PubMed]

102. Thakur, R.; VahidMohammadi, A.; Moncada, J.; Adams, W.R.; Chi, M.; Tatarchuk, B.; Beidaghi, M.; Carrero, C.A. Insights into the thermal and chemical stability of multilayered V_2CT_x MXene. *Nanoscale* **2019**, *11*, 10716–10726. [CrossRef] [PubMed]

103. Seredych, M.; Shuck, C.E.; Pinto, D.; Alhabeb, M.; Precetti, E.; Deysher, G.; Anasori, B.; Kurra, N.; Gogotsi, Y. High-temperature behavior and surface chemistry of carbide MXenes studied by thermal analysis. *Chem. Mater.* **2019**, *31*, 3324–3332. [CrossRef]

104. Naguib, M.; Mashtalir, O.; Carle, J.; Presser, V.; Lu, J.; Hultman, L.; Gogotsi, Y.; Barsoum, M.W. Two-Dimensional Transition Metal Carbides. *ACS Nano* **2012**, *6*, 1322–1331. [CrossRef] [PubMed]

105. Koyappayil, A.; Chavan, S.G.; Roh, Y.-G.; Lee, M.-H. Advances of MXenes; Perspectives on Biomedical Research. *Biosensors* **2022**, *12*, 454. [CrossRef]

106. He, T.; Liu, W.; Lv, T.; Ma, M.; Liu, Z.; Vasiliev, A.; Li, X. MXene/SnO_2 heterojunction based chemical gas sensors. *Sens. Actuators B Chem.* **2021**, *329*, 129275. [CrossRef]

107. Zhao, F.; Yao, Y.; Jiang, C.; Shao, Y.; Barceló, D.; Ying, Y.; Ping, J. Self-reduction bimetallic nanoparticles on ultrathin MXene nanosheets as functional platform for pesticide sensing. *J. Hazard. Mater.* **2020**, *384*, 121358. [CrossRef]

108. Liu, X.; Qiu, Y.; Jiang, D.; Li, F.; Gan, Y.; Zhu, Y.; Pan, Y.; Wan, H.; Wang, P. Covalently grafting first-generation PAMAM dendrimers onto MXenes with self-adsorbed AuNPs for use as a functional nanoplatform for highly sensitive electrochemical biosensing of cTnT. *Microsyst. Nanoeng.* **2022**, *8*, 35. [CrossRef]

109. Medetalibeyoglu, H.; Kotan, G.; Atar, N.; Yola, M.L. A novel and ultrasensitive sandwich-type electrochemical immunosensor based on delaminated MXene@AuNPs as signal amplification for prostate specific antigen (PSA) detection and immunosensor validation. *Talanta* **2020**, *220*, 121403. [CrossRef]

110. Laochai, T.; Yukird, J.; Promphet, N.; Qin, J.; Chailapakul, O.; Rodthongkum, N. Non-invasive electrochemical immunosensor for sweat cortisol based on L-cys/AuNPs/MXene modified thread electrode. *Biosens. Bioelectron.* **2022**, *203*, 114039. [CrossRef] [PubMed]

111. Jiang, B.; Li, C.; Dag, Ö.; Abe, H.; Takei, T.; Imai, T.; Hossain, M.; Shahriar, A.; Islam, M.; Wood, K. Mesoporous metallic rhodium nanoparticles. *Nat. Commun.* **2017**, *8*, 15581. [CrossRef] [PubMed]

112. Liu, C.; Yang, W.; Min, X.; Zhang, D.; Fu, X.; Ding, S.; Xu, W. An enzyme-free electrochemical immunosensor based on quaternary metallic/nonmetallic PdPtBP alloy mesoporous nanoparticles/MXene and conductive $CuCl_2$ nanowires for ultrasensitive assay of kidney injury molecule-1. *Sens. Actuators B Chem.* **2021**, *334*, 129585. [CrossRef]

113. Yang, X.; Zhao, L.; Lu, L.; Feng, M.; Xia, J.; Zhang, F.; Wang, Z. In Situ Reduction of Gold Nanoparticle-Decorated Ti_3C_2 MXene for Ultrasensitive Electrochemical Detection of MicroRNA-21 with a Cascaded Signal Amplification Strategy. *J. Electrochem. Soc.* **2022**, *169*, 057505. [CrossRef]

114. Mohammadniaei, M.; Koyappayil, A.; Sun, Y.; Min, J.; Lee, M.-H. Gold nanoparticle/MXene for multiple and sensitive detection of oncomiRs based on synergetic signal amplification. *Biosens. Bioelectron.* **2020**, *159*, 112208. [CrossRef] [PubMed]

115. Wang, H.; Sun, J.; Lu, L.; Yang, X.; Xia, J.; Zhang, F.; Wang, Z. Competitive electrochemical aptasensor based on a cDNA-ferrocene/MXene probe for detection of breast cancer marker Mucin1. *Anal. Chim. Acta* **2020**, *1094*, 18–25. [CrossRef]

116. Cheng, J.; Hu, K.; Liu, Q.; Liu, Y.; Yang, H.; Kong, J. Electrochemical ultrasensitive detection of CYFRA21-1 using $Ti_3C_2T_x$-MXene as enhancer and covalent organic frameworks as labels. *Anal. Bioanal. Chem.* **2021**, *413*, 2543–2551. [CrossRef]

117. Wei, Y.; Zhang, P.; Soomro, R.A.; Zhu, Q.; Xu, B. Advances in the Synthesis of 2D MXenes. *Adv. Mater.* **2021**, *33*, 2103148. [CrossRef]

118. Wu, Q.; Li, Z.; Liang, Q.; Ye, R.; Guo, S.; Zeng, X.; Hu, J.; Li, A. Ultrasensitive electrochemical biosensor for microRNA-377 detection based on MXene-Au nanocomposite and G-quadruplex nano-amplification strategy. *Electrochim. Acta* **2022**, *428*, 140945. [CrossRef]

119. Liu, J.; Tang, D. Dopamine-loaded Liposomes-amplified Electrochemical Immunoassay Based on MXene (Ti_3C_2)−AuNPs. *Electroanalysis* **2022**, *34*, 1329–1337. [CrossRef]

120. Medetalibeyoglu, H.; Beytur, M.; Akyıldırım, O.; Atar, N.; Yola, M.L. Validated electrochemical immunosensor for ultra-sensitive procalcitonin detection: Carbon electrode modified with gold nanoparticles functionalized sulfur doped MXene as sensor platform and carboxylated graphitic carbon nitride as signal amplification. *Sens. Actuators B Chem.* **2020**, *319*, 128195. [CrossRef]

121. Liu, Y.; Zeng, H.; Chai, Y.; Yuan, R.; Liu, H. Ti_3C_2/BiVO$_4$ Schottky junction as a signal indicator for ultrasensitive photoelectrochemical detection of VEGF165. *Chem. Commun.* **2019**, *55*, 13729–13732. [CrossRef] [PubMed]

122. He, Y.; Zhou, W.; Qian, G.; Chen, B. Methane storage in metal–organic frameworks. *Chem. Soc. Rev.* **2014**, *43*, 5657–5678. [CrossRef] [PubMed]

123. Baumann, A.E.; Burns, D.A.; Liu, B.; Thoi, V.S. Metal-organic framework functionalization and design strategies for advanced electrochemical energy storage devices. *Commun. Chem.* **2019**, *2*, 86. [CrossRef]

124. Li, J.; Ye, W.; Chen, C. Chapter 5—Removal of toxic/radioactive metal ions by metal-organic framework-based materials. In *Interface Science and Technology*; Chen, C., Ed.; Elsevier: Amsterdam, The Netherlands, 2019; Volume 29, pp. 217–279.

125. Kreno, L.E.; Leong, K.; Farha, O.K.; Allendorf, M.; Van Duyne, R.P.; Hupp, J.T. Metal–organic framework materials as chemical sensors. *Chem. Rev.* **2012**, *112*, 1105–1125. [CrossRef]

126. James, S.L. Metal-organic frameworks. *Chem. Soc. Rev.* **2003**, *32*, 276–288. [CrossRef]

127. Johnson, E.M.; Ilic, S.; Morris, A.J. Design strategies for enhanced conductivity in metal–organic frameworks. *ACS Cent. Sci.* **2021**, *7*, 445–453. [CrossRef]

128. Escobar-Hernandez, H.U.; Pérez, L.M.; Hu, P.; Soto, F.A.; Papadaki, M.I.; Zhou, H.-C.; Wang, Q. Thermal Stability of Metal–Organic Frameworks (MOFs): Concept, Determination, and Model Prediction Using Computational Chemistry and Machine Learning. *Ind. Eng. Chem. Res.* **2022**, *61*, 5853–5862. [CrossRef]

129. Sun, Y.; Zhou, H.-C. Recent progress in the synthesis of metal–organic frameworks. *Sci. Technol. Adv. Mater.* **2015**, *16*, 054202. [CrossRef]

130. Dourandish, Z.; Tajik, S.; Beitollahi, H.; Jahani, P.M.; Nejad, F.G.; Sheikhshoaie, I.; Di Bartolomeo, A. A Comprehensive Review of Metal-Organic Framework: Synthesis, Characterization, and Investigation of Their Application in Electrochemical Biosensors for Biomedical Analysis. *Sensors* **2022**, *22*, 2238. [CrossRef]

131. Wu, L.-Z.; Zhou, X.-Y.; Zeng, P.-C.; Huang, J.-Y.; Zhang, M.-D.; Qin, L. Hydrothermal synthesis of Ni (II) or Co (II)-based MOF for electrocatalytic hydrogen evolution. *Polyhedron* **2022**, *225*, 116035. [CrossRef]

132. Wang, F.-X.; Wang, C.-C.; Du, X.; Li, Y.; Wang, F.; Wang, P. Efficient removal of emerging organic contaminants via photo-Fenton process over micron-sized Fe-MOF sheet. *Chem. Eng. J.* **2022**, *429*, 132495. [CrossRef]

133. Menon, S.S.; Chandran, S.V.; Koyappayil, A.; Berchmans, S. Copper- Based Metal-Organic Frameworks as Peroxidase Mimics Leading to Sensitive H_2O_2 and Glucose Detection. *ChemistrySelect* **2018**, *3*, 8319–8324. [CrossRef]

134. Nazari, Z.; Taher, M.A.; Fazelirad, H. A Zn based metal organic framework nanocomposite: Synthesis, characterization and application for preconcentration of cadmium prior to its determination by FAAS. *RSC Adv.* **2017**, *7*, 44890–44895. [CrossRef]

135. Ban, Y.; Li, Y.; Liu, X.; Peng, Y.; Yang, W. Solvothermal synthesis of mixed-ligand metal–organic framework ZIF-78 with controllable size and morphology. *Microporous Mesoporous Mater.* **2013**, *173*, 29–36. [CrossRef]

136. Liu, Y.; Wei, Y.; Liu, M.; Bai, Y.; Wang, X.; Shang, S.; Chen, J.; Liu, Y. Electrochemical Synthesis of Large Area Two-Dimensional Metal–Organic Framework Films on Copper Anodes. *Angew. Chem. Int. Ed.* **2021**, *60*, 2887–2891. [CrossRef]

137. Asghar, A.; Iqbal, N.; Noor, T.; Kariuki, B.M.; Kidwell, L.; Easun, T.L. Efficient electrochemical synthesis of a manganese-based metal–organic framework for H_2 and CO_2 uptake. *Green Chem.* **2021**, *23*, 1220–1227. [CrossRef]

138. Tang, D.; Yang, X.; Wang, B.; Ding, Y.; Xu, S.; Liu, J.; Peng, Y.; Yu, X.; Su, Z.; Qin, X. One-Step Electrochemical Growth of 2D/3D Zn(II)-MOF Hybrid Nanocomposites on an Electrode and Utilization of a PtNPs@2D MOF Nanocatalyst for Electrochemical Immunoassay. *ACS Appl. Mater. Interfaces* **2021**, *13*, 46225–46232. [CrossRef]

139. Wu, W.; Decker, G.E.; Weaver, A.E.; Arnoff, A.I.; Bloch, E.D.; Rosenthal, J. Facile and Rapid Room-Temperature Electrosynthesis and Controlled Surface Growth of Fe-MIL-101 and Fe-MIL-101-NH_2. *ACS Cent. Sci.* **2021**, *7*, 1427–1433. [CrossRef]

140. Ghoorchian, A.; Afkhami, A.; Madrakian, T.; Ahmadi, M. Chapter 9—Electrochemical synthesis of MOFs. In *Metal–Organic Frameworks for Biomedical Applications*; Mozafari, M., Ed.; Woodhead Publishing: Cambridge, UK, 2020; pp. 177–195.

141. Ameloot, R.; Stappers, L.; Fransaer, J.; Alaerts, L.; Sels, B.F.; De Vos, D.E. Patterned growth of metal-organic framework coatings by electrochemical synthesis. *Chem. Mater.* **2009**, *21*, 2580–2582. [CrossRef]

142. Bazzi, L.; Ayouch, I.; Tachallait, H.; Hankari, S.E.L. Ultrasound and microwave assisted-synthesis of ZIF-8 from zinc oxide for the adsorption of phosphate. *Results Eng.* **2022**, *13*, 100378. [CrossRef]

143. Koyappayil, A.; Yeon, S.-h.; Chavan, S.G.; Jin, L.; Go, A.; Lee, M.-H. Efficient and rapid synthesis of ultrathin nickel-metal organic framework nanosheets for the sensitive determination of glucose. *Microchem. J.* **2022**, *179*, 107462. [CrossRef]

144. Lee, J.H.; Ahn, Y.; Kwak, S.-Y. Facile Sonochemical Synthesis of Flexible Fe-Based Metal–Organic Frameworks and Their Efficient Removal of Organic Contaminants from Aqueous Solutions. *ACS Omega* **2022**, *7*, 23213–23222. [CrossRef]

145. Li, J.; Liu, L.; Ai, Y.; Liu, Y.; Sun, H.; Liang, Q. Self-Polymerized Dopamine-Decorated Au NPs and Coordinated with Fe-MOF as a Dual Binding Sites and Dual Signal-Amplifying Electrochemical Aptasensor for the Detection of CEA. *ACS Appl. Mater. Interfaces* **2020**, *12*, 5500–5510. [CrossRef] [PubMed]

146. Li, W.; Yang, Y.; Ma, C.; Song, Y.; Hong, C.; Qiao, X. A sandwich-type electrochemical immunosensor for ultrasensitive detection of CEA based on core–shell Cu_2O@Cu-MOF@Au NPs nanostructure attached with HRP for triple signal amplification. *J. Mater. Sci.* **2020**, *55*, 13980–13994. [CrossRef]

147. Mulchandani, A.; Myung, N.V. Conducting polymer nanowires-based label-free biosensors. *Curr. Opin. Biotechnol.* **2011**, *22*, 502–508. [CrossRef]

148. Zhang, Y.; Kolmakov, A.; Lilach, Y.; Moskovits, M. Electronic Control of Chemistry and Catalysis at the Surface of an Individual Tin Oxide Nanowire. *J. Phys. Chem. B* **2005**, *109*, 1923–1929. [CrossRef]

149. Li, S.; Yue, S.; Yu, C.; Chen, Y.; Yuan, D.; Yu, Q. A label-free immunosensor for the detection of nuclear matrix protein-22 based on a chrysanthemum-like Co-MOFs/CuAu NWs nanocomposite. *Analyst* **2019**, *144*, 649–655. [CrossRef]

150. Feng, J.; Wang, H.; Ma, Z. Ultrasensitive amperometric immunosensor for the prostate specific antigen by exploiting a Fenton reaction induced by a metal-organic framework nanocomposite of type Au/Fe-MOF with peroxidase mimicking activity. *Microchim. Acta* **2020**, *187*, 95. [CrossRef]

151. Mehmandoust, M.; Erk, E.E.; Soylak, M.; Erk, N.; Karimi, F. Metal–Organic Framework Based Electrochemical Immunosensor for Label-Free Detection of Glial Fibrillary Acidic Protein as a Biomarker. *Ind. Eng. Chem. Res.* **2022**. [CrossRef]

152. Zhao, S.; Zhang, Y.; Ding, S.; Fan, J.; Luo, Z.; Liu, K.; Shi, Q.; Liu, W.; Zang, G. A highly sensitive label-free electrochemical immunosensor based on AuNPs-PtNPs-MOFs for nuclear matrix protein 22 analysis in urine sample. *J. Electroanal. Chem.* **2019**, *834*, 33–42. [CrossRef]

153. Joseph, J.; Iftekhar, S.; Srivastava, V.; Fallah, Z.; Zare, E.N.; Sillanpää, M. Iron-based metal-organic framework: Synthesis, structure and current technologies for water reclamation with deep insight into framework integrity. *Chemosphere* **2021**, *284*, 131171. [CrossRef] [PubMed]

154. Silva, P.; Vilela, S.M.F.; Tomé, J.P.C.; Almeida Paz, F.A. Multifunctional metal–organic frameworks: From academia to industrial applications. *Chem. Soc. Rev.* **2015**, *44*, 6774–6803. [CrossRef]

155. Wang, W.; Yu, Y.; Jin, Y.; Liu, X.; Shang, M.; Zheng, X.; Liu, T.; Xie, Z. Two-dimensional metal-organic frameworks: From synthesis to bioapplications. *J. Nanobiotechnol.* **2022**, *20*, 207. [CrossRef]

156. Cheng, T.; Li, X.; Huang, P.; Wang, H.; Wang, M.; Yang, W. Colorimetric and electrochemical (dual) thrombin assay based on the use of a platinum nanoparticle modified metal-organic framework (type Fe-MIL-88) acting as a peroxidase mimic. *Microchim. Acta* **2019**, *186*, 94. [CrossRef]

157. Sun, D.; Luo, Z.; Lu, J.; Zhang, S.; Che, T.; Chen, Z.; Zhang, L. Electrochemical dual-aptamer-based biosensor for nonenzymatic detection of cardiac troponin I by nanohybrid electrocatalysts labeling combined with DNA nanotetrahedron structure. *Biosens. Bioelectron.* **2019**, *134*, 49–56. [CrossRef]

158. Miao, J.; Du, K.; Li, X.; Xu, X.; Dong, X.; Fang, J.; Cao, W.; Wei, Q. Ratiometric electrochemical immunosensor for the detection of procalcitonin based on the ratios of SiO_2-Fc–COOH–Au and UiO-66-TB complexes. *Biosens. Bioelectron.* **2021**, *171*, 112713. [CrossRef]

159. Dong, H.; Liu, S.; Liu, Q.; Li, Y.; Li, Y.; Zhao, Z. A dual-signal output electrochemical immunosensor based on Au–MoS_2/MOF catalytic cycle amplification strategy for neuron-specific enolase ultrasensitive detection. *Biosens. Bioelectron.* **2022**, *195*, 113648. [CrossRef]

160. Wang, S.; Wang, M.; Li, C.; Li, H.; Ge, C.; Zhang, X.; Jin, Y. A highly sensitive and stable electrochemiluminescence immunosensor for alpha-fetoprotein detection based on luminol-AgNPs@Co/Ni-MOF nanosheet microflowers. *Sens. Actuators B Chem.* **2020**, *311*, 127919. [CrossRef]

161. Du, D.; Shu, J.; Guo, M.; Haghighatbin, M.A.; Yang, D.; Bian, Z.; Cui, H. Potential-Resolved Differential Electrochemiluminescence Immunosensor for Cardiac Troponin I Based on MOF-5-Wrapped CdS Quantum Dot Nanoluminophores. *Anal. Chem.* **2020**, *92*, 14113–14121. [CrossRef]

162. Dai, L.; Li, Y.; Wang, Y.; Luo, X.; Wei, D.; Feng, R.; Yan, T.; Ren, X.; Du, B.; Wei, Q. A prostate-specific antigen electrochemical immunosensor based on Pd NPs functionalized electroactive Co-MOF signal amplification strategy. *Biosens. Bioelectron.* **2019**, *132*, 97–104. [CrossRef] [PubMed]

163. Gao, X.; Liu, H.; Wang, D.; Zhang, J. Graphdiyne: Synthesis, properties, and applications. *Chem. Soc. Rev.* **2019**, *48*, 908–936. [CrossRef] [PubMed]

164. Hou, Y.-Y.; Xu, J.; Wang, F.-T.; Dong, Z.; Tan, X.; Huang, K.-J.; Li, J.-Q.; Zuo, C.-Y.; Zhang, S.-Q. Construction of an Integrated Device of a Self-Powered Biosensor and Matching Capacitor Based on Graphdiyne and Multiple Signal Amplification: Ultrasensitive Method for MicroRNA Detection. *Anal. Chem.* **2021**, *93*, 15225–15230. [CrossRef] [PubMed]

165. Pourali, A.; Rashidi, M.R.; Barar, J.; Pavon-Djavid, G.; Omidi, Y. Voltammetric biosensors for analytical detection of cardiac troponin biomarkers in acute myocardial infarction. *TrAC-Trends Anal. Chem.* **2021**, *134*, 116123. [CrossRef]

166. Adeel, M.; Rahman, M.M.; Lee, J.-J. Label-free aptasensor for the detection of cardiac biomarker myoglobin based on gold nanoparticles decorated boron nitride nanosheets. *Biosens. Bioelectron.* **2019**, *126*, 143–150. [CrossRef]

167. Cohen, M.L. Calculation of bulk moduli of diamond and zinc-blende solids. *Phys. Rev. B* **1985**, *32*, 7988–7991. [CrossRef]

168. Liu, A.Y.; Cohen, M.L. Prediction of New Low Compressibility Solids. *Science* **1989**, *245*, 841–842. [CrossRef]

169. Zhu, J.; Xiao, P.; Li, H.; Carabineiro, S.A.C. Graphitic Carbon Nitride: Synthesis, Properties, and Applications in Catalysis. *ACS Appl. Mater. Interfaces* **2014**, *6*, 16449–16465. [CrossRef]

170. Bott Neto, J.L.; Martins, T.S.; Machado, S.A.; Oliveira, O.N. Enhanced photocatalysis on graphitic carbon nitride sensitized with gold nanoparticles for photoelectrochemical immunosensors. *Appl. Surf. Sci.* **2022**, *606*, 154952. [CrossRef]

171. Wu, Z.; Lyu, Y.; Zhang, Y.; Ding, R.; Zheng, B.; Yang, Z.; Lau, S.P.; Chen, X.H.; Hao, J. Large-scale growth of few-layer two-dimensional black phosphorus. *Nat. Mater.* **2021**, *20*, 1203–1209. [CrossRef]

172. Gaufrès, E.; Fossard, F.; Gosselin, V.; Sponza, L.; Ducastelle, F.; Li, Z.; Louie, S.G.; Martel, R.; Côté, M.; Loiseau, A. Momentum-resolved dielectric response of free-standing mono-, bi-, and trilayer black phosphorus. *Nano Lett.* **2019**, *19*, 8303–8310. [CrossRef]

173. Youngblood, N.; Chen, C.; Koester, S.J.; Li, M. Waveguide-integrated black phosphorus photodetector with high responsivity and low dark current. *Nat. Photonics* **2015**, *9*, 247–252. [CrossRef]

174. Wu, Z.; Hao, J. Electrical transport properties in group-V elemental ultrathin 2D layers. *NPJ 2D Mater. Appl.* **2020**, *4*, 4. [CrossRef]

175. Li, Z.; Fu, Y.; Zhu, Q.; Wei, S.; Gao, J.; Zhu, Y.; Xue, T.; Bai, L.; Wen, Y. High-stable Phosphorene-supported Bimetallic Pt-Pd Nanoelectrocatalyst for p-Aminophenol, β-Galactosidase, and Escherichia coli. *Int. J. Electrochem. Sci* **2020**, *15*, 3089–3103. [CrossRef]

176. Shanmuganathan, K.; Ellison, C.J. Chapter 20—Layered Double Hydroxides: An Emerging Class of Flame Retardants. In *Polymer Green Flame Retardants*; Papaspyrides, C.D., Kiliaris, P., Eds.; Elsevier: Amsterdam, The Netherlands, 2014; pp. 675–707.

177. Wu, M.; Liu, S.; Qi, F.; Qiu, R.; Feng, J.; Ren, X.; Rong, S.; Ma, H.; Chang, D.; Pan, H. A label-free electrochemical immunosensor for CA125 detection based on CMK-3(Au/Fc@MgAl-LDH)n multilayer nanocomposites modification. *Talanta* **2022**, *241*, 123254. [CrossRef]

178. Ojha, R.P.; Singh, P.; Azad, U.P.; Prakash, R. Impedimetric immunosensor for the NS1 dengue biomarker based on the gold nanorod decorated graphitic carbon nitride modified electrode. *Electrochim. Acta* **2022**, *411*, 140069. [CrossRef]

179. Jalal, U.M.; Jin, G.J.; Shim, J.S. Paper-Plastic Hybrid Microfluidic Device for Smartphone-Based Colorimetric Analysis of Urine. *Anal. Chem* **2017**, *89*, 13160–13166. [CrossRef]

180. Chung, S.; Breshears, L.E.; Perea, S.; Morrison, C.M.; Betancourt, W.Q.; Reynolds, K.A.; Yoon, J.Y. Smartphone-Based Paper Microfluidic Particulometry of Norovirus from Environmental Water Samples at the Single Copy Level. *ACS Omega* **2019**, *4*, 11180–11188. [CrossRef]

biosensors

MDPI

Review

A Review on Microfluidics-Based Impedance Biosensors

Yu-Shih Chen [1], Chun-Hao Huang [1], Ping-Ching Pai [2], Jungmok Seo [1,3,*] and Kin Fong Lei [1,2,3,*]

[1] Department of Biomedical Engineering, Chang Gung University, Taoyuan 33302, Taiwan
[2] Department of Radiation Oncology, Linkou Chang Gung Memorial Hospital, Taoyuan 33305, Taiwan
[3] Department of Electrical & Electronic Engineering, Yonsei University, Seoul 120-749, Republic of Korea
*   Correspondence: jungmok.seo@yonsei.ac.kr (J.S.); kflei@mail.cgu.edu.tw (K.F.L.)

Abstract: Electrical impedance biosensors are powerful and continuously being developed for various biological sensing applications. In this line, the sensitivity of impedance biosensors embedded with microfluidic technologies, such as sheath flow focusing, dielectrophoretic focusing, and interdigitated electrode arrays, can still be greatly improved. In particular, reagent consumption reduction and analysis time-shortening features can highly increase the analytical capabilities of such biosensors. Moreover, the reliability and efficiency of analyses are benefited by microfluidics-enabled automation. Through the use of mature microfluidic technology, complicated biological processes can be shrunk and integrated into a single microfluidic system (e.g., lab-on-a-chip or micro-total analysis systems). By incorporating electrical impedance biosensors, hand-held and bench-top microfluidic systems can be easily developed and operated by personnel without professional training. Furthermore, the impedance spectrum provides broad information regarding cell size, membrane capacitance, cytoplasmic conductivity, and cytoplasmic permittivity without the need for fluorescent labeling, magnetic modifications, or other cellular treatments. In this review article, a comprehensive summary of microfluidics-based impedance biosensors is presented. The structure of this article is based on the different substrate material categorizations. Moreover, the development trend of microfluidics-based impedance biosensors is discussed, along with difficulties and challenges that may be encountered in the future.

Keywords: microfluidic; impedance biosensor; electrical impedance flow cytometer; electrochemical impedance spectroscopy

Citation: Chen, Y.-S.; Huang, C.-H.; Pai, P.-C.; Seo, J.; Lei, K.F. A Review on Microfluidics-Based Impedance Biosensors. *Biosensors* **2023**, *13*, 83. https://doi.org/10.3390/bios13010083

Received: 21 November 2022
Revised: 20 December 2022
Accepted: 28 December 2022
Published: 3 January 2023

1. Introduction

Biosensors are mainly used to measure or perceive signals from biological responses. Electrical biosensors can be generally classified into potential, current, and impedance sensors [1]. H. E. Ayliffe pioneered the measurement of single-cell impedance in a microchannel in 1999 [2]. Biological substances can be detected using a pair of microelectrodes with a gap of several μm, in a microchannel 10 μm in width. This narrow microchannel allowed for a more accurate impedance measurement of human polymorphonuclear leukocytes and teleost fish red blood cells. Subsequently, the electrical and equivalent circuit models of single cells were established [3,4].

As early as 1984, Giaever et al. designed a device on a Petri dish that could monitor impedance in order to investigate the density and cell migration of fibroblasts [5]. The impedance signal was obtained by gold electrodes 130 μm in width, which were deposited by a metal evaporator using a mask. Fibroblasts were attached to the gold electrodes, and their cellular response could be represented by the impedance measured across the electrodes [6]. The measured impedance values could differentiate normal fibroblasts from transformed cells. When fibronectin and gelatin were coated on the gold electrodes, the fibroblasts showed a better response. Based on the results of electrical signal monitoring, cell movement was observed. Later, Giaever published an article proposing that the phenomenon of cell movement is called micromotion [7], following which the authors put forward a mathematical model establishing the theory of cell movement analysis. Experimentally, small fluctuations in impedance electrical signals were measured as direct evidence of cell movement. Unlike cells observed by light microscopy, impedance biosensors are designed to continuously track the vertical movement of detected cells [8]. The range of cell motion can be as small as 1 nm. The impedance value and fluctuation of each type of cell differs and, so, may be used as a cell fingerprint. Regarding electrical impedance spectroscopy, Grossi et al. carried out a series of literature reviews [9]. The impedance spectrum is usually matched with an equivalent circuit model, and the impedance spectrum obtained by each test object can be expressed as the electrical fingerprint of the test object. Sun et al. used electrochemical impedance spectroscopy to explain the dielectric properties of individual polyelectrolyte microcapsules with different shell thicknesses [10]. The authors built a complete equivalent circuit model of a single solid spherical particle in suspension, and the resistance of the shell and the capacitance of the inner core were used to determine the permittivity and conductivity of individual capsule shells. The study indicated that the conductivity of the six- and nine-layer microcapsule shells could be estimated as 28 ± 6 and 3.3 ± 1.7 mS m^{-1}, respectively.

On the other hand, the first microchannel-based flow cytometer was designed by Kamentsky et al. in 1965 [11], with a cell throughput of 500 cells per second. Initially, optical detection was used, using wavelengths of 253 nm and 410 nm. Later on, S. Gawad developed an impedance spectroscopy flow cytometer using Pt as an electrode, which could reduce the electrode impedance and allow the measurement frequency to be as low as 10 kHz [12]. By using a three-electrode design, the impedance value when the cells pass through can be easily measured. The peak impedance value is related to the cell size, and the electrode spacing and the time difference between two impedance signals could also be used to determine the velocity of the measured particle. Gawad et al. established a model to discriminate cell size, membrane capacitance, and cytoplasmic conductivity using a miniature cytometer [13]. Moreover, Cheung et al. obtained amplitude, opacity, and phase information using a microfluidic impedance cytometer, which can be used to distinguish different cells [14], where the measured amplitude can determine the cell size. Opacity was used to distinguish polystyrene beads from red blood cells (RBCs), while phase information was used to distinguish between RBCs and ghosts. RBCs and RBCs fixed in glutaraldehyde could also be distinguished by opacity. The definition of opacity was published in an article by R. A. Hoffman [15]; in particular, it is the ratio of the high-frequency impedance to low-frequency impedance of a particle. Some scholars have studied the development of fluidic impedance cytometers for micro-single cells [16].

Impedance biosensors integrated with microfluidic technology can be categorized into two major technologies: electrical impedance flow cytometry and electrochemical impedance spectroscopy. In the design of a microchannel, an electrical impedance flow cytometer can measure dynamic objects using impedance technology. On the other hand, electrochemical impedance spectroscopy is usually combined with microcavity structures. Static objects can be measured using impedance measurement techniques. In this review article, microfluidics-based impedance biosensors are comprehensively summarized and classified according to the different substrate materials. The development trend, difficulties, and challenges associated with microfluidics-based impedance biosensors, which are expected to be encountered in the future, are also discussed.

2. Silicon-Based Impedance Biosensors

Silicon-based impedance biosensors are only possible with the fabrication of the electrodes by micromachining. Therefore, the development of silicon impedance biosensors began earlier; however, a number of articles have also been published recently, due to relevant special processes.

An impedance biosensor with a nanometer-wide interdigitated electrode array was developed [17], with electrode widths and spacings of 250–500 nm microfabricated using deep UV lithography. The same article also verified the binding of biomolecular structures to nanoscale electrode surfaces. The impedance signal results indicate that the immobilization of glucose oxidase on the electrode can be monitored. A silicon-based microfabricated biochip was designed to measure the electrical impedance spectrum [18], which could measure the conductance change in a 30 nl volume of bacterial suspension and showed the viability of the bacteria. The same research demonstrated electrical impedance values for the live micro-organism *Listeria innocuous*. By-products after bacterial metabolism have also been shown to change the electrical impedance value; for example, Radke et al. developed an impedance biosensor for bacterial detection using immobilized antibodies. The interdigitated electrode arrays were designed on silicon-based biosensors [19]. *Escherichia coli*-specific antibodies were immobilized on the electrodes, and impedance changes due to bacteria immobilized on the interdigitated gold electrodes were observed. Impedance signals at low frequencies showed that bacteria bound to the sensor electrode surface within 5 min. The rate of binding was the most pronounced in the first 20 min but slowed down significantly after 35 min. At high frequencies, impedance does not change over time. Test concentrations can be as low as 10 CFU/mL of bacterial suspension. The same research team used an impedance biosensor with immobilized antibodies and interdigitated electrode arrays to detect pathogenic *Escherichia coli* O157:H7 and Salmonella infants [20]. It mainly detects bacteria in food or water. P-type (100) silicon wafers were used in that article. Coplanar impedance sensors were designed on glass and fabricated by photolithography [21]. They verified that the spacing of the set of coplanar electrodes is a more important parameter than the electrode area. As the spacing of the electrode design increases, the impedance value increases accordingly. An impedance sensor was used to monitor drug-affected spheroids in a microcavity [22]. Silicon wafers were microfabricated into microcavities through wet anisotropic etching. The impedance sensor was designed with 15 microcavities, and 4 electrodes in each cavity were used to sense the impedance of the spheroids. OLN93 cell spheroids were most loosely organized and peaked at around 180 kHz, while Bro cell spheroids had a more compact structure and showed a peak at 100 kHz. They also observed the impedance of the spheroids 8 h after drug administration. Impedance increased with forskolin, camptothecin, and staurosporine and decreased with doxorubicin and tamoxifen. Single-cell impedance and optical sensing were integrated into a single chip for real-time viability assessment [23]. Single-cell capture microwell chips were obtained by etching silicon wafers with KOH. The induction chip is composed of two cavities (upper and lower). The adhesion changes of RAW264.7 macrophages can be assessed through the impedance value of the sensing chip. In an interdigitated electrode array, the gap between the electrodes is an important factor to improve the sensitivity of the biosensor [24]. Three-dimensional interdigitated electrode arrays were fabricated to sense the impedance signals of proteins. C-reactive protein-specific antibodies were immobilized on the electrode surface. The results showed that electrochemical impedance spectroscopy can be used to monitor the concentration of C-reactive protein in human serum. An impedance biosensor that can sense the concentration of picesterol was developed [25]. The electrodes on the silicon wafer were designed with interdigitated electrodes, where the distance between two electrodes was 15 μm. Cortisol-specific monoclonal antibodies were immobilized on the surface of the microelectrodes. Cortisol binding to antibodies can be signaled by changing the impedance values. The experimental results showed that the impedance biosensor can accurately detect cortisol in the range of 1 pM to 10 nM in saliva. An electrical impedance biosensor was constructed through a CMOS-process, using high-density sensing electrodes for the detection of breast tumor cells (MCF-7) [26]. A total of 96 × 96 gold microelectrodes were designed with a sensing area of 3.5 mm × 3.5 mm. The impedance signal is read out by an integrated circuit fabricated with 0.18 μm CMOS technology. The results showed that the increase in impedance was associated with cell binding to the electrode surface. Ma et al. designed an impedance biosensor that can detect suspended DNA, as shown in Figure 1a [27], where the sensing electrodes are fabricated using 0.35 μm CMOS technology. They proposed that the impedance of a solution is highly dependent on the concentration. Moreover, the impedance value of the sample solution is also highly correlated with the length of the DNA fragment. In other words, the biological samples obtained after PCR can be tested using the biosensors designed in that article. Lisandro Cunci developed electrochemical impedance biosensors that can detect telomerase in cancer cells [28]. The flexible heater and temperature sensor were designed together in the biosensor. Single-stranded DNA probes were immobilized on the surface of the interdigitated gold electrode array. Jurkat cells were tested for telomerase and showed a 14-fold increase in electrical resistance. The sensitivity of electrochemical impedance spectroscopy biosensors was enhanced using electric field focusing of magnetic beads in microwells [29], where the antibody is immobilized on the surface of the magnetic beads. Microwells are fabricated on silicon wafers using high aspect ratio SU-8 microstructures. In the same article, prostate-specific antigen in a PBS buffer and human plasma were used to validate the argument that focused magnetic beads can improve the sensitivity. The experimental results showed that prostate-specific antigens at low concentrations (i.e., tens or hundreds of fg/mL) could be detected. Pursey et al. integrated surface plasmon resonance and electrochemical impedance spectroscopy into a microfluidic chip, targeting bladder cancer-associated DNA sequences, as shown in Figure 1b [30]. Gold electrodes are microfabricated on the silicon layer, and 20 sensors on the same wafer simultaneously detect three different DNA markers for bladder cancer. Signals were measured within a short period of 20 min. Impedance biosensors that can monitor bacteria have also been developed. In one study, 216 three-dimensional interdigitated electrodes of 3 μm width and with 3 μm gap were microfabricated on a silicon wafer [31]. To improve the sensitivity, the three-dimensional electrodes were separated by an insulating layer. In addition to sensing the impedance signals of bacteria, the three-dimensional electrode can also concentrate bacteria. An impedance sensor fabricated in microgrooves was developed, where silicon substrates were microfabricated to create the microgrooves, as shown in Figure 1c [32]. Two gold electrodes were microfabricated in the microgrooves, which can be used to sense the impedance value of three-dimensional cancer cells, and changes in impedance can reflect the proliferation and apoptosis of three-dimensional cancer cells. Impedance sensors can also identify cancer cells that are affected by drugs. A biosensor combining impedance and photoelectrochemical analysis of cancer cells was designed, as shown in Figure 1d [33], where monocrystalline silicon was used as a substrate for the biosensor. The serrated interdigitated electrodes not only can focus cells but also sense impedance signals. They identified four different types of cancer cells: esophageal cancer (CE81T cell), esophageal cancer (OE21 cell), lung adenocarcinoma (A549 cell), and bladder cancer (TSGH-8301 cell).

The development of silicon chips for microfluidic impedance sensors depends on the development of the microelectromechanical system. Many silicon-based microfabrication processes and metal microelectrode processes are derived from microelectromechanical system microfabrication. Therefore, this is discussed in the first part of this article. Due to the special process requirements, several articles have focused on silicon-based microfluidic impedance sensors. In recent years, the COVID-19 virus has spread throughout the world, and this multi-mutated virus needs to be sensed through the use of biosensors. The MEMS process developed by Taiwan's Taiwan Semiconductor Manufacturing Company may provide the direction for such a technology to be commercialized.

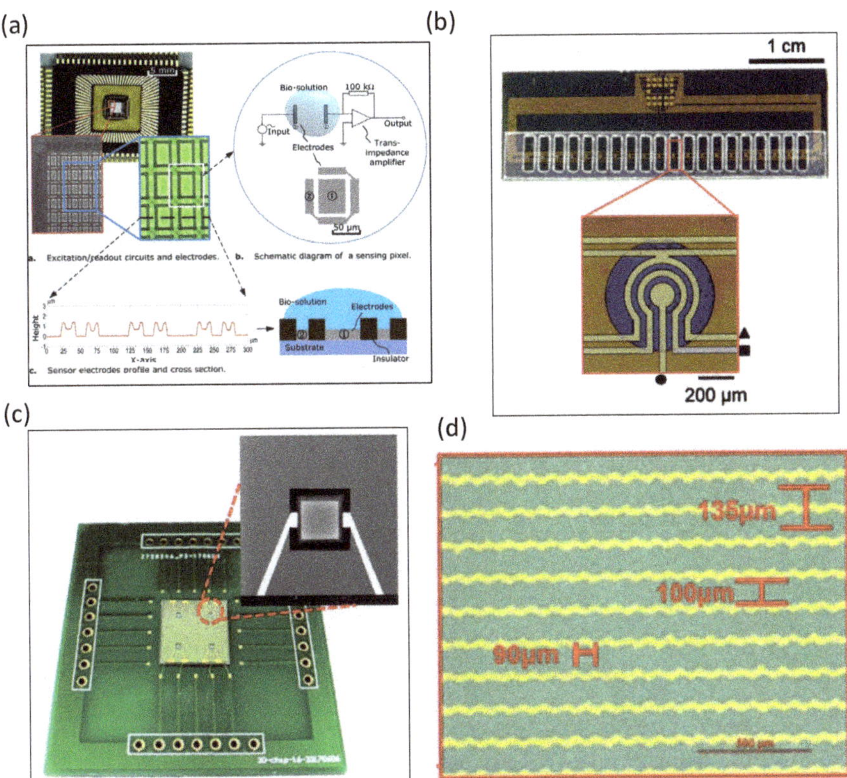

Figure 1. The silicon-based impedance biosensors. (**a**) A low-cost 0.35 mm CMOS technology by TSMC (Taiwan) was used to fabricate the micro-array chip that sensed DNA characterization. Reproduced with permission from [27]. Copyright Scientific Reports 2013. (**b**) Integrated surface plasmon resonance and electrochemical impedance spectroscopy in a microfluidic chip [30]. Reproduced with permission from [30]. Copyright Sensors and Actuators B: Chemical 2017. (**c**) Silicon substrates were microfabricated to create microgrooves. Reproduced with permission from [32]. Copyright Microsystems and Nanoengineering 2020. (**d**) A biosensor combining impedance and photoelectrochemical was designed. Reproduced with permission from [33]. Copyright Biosensors 2022.

3. Printed Circuit Board (PCB)-Based Impedance Biosensors

Printed circuit boards (PCBs) are flat plates that were originally used to make circuits, which are very commercialized. Their electrode width and electrode spacing are not very small; therefore, the sensitivity limit of the sensor is not low either.

Electrochemical impedance spectroscopy biosensors consist of PCBs with gold-coated electrodes, which are mainly used to detect plant pathogens [34], where the antibody is first bound to the surface of the electrochemical sensor. For papaya ring spot virus, the biosensor was shown to be capable of detecting papaya ring spot virus coat protein with high sensitivity. A chip using a microfluidic impedance sensing system to detect the transgenic protein Cry1Ab was designed, as shown in Figure 2a [35]. Gold electrodes were printed on commercial printed circuit boards, with the spacing between two printed electrodes being 250 µm. The impedance signal at the optimal test frequency (358.3 Hz) presented a good linear relationship with the concentration of the transgenic protein Cry1Ab in the range of 0–0.2 nM. Clinically, the degree of red blood cell agglutination is divided into five grades by visual inspection, which is routinely conducted in hospitals [36]. An electrical impedance blood-sensing chip was designed by Chang et al. to distinguish the degree of blood agglutination. The interdigitated electrode array was designed on a PCB.

ZnO nanowires were synthesized on the surface of the electrode array in order to improve the sensitivity of impedance measurement. An electrical impedance sensor was used to detect the degree of fibrosis in liver tissue, as shown in Figure 2b [37], where liver tissue was detected by a pair of gold electrodes on the PCB. The experimental results indicate that the maximum resistance difference between healthy and fibrotic tissue was about 2 kΩ at day 8.

(a)   (b)

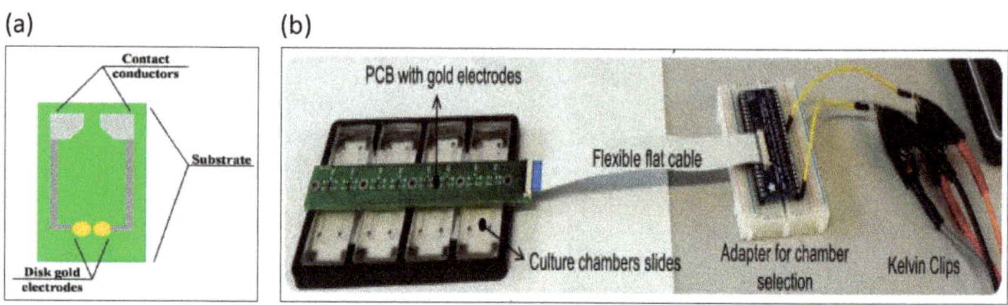

Figure 2. The PCB-based impedance biosensors. (**a**) Gold electrodes were printed on commercial printed circuit boards. Reproduced with permission from [35]. Copyright Scientific Reports 2013. (**b**) An electrical impedance sensor was used to detect the degree of fibrosis in liver tissue. Reproduced with permission from [37]. Copyright Biosensors 2022.

The development of PCB chips is limited by the electrode line width in traditional PCB processes. Microelectrodes that are not tiny enough will lead to an inability to improve the sensitivity. Therefore, PCB chips are more suitable for the production or commercial design where the induction is fast and the sensitivity requirements are not high.

4. Polymer-Based Impedance Biosensors

In addition to glass, the most commonly used substrates in the field of microfluidics are polymer chips. Different from the stretchable chips mentioned in later chapters, the polymer chips mentioned in this chapter are formed of hard and inflexible materials.

A study considering electrochemical impedance spectroscopy on polymer substrates was developed [38], in which the authors designed a nanoscale interdigitated electric shock array, in which the electrode width was only 200 nm and the electrode spacing was 500 nm. Gold nanometer interdigitated electrode arrays were patterned on cyclic olefin copolymer substrates. Experiments have demonstrated selective iDEP capture and impedance detection on polystyrene microspheres and *Bacillus subtilis* spores [39]. The authors used oxides to passivate the sensing electrode of the sensor in order to avoid the metal and electrolyte having adverse effects on the electrode surface. Cyclic olefin copolymer was used as the substrate of the sensor. Studies have used all-polymer electrochemical microfluidic biosensors for electrochemical impedance spectroscopy, as shown in Figure 3a [40], where polymer materials from Topas Corporation were used as substrates and conductive polymer bilayers were used as electrode materials. Electrochemical impedance spectroscopy was able to detect ampicillin in a concentration range from 100 pM to 1 μM and kanamycin A from 10 nM to 1 mM. Figure 3b shows how Pires et al. combined an impedance sensor and a current sensor to detect biofilms in water [41]. Two microfluidic chambers were designed on a cyclic olefin copolymer substrate with four impedance sensors and three current sensors in each cavity. A conductive polymer (PEDOT:TsO) was fabricated as an interdigitated electrode array for impedance biosensors [42], where the cyclic olefin copolymer produced by TOPAS was used as the base material for the conductive polymer electrodes. A microfluidic impedance sensor was used to detect the food additive clenbuterol hydrochloride [43], where the electrodes were patterned on poly (ethylene terephthalate) films. Polyaniline@graphene oxide nanocomposites were used to functionalize the sensing electrodes, and the microfluidic impedance sensors could detect down to 0.12 ppb. Sharif et al. integrated a microfluidic system and a magnetic separation procedure to develop a novel impedance sensor for the detection of various foodborne pathogens [44]. Their results showed that the impedance sensor was effective for the detection of various foodborne pathogens, including *Escherichia coli* (*E. coli* O157:H7), *Vibrio parahaemolyticus* (*V. parahaemolyticus*), *Staphylococcus aureus* (*S. aureus*), and *Listeria monocytogenes* (*L. monocytogenes*). Polymethyl methacrylate (PMMA) was used as the substrate. D-dimer is a biomarker in the blood that can be used to diagnose deep vein thrombosis and pulmonary embolism [45]. Lakey et al. designed a polymer microfluidic impedance sensor for the detection of D-dimer, where interdigitated electrode arrays were patterned on polyethylene naphthalate (PEN) substrates. Ma et al. developed an electrochemical impedance spectroscopy approach for the detection of endotoxins [46]. The electrodes were screen-printed on a polyethylene terephthalate (PET) substrate, which contained carbon for the working and auxiliary electrodes and Ag/AgCl for the reference electrode. The sensitivity of the impedance biosensors could be as low as 500 pg mL^{-1}, and the total measurement time was only half an hour. Niaraki et al. used graphene microelectrodes to monitor neuronal growth and detachment after death, as shown in Figure 3c [47]. Kapton Polyimide (PI) was used as a substrate for patterned graphene electrodes. The microelectrodes fabricated by this research team feature a wrinkled surface morphology, which allows for a fast response time to be achieved. Chmayssem et al. integrated a cell culture chamber with electrochemical impedance spectroscopy, as shown in Figure 3d [48]. The researchers used a screen-printing technique to fabricate the microelectrodes on polyethylene terephthalate (PET) sheets. The electrode material was selected, and Ag/AgCl was used as the low interface impedance electrode on the PET sheet. On the top layer of electrodes was carbon-biocompatible ink rich in IrOx particles. Hantschke et al. integrated electrophoretic and electrical impedance sensors for point-of-care (POC) diagnostics [49], where the electrodes and microchannels were fabricated on polymethylmethacrylate (PMMA) substrates.

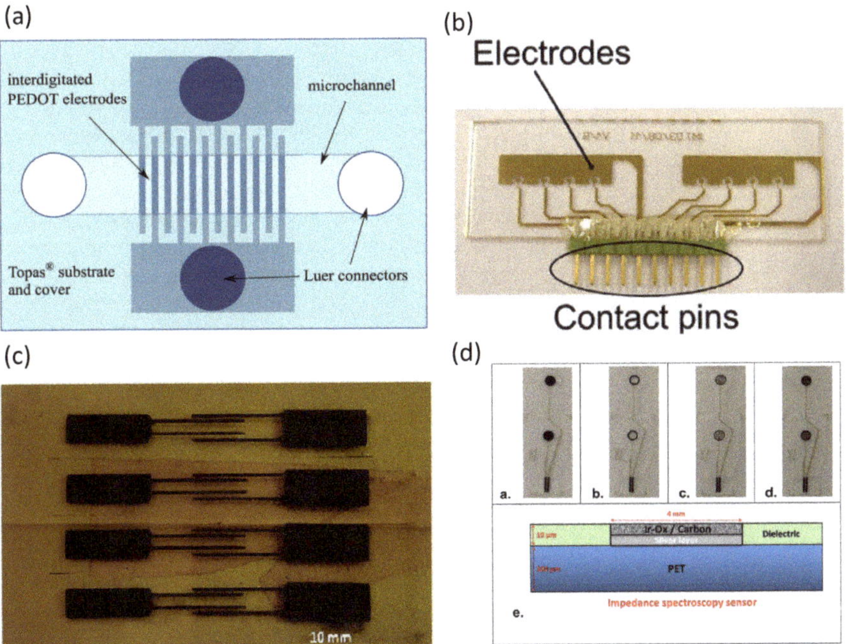

Figure 3. The polymer-based impedance biosensors. (**a**) All-polymer electrochemical microfluidic

biosensors are used. Reproduced with permission from [40]. Copyright Biosensors and Bioelectronics 2013. (**b**) Two microfluidic chambers were designed on a cyclic olefin copolymer substrate. Reproduced with permission from [41]. Copyright Biosensors and Bioelectronics 2013. (**c**) Kapton Polyimide (PI) was used as a substrate. Reproduced with permission from [47]. Copyright Biosensors and Bioelectronics 2022. (**d**) Microelectrodes were microfabricated on polyethylene terephthalate (PET) sheets. Reproduced with permission from [48]. Copyright Biosensors 2022.

The key to the use of polymer chips in microfluidic impedance biosensors is the combination of microelectrodes and polymer substrates. The sensitivity of the biosensors were determined by the size of the microelectrodes. Polymer chips are used similarly to glass chips, with both being hard and inelastic materials. On the other hand, polymer materials are suitable for mass production. Therefore, polymer chips can be one of the options for commercialization.

5. Glass-Based Impedance Biosensors

Microfluidic chips with glass as the substrate are very common. There have been some articles on materials that can replace glass, such as ITO glass, Pyrex glass, and SiO_2 glass. Therefore, the classification in this chapter is based on what the biosensors on glass can monitor. In this line, the subsections are classified regarding the detection of bacteria, blood, cells, DNA, proteins, toxins, and viruses.

5.1. Detection of Bacteria

Ruan et al. used electrochemical impedance spectroscopy to detect *E. coli* O157:H7 [50]. An anti-*E. coli* O157:H7 antibody was immobilized on the surface of an indium tin oxide (ITO) electrode. The binding of the antibody to the antigen changed the impedance signal. The limit of the sensor to detect bacteria was as low as 6×10^3 cells/mL. Yang et al. used an impedance sensor to detect Salmonella typhimurium and observed impedance changes during bacterial growth [51]. The material of the interdigitated electrode array was ITO. Four frequencies (10 Hz, 100 Hz, 1 kHz, and 10 kHz) were used to record impedance growth curves in the experiment. The impedance changes only when the bacterial count reaches 10^5–10^6 CFU/mL. Experimental data indicated that the greatest impedance change was observed at 10 Hz. $[Fe(CN)6]^{3-/4-}$ was used as a redox probe [52], which increases the electron transfer resistance on the antibody-immobilized microelectrode surface. Impedance immunosensors for the detection of Listeria were developed [53], for which TiO_2 nanowires were immobilized on gold electrodes, and monoclonal antibodies were immobilized on the nanowires. Listeria was then specifically captured using the antibodies. The impedance immunosensors sense impedance changes induced by the nanowire–antibody–bacteria system. Tan et al. designed a microfluidic impedance immunosensor for the detection of *Escherichia coli* and *Staphylococcus aureus* [54]; in particular, specific antibodies were immobilized on alumina nanoporous membranes, and bacteria were captured on the nanoporous membranes by antibodies. In a 2 h rapid assay, the sensitivity was as high as 10^2 CFU/mL. J. Lum designed an impedance biosensor, mainly for avian influenza virus subtype H5N1, as shown in Figure 4a [55]. Immunomagnetic nanoparticles and interdigitated microelectrodes were designed on a microfluidic chip. The polyclonal antibody was immobilized on the surface of the microelectrode, which binds to the avian influenza virus H5N1 to generate an impedance signal. Chicken red blood cells were used as biomarkers attached to the interdigital microelectrodes. The results showed that the impedance signal was increased by more than 100%. An impedance biosensor with focusing and sensing electrodes was designed to detect *E. coli* O157:H7 [56]. The sensitivity lower limit of detection for impedance sensor measurements was 3×10^2 CFU/mL. By focusing on p-DEP, the measurement sensitivity can be increased by 2.9 to 4.5 times. Couniot et al. integrated a CMOS process to design an impedance sensor for bacterial detection in urine, as shown in Figure 4b [57], where the detection of impedance spectroscopy was mainly directed at *Staphylococcus epidermidis*. Pyrex wafers were used as substrates for the impedance sensors. Liu et al. developed impedance-based biosensors that can be used to simultaneously detect Salmonella Serotypes B, D, and E by relying on three sensing IDE arrays, as shown in Figure 4c [58]. The sensor was designed with focusing electrodes to generate dielectrophoretic force, where the focusing force can increase the sensitivity by 4–4.5 times, and the detection limit is as low as 8 cells/mL. Other results indicated that the sensor could also distinguish between dead and live cells. Dastider et al. designed a microfluidic impedance sensor that can sense low concentrations of *E. coli* O157:H7 [59]. Upstream of the microchannel, the authors designed interdigitated focusing electrodes, in which the electrodes were arranged in a 45-degree-inclined manner. Positive DEP forces were used to focus cells in the center of the microchannel. Then, downstream of the microchannel, three sets of interdigital electrode arrays (IDEAs) were designed to sense impedance. An *E. coli* antibody was functionalized on the sensing electrode, which captured *E. coli* and resulted in a change in impedance. Experiments have shown that the microfluidic impedance sensor can detect coliform bacteria at a concentration of 39 CFU/mL. A microfluidic impedance sensor was used to detect Salmonella Serotypes B and D in food, as shown in Figure 4d [60]. There are two sensing areas in the chip, and the sensing electrodes are composed of interdigitated electrodes. The shocks were coated with antibodies against Salmonella. The experimental results demonstrate that the impedance sensor can detect Salmonella as low as 300 cells/mL.

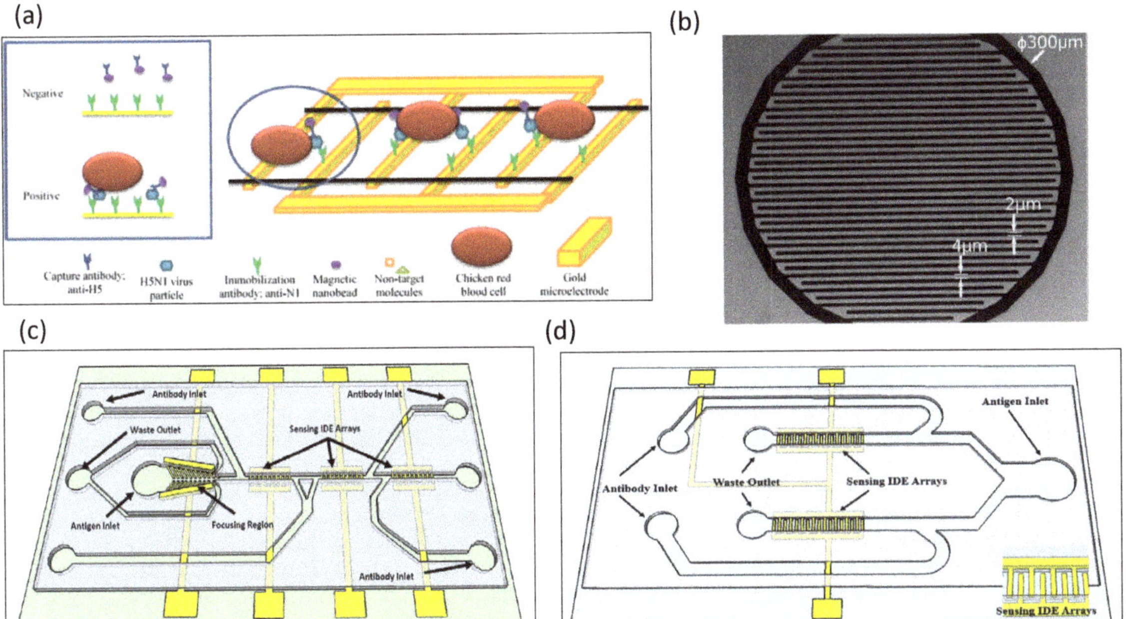

Figure 4. Glass-based impedance biosensors for detecting bacteria. (**a**) A polyclonal antibody was immobilized on the surface of a microelectrode and bound to avian influenza virus H5N1 to generate an impedance signal. Reproduced with permission from [55]. Copyright Biosensors and Bioelectronics 2012. (**b**) Pyrex wafers were used as substrates for impedance sensors. Reproduced with permission from [57]. Copyright Biosensors and Bioelectronics 2015. (**c**) An impedance-based biosensors were used to simultaneously detect Salmonella Serotypes B, D, and E. Reproduced with permission from [58]. Copyright Scientific Reports 2018. (**d**) There were two sensing areas in the chip, and the sensing electrodes were composed of interdigitated electrodes. Reproduced with permission from [60]. Copyright PLoS ONE 2019.

Many of the keys aspects in the design of impedance biosensors used to sense bacteria are related to improving the sensitivity. As the combination of antigen and antibodies can lead to changes in the impedance signal, many studies on antibodies immobilized on electrodes have been carried out.

5.2. Detection of Blood Samples

The manufacturing of electrical impedance measurement systems became possible with microfabrication technology. For example, wet etching was used to obtain microchannels on a coverslip [2], and gold electrodes were plated on both sides of the microchannel. The authors scanned the electric impedance spectroscopy of ionic salt solutions, air, and deionized (DI) water in the frequency range from 100 Hz to 2 MHz. Their experimental results showed the efficacy of electrical impedance spectroscopy for human polymorphonuclear leukocytes and teleost fish red blood cells. Mishra et al. used three microelectrodes on a chip to sense the impedance of human CD4(+) cells in blood, as shown in Figure 5a [61]. The reference electrode, working electrode, and counter electrode were microfabricated on glass wafers. As the protein adsorbed onto the microelectrode surface, the detected impedance value increased even more. The impedance also increased with the number of captured cells. Kuttel et al. used impedance spectroscopy to detect red blood cells infected with *Babesia bovis* [62]. The change in impedance was mainly due to the presence of the parasite in the cell changing the impedance value of the original red blood cells, white blood cells, or platelets. Therefore, infected cells could be easily and quickly distinguished from healthy cells. In areas without good infrastructure or in very remote areas, the use of impedance spectroscopy to detect parasites in whole blood samples can greatly reduce the time of diagnosis for medical personnel. Holmes et al. developed a high-speed microfluidic single-cell impedance cytometer using dual frequency for whole blood analysis [63], which is mainly used for the impedance measurement and identification of T lymphocytes, monocytes, and neutrophils. The experiments showed that, at low frequencies, T lymphocytes and neutrophils can be distinguished. The cells were conjugated with fluorescently labeled antibodies, allowing the system to analyze fluorescence and impedance simultaneously. Han et al. developed a microfluidic chip that integrates red blood cell lysis with a microfluidic impedance cytometer, as shown in Figure 5b [64]. Their laboratory developed a buffer that not only lyses red blood cells but also increases the identification of monocytes and neutrophils. The method for multi-step cell lysis described in this paper is of great help for the microfluidic system, in terms of whole blood analysis. Lei et al. designed an electrical impedance to monitor the blood coagulation process in a microfluidic chip, which can obtain impedance results consistent with clinical reports at different temperatures and blood cell counts, as shown in Figure 5c [65]. This device provides a new analytical method for the sensitive and real-time monitoring of coagulation in whole blood samples. Song et al. developed a microfluidic impedance flow cytometer to identify undifferentiated and differentiated mouse embryonic stem cells [66], where two micropores and three electrodes were designed in the chip. The experimental results indicated that undifferentiated stem cells and polystyrene spheres could be distinguished at any frequency, while undifferentiated and differentiated stem cells require higher frequency and opacity to be distinguished. Du et al. designed an electrical impedance flow cytometer targeting red blood cells infected with *Plasmodium falciparum* [67]. The physiological and electrical properties of erythrocytes were altered 48 h after *P. falciparum* infection. In addition to the cytometer, the authors incorporated new offset parameters to make it easier to distinguish infected erythrocytes from uninfected erythrocytes. Spencer et al. used a microfluidic impedance cytometer for the detection of a representative circulating tumor cell (the MCF7 tumor cell line) [68]. The red blood cells were removed by lysis, and the buffer did not affect the dielectric properties of the MCF7 cells. Through impedance analysis, MCF7 cells were shown to have a larger size and membrane capacitance. The experimental results indicated that 100 MCF7 cells could be detected in 1 mL of whole blood. The average recovery rate was as high as 92%. Liu et al. developed an electrical impedance microflow cytometer that can control oxygen flow for the analysis of sickle blood cells, as shown in Figure 5d [69]. The two-layer microfluidic channel was separated by a 150 µm thick PDMS film. The upper layer is a serpentine gas channel that controls oxygen, while the lower layer is a microchannel through which sickle cells and red blood cells flow. Ti/Au electrodes were designed to measure the impedance of sickle cells in the lower microchannel. Under normoxic conditions, the authors distinguished between normal and sickle cells using impedance signals measured at intermediate frequencies. The same research demonstrated that impedance signals can be obtained without the need for hemolysis. Their experimental results also proved that the impedance signal of the microflow cytometer can be used as an indicator of red blood cell disease and sickle cell disease.

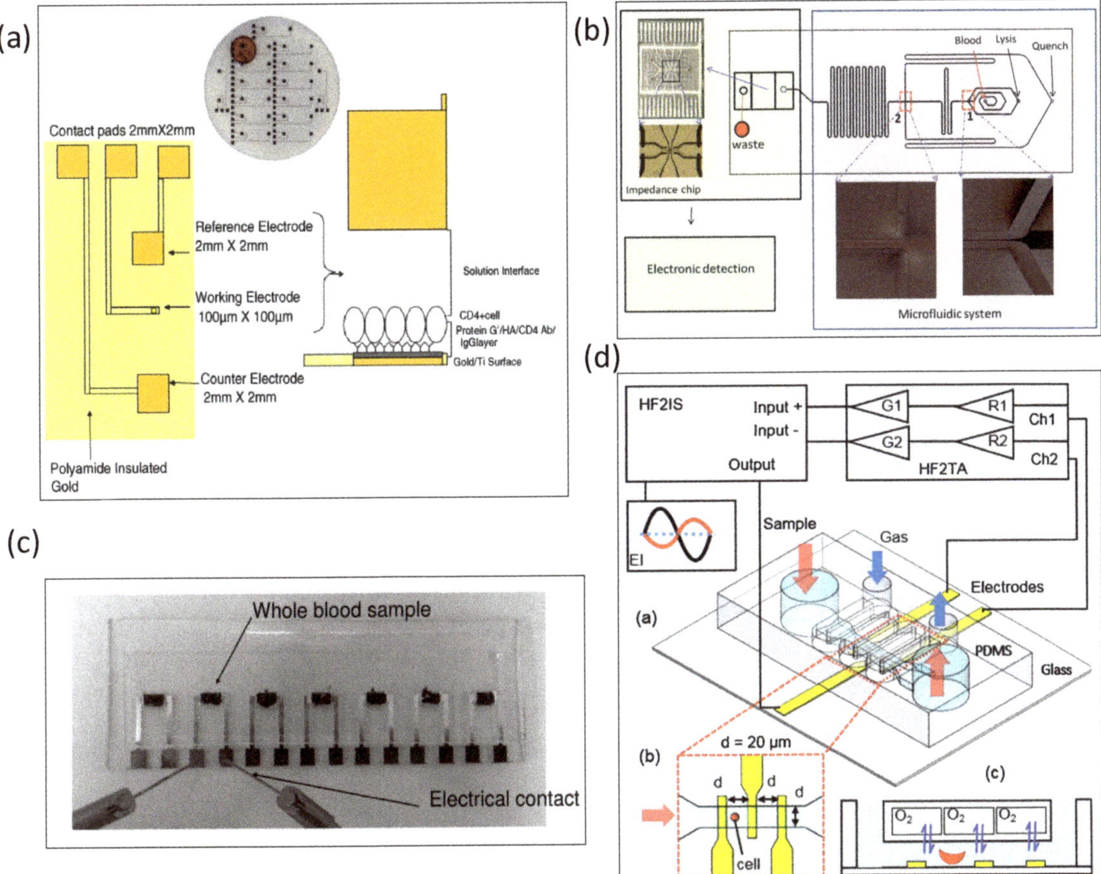

Figure 5. Glass-based impedance biosensors for detecting blood. (**a**) Three microelectrodes were used to sense the impedance of human CD4(+) cells in blood. Reproduced with permission from [61]. Copyright Biosensors and Bioelectronics 2005. (**b**) A microfluidic impedance cytometer was for red blood cell lysis. Reproduced with permission from [64]. Copyright Analytical Chemistry 2011. (**c**) A microfluidic chip was designed to detect electrical impedance for monitoring the blood coagulation process. Reproduced with permission from [65]. Copyright PLoS ONE 2013. (**d**) An electrical impedance microflow cytometer with controlled oxygen flow for the analysis of sickle blood cells. Reproduced with permission from [69]. Copyright Sensors and Actuators B: Chemical 2018.

5.3. Static Cell Analyzed by Electrical Impedance Spectroscopy

A microfluidic impedance sensor was designed to measure the cell migration of cancer cells in a three-dimensional extracellular matrix [70]. A total of 16 sets of sensing electrode arrays and cell grabbing arrays were designed in the microchannel. Under continuous monitoring, the migration of MDA-MB-231 cells allowed for a rapid change in impedance amplitude (of about 10 Ω/s). Liu et al. designed a microfluidic chip with embedded measurement electrodes to monitor the cell migration process using impedance measurement technology [71]. Cells were measured and recorded in the microfluidic channel as they pass through multiple parallel electrodes. This method enables the accurate and objective recording of cell migration activity and the calculation of migration rates among different stimulating drugs. Huang et al. designed a microchannel filled with Matrigel to quantify cell migration velocity as an assay tool, as shown in Figure 6a [72]. The successful measurement of cells suspended in a 3D environment and the induction of cell migration by stimulatory factors were used to record the migration speed of cells. The measurement sensitivity was better than that of a traditional trans-well assay.

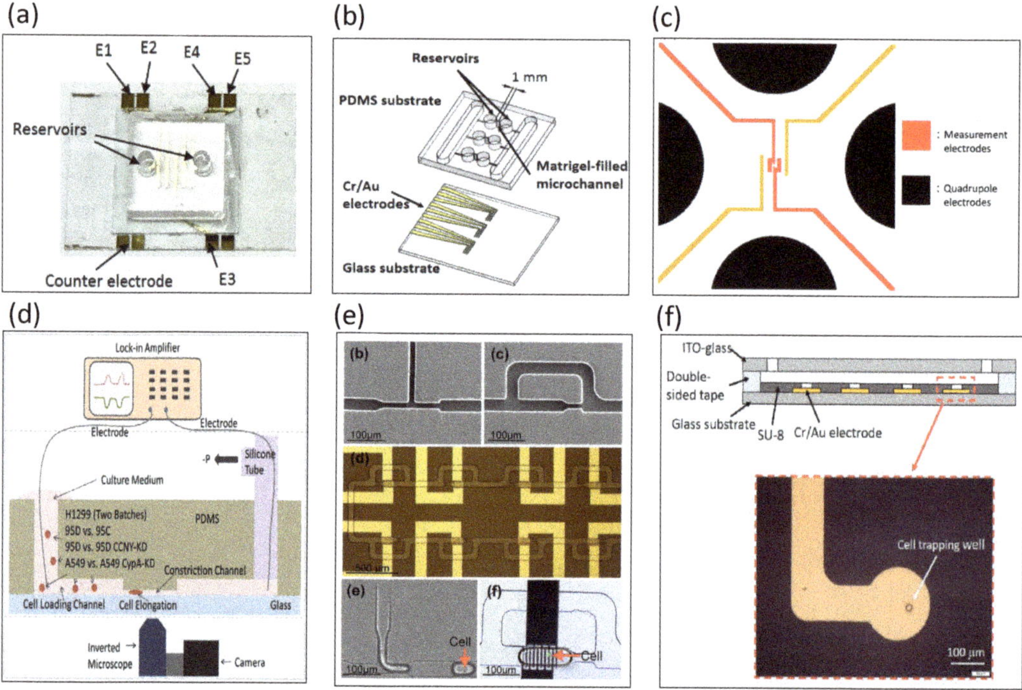

Figure 6. Static cell analyzed with electrical impedance spectroscopy. (**a**) A microchannel was designed filled with Matrigel to quantify cell migration velocity. Reproduced with permission from [72]. Copyright Analytica Chimica Acta 2020. (**b**) This chip was designed to induce angiogenesis to extend into the microchannel. Reproduced with permission from [73]. Copyright Sensors and Actuators B: Chemical 2022. (**c**) The relationship between the electrochemical impedance spectroscopy and withstand voltage was established with four different cells (HeLa, A549, MCF-7, and MDA-MB-231). Reproduced with permission from [74]. Copyright Sensors and Actuators B: Chemical 2012. (**d**) Because of the constriction of the microchannel, tumor cells, thus, are elongated for sensing impedance. Reproduced with permission from [75]. Copyright Biosensors and Bioelectronics 2014. (**e**) A microfluidic impedance sensing chip with droplets and microelectrode arrays was used to monitor the osteogenic differentiation of bone marrow mesenchymal stem cells. Reproduced with permission from [76]. Copyright Biosensors and Bioelectronics 2019. (**f**) Single stem cells were captured in a 20 µm chamber by dielectrophoresis. Reproduced with permission from [77]. Copyright Talanta 2021.

Lei et al. developed a perfusion three-dimensional (3D) cell culture microfluidic chip combined with real-time and non-invasive impedance monitoring [73]. This device can simulate complex 3D biological microenvironments to culture cells and monitor the impedance changes under different concentrations of drug stimulation through impedance measurements. The impedance results are analyzed to determine the cell proliferation and chemosensitivity of 3D cell cultures. Lei et al. designed an impedance measurement device for cell colonies cultured on hydrogels [74–76]. Huang et al. constructed a 3D biological barrier using Matrigel and induced angiogenesis to extend into the microchannel, as shown in Figure 6b [77]. The angiogenesis process could be monitored by label-free impedance, using electrodes at the bottom of the microchannel. The device can also successfully quantify the time and distance of angiogenesis, thereby providing a reliable and quantitative method for the assay of angiogenesis.

Bieberich et al. developed electrical cell impedance spectroscopy to monitor the impedance response of PC12 and embryonic stem cells forming synapses [78]. Jang et al. published a study combining a cell capture method with microfabricated impedance spectroscopy [79], in which three micropillars were designed in the center of the microchannel to capture single HeLa cells. Cho et al. detailed the integrated microfluidic capture of single-cell technology with electrical impedance spectroscopy [80]. Hildebrandt et al. developed electrochemical impedance spectroscopy to distinguish the osteogenic differentiation of human mesenchymal stem cells [81]. In the application of cellular impedance to infectious parasites, Houssin et al. designed an electrochemical impedance spectroscopy approach to detect the presence of oocysts [82]. Dalmay et al. developed impedance spectroscopy to distinguish cancer stem cells and U87 glial cells (differentiated cells) [83]. The impedance spectrum designed by Bagnaninchi et al. can instantly monitor adipose stem cell (ADSC) differentiation [84]. Hong et al. established the relationship between electrochemical impedance spectroscopy and the withstand voltage of four different cells (HeLa, A549, MCF-7, and MDA-MB-231), as shown in Figure 6c [85]. Under a strong electric field, the cytoplasmic resistance decreases due to the opening of ion channels. The experimental results showed that different cells have not only different impedance spectra but also different withstand voltages. Chen et al. designed a microfluidic chip for capturing single cells and measuring impedance values [86]. Zhao et al. proposed to convert the measured impedance values into membrane capacitance (C-specific membrane) and cytoplasmic conductivity (σ cytoplasm), as shown in Figure 6d [87]. Due to the constriction of the microchannel, tumor cells become elongated. The experimental results demonstrated that tumor cells can be distinguished using two parameters: C-specific membrane and sigma cytoplasm. Ruan et al. integrated dielectrophoretic force and impedance sensors to detect lung circulating tumor cells [88]. Fan et al. designed a microfluidic impedance sensing chip with droplets and microelectrode arrays to monitor the osteogenic differentiation of bone marrow mesenchymal stem cells, as shown in Figure 6e [89]; the authors also proposed a model of cellular droplets. Lei et al. captured single stem cells in a 20 μm chamber by dielectrophoresis, as shown in Figure 6f [90]; the cells were measured in the impedance spectrum range of 2–20 kHz.

5.4. Dynamic Cell Analyzed by Microfluidic Impedance Cytometry

The development of microfluidic impedance flow cytometry is important for cell analysis. In this subsection, microfluidic impedance flow cytometry is divided into two parts for discussion: one is the principle of microfluidic impedance flow cytometry, and the other includes the sensing applications of microfluidic impedance flow cytometry.

Sun et al. designed two microfluidic impedance cytometers with parallel facing electrodes and coplanar electrodes [91]. For impedance measurements, parallel facing electrode designs are more sensitive than coplanar electrode designs. Holmes et al. used a microfabricated flow cytometer for the discrimination of micron-sized polymer beads [92]. Fluorescently labeled proteins were immobilized on the beads, which can be used to analyze the immune response. Negative dielectrophoretic force was used to focus the polymer beads by the focusing electrodes. An electrical impedance flow cytometer for the high-speed analysis of particles was developed [93]. The impedance signal of polystyrene beads could be obtained in as little as 1 ms. Compared with microchannels fabricated by soft lithography, Kummrow et al. used ultraprecision milling technology to design 3D microchannels with horizontal and vertical focusing capabilities [94]. Fiber optics, mirrors, and electrodes were integrated into a flow cytometer for blood cells. Spencer et al. conducted a study on how the position of particles in a microchannel affects impedance measurements [95]. Impedance is related to the position of the particle in the vertical direction. A flow cytometer was designed by Barat et al. in order to measure both the optical and electrical properties of particles [96]. Daniel Spencer integrated optical fibers and waveguides into an impedance flow cytometer to measure electrical impedance (electrical volume and opacity), fluorescence, and large-angle side scatter without the need for particle focusing [97]. Haandbæk et al. published an article on microfabricated flow cytometry at high frequencies [98]. The experimental results indicated the ability at high frequencies to distinguish not only wild-type yeast and mutant strains but also opacity values at frequencies above 50 MHz.

David et al. used impedance flow cytometry to measure the viability and membrane potential of *Bacillus megaterium* cells [99]. A microfluidic impedance cytometer was designed for platelet analysis by Evander et al. [100], where the focusing electrodes allow for secondary focusing of the sample through the dielectrophoretic force. Lin et al. used a microfluidic impedance cytometer to detect quantified protein biomarkers [101]. For whole blood analysis, Simon et al. developed a microfabricated AC impedance cytometer with multi-frequency AC impedance and side scatter analysis capabilities [102]. Xie et al. proposed the concept of using a microfabricated impedance cytometer to detect electronic biomarkers [103]. By changing the dielectric properties of the particle, the authors designed a nanoelectronic barcode particle as an electronic biomarker. McGrath et al. used a microfluidic impedance flow cytometer to distinguish the oocysts of protozoan parasites, as shown in Figure 7a [104]. This chip can distinguish between *Cryptosporidium parvum*, *Cryptosporidium muris*, and *Giardia lamblia* within minutes. A microfluidic impedance cytometer integrating inertial focusing and liquid electrodes was developed for the high-throughput measurement of human breast tumor cells and leukocytes, as shown in Figure 7b [105]. The purpose of inertial focusing is to reduce cell adhesion and ensure that single cells pass through the sensing area. An interesting study involved the use of two microneedles, placed on either side of a microchannel to sense impedance values [106]. For clinical analysis and judgment, Sun et al. used multi-frequency impedance spectroscopy and machine learning to rapidly distinguish the survival of cancer cells under the action of anti-matriptase-conjugated drugs [107]. Sui et al. developed an impedance flow cytometer to detect spheroid green algae cells (Picochlorum SE3) at different salt concentrations, as shown in Figure 7c [108]. Mahesh et al. published a study on the observed "double-peak" characteristics of individual cells with high sensitivity to changes in cell membrane capacitance [109]. This phenomenon has limitations: it operates at the lower frequencies (400–800 kHz) of the beta-dispersion mechanism, and the microelectrodes must be coplanar and paired. The authors pointed out that changes in cell size and membrane capacitance can be resolved using a single frequency. A microfluidic impedance cytometer was used for the analysis of antigen-specific T lymphocytes, as shown in Figure 7d [110]. The experimental results demonstrated that differences in impedance can be observed among dead, healthy, and activated lymphocytes. Caselli et al. used artificial intelligence methods to decipher signals from microfluidic impedance cytometers [111]. The authors demonstrated two advances: (i) the use of a neural network to determine the dielectric properties of single cells in raw impedance data streams and (ii) resolving the impedance signatures of coincident cells. The results demonstrated that the neural network could increase the signal processing capability of the microfluidic impedance cytometer.

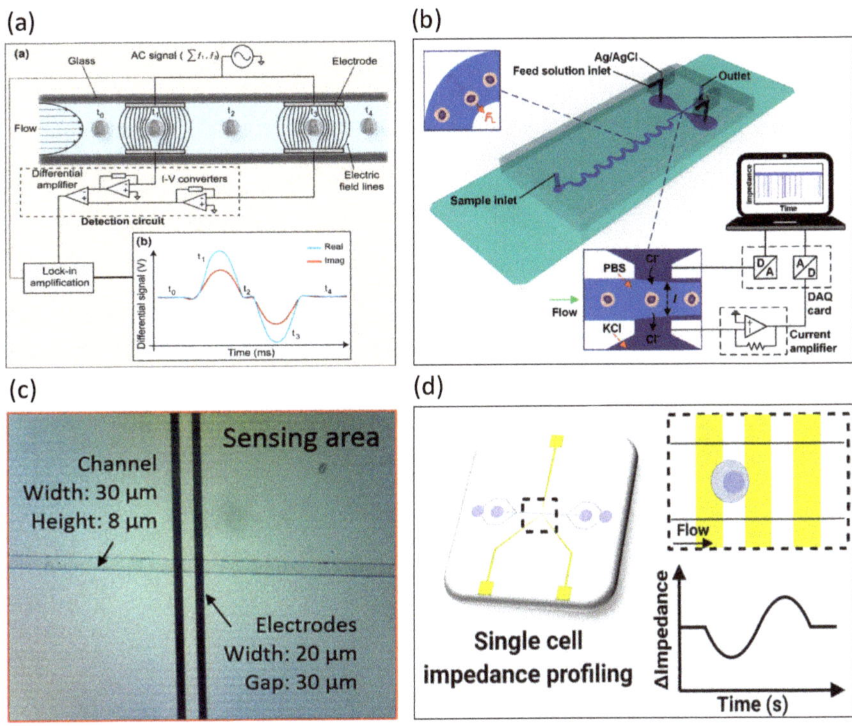

Figure 7. Dynamic cell analyzed by microfluidic impedance cytometry. (**a**) A model of the impedance flow cytometer was established to distinguish Cryptosporidium parvum, Cryptosporidium muris, and Giardia lamblia in minutes. Reproduced with permission from [104]. Copyright Scientific Reports 2017. (**b**) A microfluidic impedance cytometer integrating inertial focusing and liquid electrodes was developed. Reproduced with permission from [105]. Copyright Analytical Chemistry 2017. (**c**) An impedance flow cytometer was developed to detect spheroid green algae cells (Picochlorum SE3) at different salt concentrations. Reproduced with permission from [108]. Copyright Scientific Reports 2020. (**d**) A microfluidic impedance cytometer was used for the analysis of antigen-specific T lymphocytes. Reproduced with permission from [110]. Copyright Sensors and Actuators B: Chemical 2021.

5.5. Detection of Viruses

The key to the use of impedance biosensors for virus sensing is the immobilization of antibodies, where the antibody must be immobilized on the surface of the electrode first. However, the impedance signal will be altered due to the binding of both antibodies and antigens.

Pathogenic avian influenza viruses have been detected by impedance sensors [112]. The polyclonal antibody was immobilized on the surface of a gold microelectrode by protein A. The antibody–antigen binding reaction in the same research could be amplified by red blood cells, and the proposed impedance immunosensor could be completed in as little as 2 h. A portable impedance biosensor for avian influenza virus H5N2 was developed by Wang et al. [113]. Magnetic nanobeads and interdigitated array microelectrodes were integrated on a microfluidic chip, where the magnetic nanobeads captured avian influenza virus subtype-specific antibodies. The entire detection time was less than 1 h, which was greatly reduced compared with traditional avian influenza virus detection. The experimental results indicate that the sensitivity of the impedance biosensor is comparable to that of real-time reverse transcriptase PCR. Jacob Lum designed an impedance biosensor, mainly for avian influenza virus subtype H5N1 [55]. For this purpose, immunomagnetic nanoparticles and interdigitated microelectrodes were designed on a microfluidic chip. The polyclonal antibody was immobilized on the surface of the microelectrode, which binds to avian influenza virus H5N1 to generate an impedance signal. Chicken red blood cells were used as biomarkers attached to interdigital microelectrodes. The results showed that the impedance signal can be increased by more than 100%. For the detection of human immunodeficiency virus, Shafiee et al. designed a microfluidic sensing chip to measure the impedance spectrum of virus nanolysates [114]. Two gold electrodes were designed on both sides of the microchannel cavity. The lysis of the virus results in the release of the ions and charged molecules of the virus into a non-ionic solution. That research demonstrated that impedance spectroscopy provides a convenient and rapid tool for the detection of multiple pathogens. Singh et al. developed an electrical impedance sensing chip to detect influenza H1N1 virus, as shown in Figure 8a [115,116], for which three microelectrodes were fabricated on a glass substrate. The authors used reduced graphene oxide and monoclonal antibodies as electrode modifications.

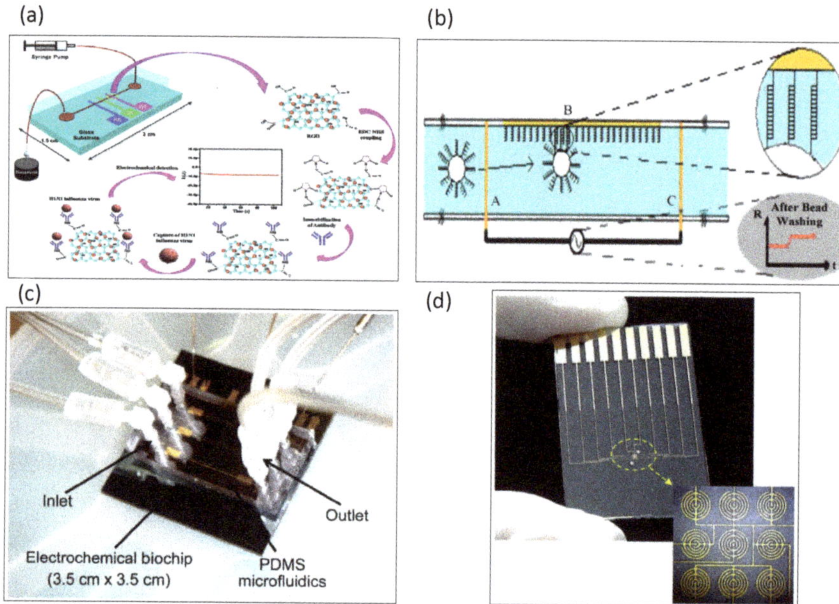

Figure 8. Glass-based impedance biosensors for detecting other organisms and chemicals. (**a**) An electrical impedance sensing chip was developed to detect influenza H1N1 virus. Reproduced with permission from [115]. Copyright Scientific Reports 2017. (**b**) Impedance sensors were developed for DNA hybridization. Reproduced with permission from [117]. Copyright Sensors and Actuators B: Chemical 2011. (**c**) Single-stranded DNA probes were functionalized onto electrodes. Complementary DNA hybridization was then induced using electrochemical impedance spectroscopy. Reproduced with permission from [118]. Copyright Biosensors and Bioelectronics 2012. (**d**) Galectin-1 protein is a biomarker of bladder cancer. An impedance immunosensor was developed to detect Galectin-1 protein in urine. The dielectrophoretic force was used to capture Galectin-1 antibody to improve the sensitivity of the sensor. Reproduced with permission from [119]. Copyright Biosensors and Bioelectronics 2016.

5.6. Detection of Other Analytes and Chemicals

Berdat et al. used an impedance sensor based on an interdigitated microelectrode array to sense DNA [116], for which 5 μm wide microelectrodes were fabricated using a lift-off process method. The complementary probe was first immobilized on the electrode and hybridized to the target ssDNA. Finally, impedance sensors were used to detect the pathogen *Salmonella choleraesuis* in dairy products. Javanmard et al. developed a set of impedance sensors for DNA hybridization, as shown in Figure 8b [117]. Oligonucleotide probes were immobilized on the surface of the microchannel, and target DNA strands were immobilized on the surface of polystyrene beads. Contact between the probe and the target DNA strand results in the hybridization of the DNA, leading to the capture of the polystyrene beads on the surface of the microchannel. An impedance chip for sensing DNA hybridization was developed by Hadar et al., as shown in Figure 8c [118]. Single-stranded DNA probes were functionalized onto electrodes, and complementary DNA hybridization was then induced using electrochemical impedance spectroscopy. DC-biased AC electro-osmotic vortex was utilized to design a label-free electrochemical impedance spectroscopy (EIS)-based DNA biosensing chip [119]. Based on the electro-osmotic vortex, 20-base target DNA fragments were hybridized to achieve 90% within 141 s. The ultrasensitive detection limit was 0.5 aM. Another study indicated that the electric field was manipulated by alternating current (AC) electrokinetics to improve hybridization efficiency and reduce hybridization time [120]. Thus, the chip was realized for faster and more efficient detection.

Galectin-1 protein is a biomarker of bladder cancer. An impedance immunosensor for detecting bladder cancer in urine was developed, as shown in Figure 8d [121]. Before measuring the impedance signal, the authors used dielectrophoretic force to capture nanoprobes (Gal-1 antibody) on the surface of the microelectrode, in order to improve the sensitivity of the sensor. Alsabbagh et al. designed a microfluidic impedance biosensor for the detection of myocardial infarction proteins by electrochemical impedance spectroscopy [122]; in particular, Troponin I, which is a biomarker for the diagnosis of myocardial infarction, was targeted. Self-assembled thiolated oligonucleotides tested on gold electrodes were found to perform better, as they improved the performance of the impedance signal. Fluorescence analysis and electrochemical impedance spectroscopy were integrated to measure aggregated C-reactive proteins [123]. The circular array of electrodes was designed to create electrokinetic flow for C-reactive protein aggregation. Interdigitated microelectrode arrays were modified by the self-assembled monolayers of mercaptocaproic acid for detecting the arthritis anti-CCP-ab biomarker [124]. The experimental results showed that the sensor response increased linearly with the stepwise increase of the biomarker concentration. A polyaniline (PANI)/MoS2-modified screen-printed electrode was detected for anti-cyclic citrullinated peptide [125]. Among them, the polymerized PANI-Au nanomatrix was utilized to entrap the aCCP antibodies for amplification of the higher signal. A peptide-based electrochemical sensor was used to detect autoantibodies for the diagnosis of rheumatoid arthritis [126]. The developed peptides were modified on the gold surface of the working electrode by a self-assembled monolayer method. Subsequently, the sensor was clinically demonstrated to have better sensitivity using 10 clinically validated samples from rheumatoid arthritis patients and 5 healthy control samples.

Chiriaco et al. designed a microfluidic electrochemical impedance biosensor to detect cholera toxin [127] with sensitivity less than 10 pM. Liu et al. developed a biosensor for the rapid screening of toxic substances in drinking water [128]. Cells are damaged due to toxins in water, resulting in decreased impedance and an increase in resonance frequency. Impedance and mass sensing measurements can help to improve sensor accuracy. An impedance biosensor was used to detect the cytotoxicity of tamoxifen in cervical cancer cells [129]. The experimental results showed that the dose of tamoxifen resulted in a significant reduction in the number of HeLa cells. The same article demonstrated that impedance biosensors can be used for the evaluation of novel drugs and cytotoxicity.

A microfluidic impedance sensor was designed for pesticide detection in vegetables [130]. Anti-chlorpyrifos monoclonal antibody was immobilized on an interdigitated electrode array. The capture of chlorpyrifos produced a change in impedance. Zeng et al. integrated magnetic focusing into impedance microsensors for oil monitoring [131]. The highly focused magnetic field was derived from two electromagnetic coils and eight silicon steel tips, where the silicon steel tips greatly improved the sensitivity of the sensor.

5.7. Conclusions of the Glass Chips

Microfluidic impedance sensor glass chips have two great development directions: electrochemical impedance spectroscopy and electrical impedance flow cytometry. The difference between the two lies in the state of the measured object. For example, in an electrical impedance flow cytometer, as shown in Figure 6, the cells are stationary. The cells may be attached to the bottom of a microgroove or a microchamber, where the culture medium is quiescent. On the other hand, cells may be trapped by the microstructures and focused by the dielectrophoretic force, thus, fixing them in one position. In an electrical impedance flow cytometer, the cells are in a dynamic state, as shown in Figure 7. For the impedance sensing of other organisms, the main technology relies on the modification of electrodes and the combination of antigens and antibodies.

6. Paper-Based Impedance Biosensors

Lei et al. combined the electrical impedance measurement technique with the hydrophilic properties of a paper base into a system for recording and analysis [132]. Through impedance analysis, a trend proportional to cell proliferation was observed. A paper-based electrochemical impedance DNA sensor for tuberculosis detection was developed by Teengam et al., as shown in Figure 9a [133]. Carbon graphene inks were printed as working and counter electrodes, while reference electrodes and conductive pads were screen-printed with silver/silver chloride ink. Pyrrolidine peptide nucleic acid (acpcPNA) was immobilized on cellulose paper, and changes in impedance were induced in the presence of *Mycobacterium tuberculosis*. Rengaraj et al. developed paper-based electrodes for the impedance detection of bacteria in water, as shown in Figure 9b [134]. The paper-based electrodes were made of carbon on hydrophobic paper formed by screen printing. The cellulose was cross-linked before use in order to enhance the strength of the paper substrate and the electrical properties of the screen-printed electrodes. The same article was the first to combine the hydrophobicity of paper substrates with the electrochemical functionalization of electrodes. A paper-based microfluidic impedance chip was developed to sense alpha-fetoprotein in human serum using peptide-modified plastic paper [135]. The sensor included two layers, where the upper layer was cellulose chromatography paper, and the lower layer was plastic paper. Among them, the sensing electrodes were printed with Ag-20 wt % graphene. The limit value of alpha-fetoprotein in PBS detected by the sensor was 1 ng/mL. Lei et al. used hydrogel to encapsulate cells and then cultured them on top of paper substrates [136]. The resulting analysis could more accurately distinguish the impedance differences between two cells and under the action of a drug. Some studies have used electrochemical impedance spectroscopy to detect miRNA-34a, which is a biomarker of cancer and Alzheimer's disease [137]. PAMAM dendrimers were modified on the surface of screen-printed electrodes. The results indicated that the difference between miRNA-34a and miRNA-15 or miRNA-660 could be distinguished through electrochemical impedance spectroscopy. Cardiac troponin I is a biomarker for the early diagnosis of acute myocardial infarction. A paper-based impedance immunosensor was developed to detect cardiac troponin I [138], for which multi-walled carbon nanotubes were immobilized on carbon ink electrodes. Then, cardiac troponin I antibody was immobilized on multi-walled carbon nanotubes. Finally, cardiac troponin I was captured by the antibody, affecting the impedance signal. Li et al. designed paper-based electrochemical impedance spectroscopy to detect coronavirus (COVID-19), as shown in Figure 9c [139]. The carbon ink was first printed on paper, and a layer of zinc oxide nanowires was grown on the carbon ink. Then, probes and blocking molecules were immobilized on its surface. Finally, the target molecules were captured for impedance sensing. The results showed that the enhanced ZnO nanowires can improve impedance sensitivity. Electrochemical impedance spectroscopy was used to analyze artificial sweat using hand-painted electrodes [140]. Electrodes were drawn on the opposite side of the paper in order to reduce the double-layer capacitance. The silver electrode pattern paper chip had stronger impedance stability than the graphite electrode paper chip. Research and development of paper-based electrochemical impedance spectroscopy was conducted for the detection of microRNA 155 [141], for which gold nanoparticles changed the properties of paper-based electrodes. S. Karuppiah developed a paper-based impedance biosensor for monitoring bacteria in water [142]. The electrodes of the biosensor were screen-printed with graphene (G) and then surface-modified with graphene oxide (GO). The lectin concanavalin A was then immobilized on the modified electrode described above. A paper-based impedance sensor was used to sense miRNA-155 and miRNA-21 for the early diagnosis of lung cancer, as shown in Figure 9d [143]. The authors used MoS2 crystals and MoS2 nanosheets to modify the paper-based electrodes. Paper-based electrochemical impedance spectroscopy was used to detect foodborne pathogens (Listeria) [144]. Tungsten disulfide nanostructures were used as paper-based electrodes for impedance sensors.

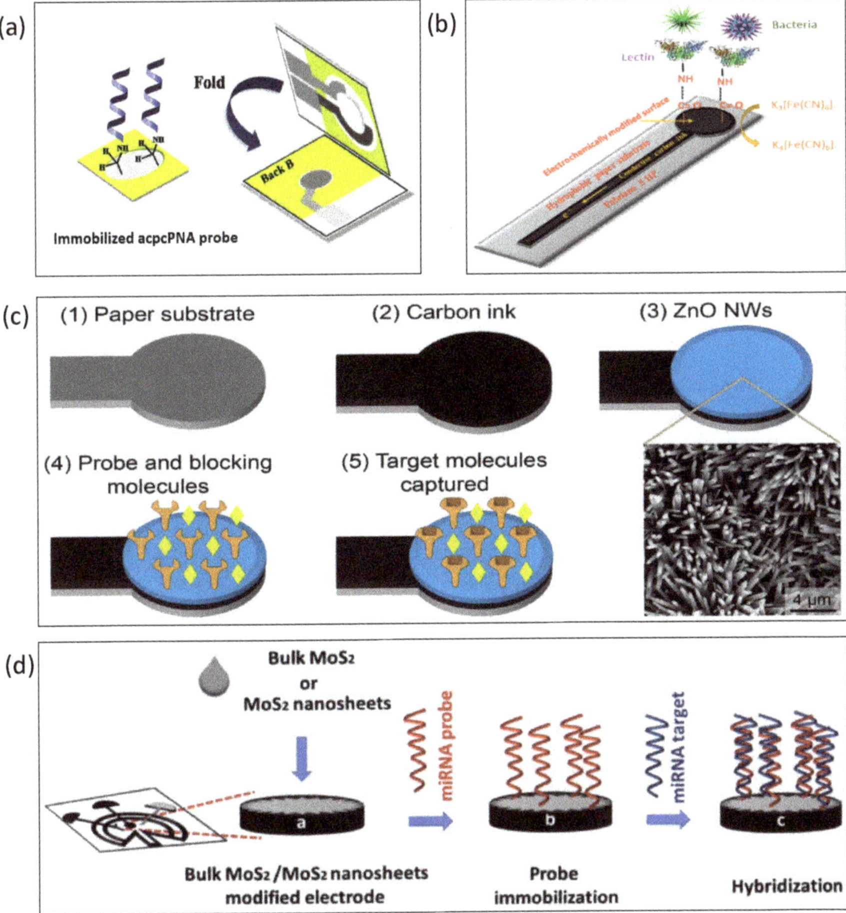

Figure 9. Paper-based impedance biosensors. (**a**) A paper-based electrochemical impedance DNA sensor was developed for tuberculosis detection. Reproduced with permission from [133]. Copyright Analytica Chimica Acta 2018. (**b**) Paper-based electrodes for impedance sensors were used to detect the bacteria in water. Reproduced with permission from [134]. Copyright Sensors and Actuators B: Chemical 2018. (**c**) A paper-based electrochemical impedance spectroscopy was designed to detect coronavirus (COVID-19). Reproduced with permission from [139]. Copyright Biosensors and Bioelectronics 2021. (**d**) Paper-based impedance sensor was used to sense miRNA-155 and miRNA-21 for early diagnosis of lung cancer. Reproduced with permission from [143]. Copyright Talanta 2022.

Paper-based chips have great advantages due to their low material cost. Microelectrodes with small line widths on paper-based chips and antibodies combined onto the microelectrodes will be key technologies for the development of paper-based chips.

7. Stretchable Biosensors

Furniturewalla et al. designed a microfluidic impedance cytometer on a flexible circuit board in the form of a portable wristband [145]. Lock-in amplification, a microfluidic biosensor, a microcontroller, and a Bluetooth module are integrated into the wristband. Flexible and stretchable biosensors for skin physiological parameter monitoring have been developed [146], where screen printing was used to fabricate sensing electrodes in flexible and stretchable conductive materials originally intended for epidermal tattooing. A retractable body biosensor for sensing the biomarker cortisol in sweat was published, as shown in Figure 10a [147]. A pullable body impedance biosensor was designed at the bottom layer and attached to the skin, following which microfluidic microvalves and microchannels were applied to this wearable patch. A stretchable microfluidic immunobiosensor patch was used for sensing neuropeptide Y in human sweat [148]. Conductive microfibers that can be stretched help to improve the sensitivity of the biosensor patches. Sensors attached to the skin can detect biomarker concentrations in human sweat at levels as low as fm. A wearable microfluidic impedance immunosensor for sweat cortisol detection was designed, as shown in Figure 10b [149], where microfluidic channels and chambers were integrated into the wearable patch, and Ti3C2Tx MXene nanosheets were incorporated in the porous structure of graphene. The wearable microfluidic impedance immunosensor could detect cortisol down to 88 pM.

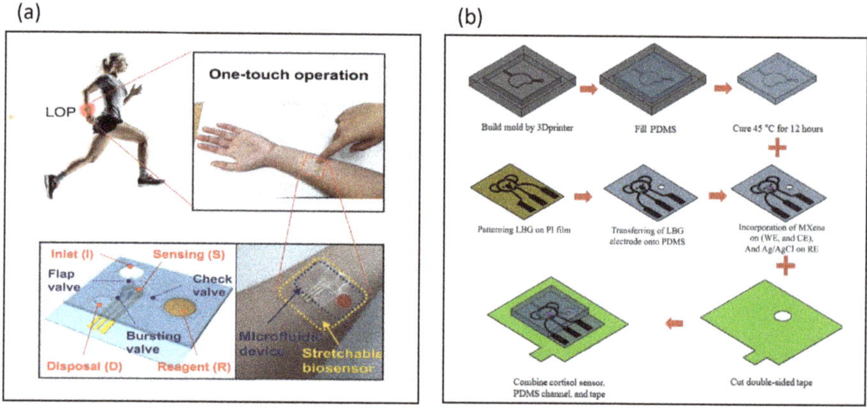

Figure 10. Stretchable impedance biosensors. (**a**) All-polymer electrochemical microfluidic biosensors were used for sensing the biomarker cortisol in sweat. The pullable body impedance biosensor was designed at the bottom layer and attached to the skin. Reproduced with permission from [147]. Copyright Biosensors and Bioelectronics 2020. (**b**) A wearable microfluidic impedance immunosensor was designed for sweat cortisol detection. Microfluidic channels and chambers were integrated into the wearable patch. Reproduced with permission from [149]. Copyright Sensors and Actuators B: Chemical 2021.

Stretchable chips for microfluidic impedance sensors comprise a novel research direction. At present, impedance sensing signals are mostly measured on the skin. The stretchability and bendability of such materials are key features, especially the flexibility of the electrode materials. In this line, conductive hydrogels may be a helpful technology.

8. Conclusions

Impedance biosensors integrated with microfluidic technology are powerful tools for understanding electrical information in microscopic and sub-microscopic organisms. The integrated sensors have the key characteristics of improved sensitivity, reduced reagent consumption, short analysis time, reduced instrument size, and simple operation. In this paper, the developed microfluidic impedance sensors were classified into six categories: silicon chip-, PCB chip-, polymer chip-, glass chip-, paper chip-, and stretchable chip-based. No matter which type of chip is to be developed, the microfabrication of microelectrodes and the bonding of chemicals are key technologies.

Silicon-based chips are made by the MEMS process developed by Taiwan Semiconductor Manufacturing Co., Ltd., which can become the commercialization direction of electrical impedance biosensors. The development of PCB chips was limited by the line width of the electrodes, which makes it impossible to improve sensitivity. Therefore, PCB chips are more suitable for production or commercial designs with a fast sensing speed and low sensitivity requirements. Polymer materials are suitable for mass production. Therefore, polymer chips can be one of the options for commercialization. Microfluidic impedance sensor glass chips have two great development directions: electrochemical impedance spectroscopy and electrical impedance flow cytometry. The difference between the two lies in the state of the measured object. In an electrical impedance flow cytometer, the cells are stationary. In an electrical impedance flow cytometer, the cells are in a dynamic state. For the impedance sensing of other organisms, the main technology relies on the modification of electrodes and the combination of antigens and antibodies.

Since the development of microfluidic impedance sensors, the development of the principle has mostly stagnated since 2014, especially electrical impedance flow cytometers. Therefore, with regard to the development of sensing applications, specific antigen–antibody binding may not easily to become a break-point in relevant research. Instead, paper-based chips and stretchable chips are expected to become the focus of future development related to microfluidic impedance sensors. As the COVID-19 virus has spread globally in recent years, this variable virus needs to be sensed by electrical impedance biosensors.

Author Contributions: Writing—original draft preparation, Y.-S.C. and C.-H.H.; writing—review and editing, P.-C.P., J.S. and K.F.L.; funding acquisition, J.S. and K.F.L. All authors have read and agreed to the published version of the manuscript.

Funding: This work was supported by the National Science and Technology Council, Taiwan (Project No. MOST111-2221-E-182-006-MY3) and the National Research Foundation of Korea (NRF) grant funded by the Korean Government (MSIT) (Project No. 2022R1A2C4001652).

Institutional Review Board Statement: Not applicable.

Informed Consent Statement: Not applicable.

Data Availability Statement: Not applicable.

Conflicts of Interest: The authors declare no conflict of interest.

References

1. Kim, S.; Song, H.; Ahn, H.; Kim, T.; Jung, J.; Cho, S.K.; Shin, D.M.; Choi, J.R.; Hwang, Y.H.; Kim, K. A Review of Advanced Impedance Biosensors with Microfluidic Chips for Single-Cell Analysis. *Biosensors* **2021**, *11*, 412. [CrossRef]
2. Ayliffe, H.E.; Frazier, A.B.; Rabbitt, R.D. Electric impedance spectroscopy using microchannels with integrated metal electrodes. *J. Microelectromech. Syst.* **1999**, *8*, 50–57. [CrossRef]
3. Zhang, Z.; Huang, X.W.; Liu, K.; Lan, T.C.; Wang, Z.X.; Zhu, Z. Recent Advances in Electrical Impedance Sensing Technology for Single-Cell Analysis. *Biosensors* **2021**, *11*, 470. [CrossRef]
4. Lei, K.F. Review on Impedance Detection of Cellular Responses in Micro/Nano Environment. *Micromachines* **2014**, *5*, 1–12. [CrossRef]
5. Giaever, I.; Keese, C.R. Monitoring fibroblast behavior in tissue culture with an applied electric field. *Proc. Natl. Acad. Sci. USA* **1984**, *81*, 3761–3764. [CrossRef]

6. Giaever, I.; Keese, C.R. Use of electric fields to monitor the dynamical aspect of cell behavior in tissue culture. *IEEE Trans. Biomed. Eng.* **1986**, *33*, 242–247. [CrossRef]
7. Giaever, I.; Keese, C.R. Micromotion of mammalian-cells measured electrically. *Proc. Natl. Acad. Sci. USA* **1991**, *88*, 7896–7900. [CrossRef]
8. Giaever, I.; Keese, C.R. A morphological biosensor for mammalian-cells. *Nature* **1993**, *366*, 591–592. [CrossRef]
9. Grossi, M.; Riccò, B. Electrical impedance spectroscopy (EIS) for biological analysis and food characterization: A review. *J. Sens. Sens. Syst.* **2017**, *6*, 303–325. [CrossRef]
10. Sun, T.; Bernabini, C.; Morgan, H. Single-Colloidal Particle Impedance Spectroscopy: Complete Equivalent Circuit Analysis of Polyelectrolyte Microcapsules. *Langmuir* **2010**, *26*, 3821–3828. [CrossRef]
11. Kamentsky, L.A.; Melamed, M.R.; Derman, H. Spectrophotometer: New instrument for ultrarapid cell analysis. *Science* **1965**, *150*, 630–631. [CrossRef] [PubMed]
12. Gawad, S.; Schild, L.; Renaud, P. Micromachined impedance spectroscopy flow cytometer for cell analysis and particle sizing. *Lab Chip* **2001**, *1*, 76–82. [CrossRef]
13. Gawad, S.; Cheung, K.; Seger, U.; Bertsch, A.; Renaud, P. Dielectric spectroscopy in a micromachined flow cytometer: Theoretical and practical considerations. *Lab Chip* **2004**, *4*, 241–251. [CrossRef] [PubMed]
14. Cheung, K.; Gawad, S.; Renaud, P. Impedance spectroscopy flow cytometry: On-chip label-free cell differentiation. *Cytom. Part A* **2005**, *65A*, 124–132. [CrossRef] [PubMed]
15. Hoffman, R.A.; Johnson, T.S.; Britt, W.B. Flow cytometric electronic direct current volume and radiofrequency impedance measurements of single cells and particles. *Cytometry* **1981**, *1*, 377–384. [CrossRef]
16. Sun, T.; Morgan, H. Single-cell microfluidic impedance cytometry: A review. *Microfluid. Nanofluid.* **2010**, *8*, 423–443. [CrossRef]
17. Van Gerwen, P.; Laureyn, W.; Laureys, W.; Huyberechts, G.; De Beeck, M.O.; Baert, K.; Suls, J.; Sansen, W.; Jacobs, P.; Hermans, L.; et al. Nanoscaled interdigitated electrode arrays for biochemical sensors. *Sens. Actuator B Chem.* **1998**, *49*, 73–80. [CrossRef]
18. Gomez, R.; Bashir, R.; Sarikaya, A.; Ladisch, M.R.; Sturgis, J.; Robinson, J.P.; Geng, T.; Bhunia, A.K.; Apple, H.L.; Wereley, S. Microfluidic Biochip for Impedance Spectroscopy of Biological Species. *Biomed. Microdevices* **2001**, *3*, 201–209. [CrossRef]
19. Radke, S.M.; Alocilja, E.C. Design and fabrication of a microimpedance biosensor for bacterial detection. *IEEE Sens. J.* **2004**, *4*, 434–440. [CrossRef]
20. Radke, S.M.; Alocilja, E.C. A microfabricated biosensor for detecting foodborne bioterrorism agents. *IEEE Sens. J.* **2005**, *5*, 744–750. [CrossRef]
21. Hong, J.; Yoon, D.S.; Kim, S.K.; Kim, T.S.; Kim, S.; Pak, E.Y.; No, K. AC frequency characteristics of coplanar impedance sensors as design parameters. *Lab Chip* **2005**, *5*, 270–279. [CrossRef]
22. Kloss, D.; Fischer, M.; Rothermel, A.; Simon, J.C.; Robitzki, A.A. Drug testing on 3D in vitro tissues trapped on a microcavity chip. *Lab Chip* **2008**, *8*, 879–884. [CrossRef]
23. James, C.D.; Reuel, N.; Lee, E.S.; Davalos, R.V.; Mani, S.S.; Carroll-Portillo, A.; Rebeil, R.; Martino, A.; Apblett, C.A. Impedimetric and optical interrogation of single cells in a microfluidic device for real-time viability and chemical response assessment. *Biosens. Bioelectron.* **2008**, *23*, 845–851. [CrossRef]
24. Singh, K.V.; Whited, A.M.; Ragineni, Y.; Barrett, T.W.; King, J.; Solanki, R. 3D nanogap interdigitated electrode array biosensors. *Anal. Bioanal. Chem.* **2010**, *397*, 1493–1502. [CrossRef]
25. Arya, S.K.; Chornokur, G.; Venugopal, M.; Bhansali, S. Antibody functionalized interdigitated mu-electrode (ID mu E) based impedimetric cortisol biosensor. *Analyst* **2010**, *135*, 1941–1946. [CrossRef]
26. Chen, Y.; Wong, C.C.; Pui, T.S.; Nadipalli, R.; Weerasekera, R.; Chandran, J.; Yu, H.; Rahman, A.R.A. CMOS high density electrical impedance biosensor array for tumor cell detection. *Sens. Actuator B Chem.* **2012**, *173*, 903–907. [CrossRef]
27. Ma, H.B.; Wallbank, R.W.R.; Chaji, R.; Li, J.H.; Suzuki, Y.; Jiggins, C.; Nathan, A. An impedance-based integrated biosensor for suspended DNA characterization. *Sci. Rep.* **2013**, *3*, 7. [CrossRef]
28. Cunci, L.; Vargas, M.M.; Cunci, R.; Gomez-Moreno, R.; Perez, I.; Baerga-Ortiz, A.; Gonzalez, C.I.; Cabrera, C.R. Real-time detection of telomerase activity in cancer cells using a label-free electrochemical impedimetric biosensing microchip. *RSC Adv.* **2014**, *4*, 52357–52365. [CrossRef]
29. Shin, K.S.; Ji, J.H.; Hwang, K.S.; Jun, S.C.; Kang, J.Y. Sensitivity Enhancement of Bead-based Electrochemical Impedance Spectroscopy (BEIS) biosensor by electric field-focusing in microwells. *Biosens. Bioelectron.* **2016**, *85*, 16–24. [CrossRef]
30. Pursey, J.P.; Chen, Y.; Stulz, E.; Park, M.K.; Kongsuphol, P. Microfluidic electrochemical multiplex detection of bladder cancer DNA markers. *Sens. Actuator B Chem.* **2017**, *251*, 34–39. [CrossRef]
31. Brosel-Oliu, S.; Mergel, O.; Uria, N.; Abramova, N.; van Rijn, P.; Bratov, A. 3D impedimetric sensors as a tool for monitoring bacterial response to antibiotics. *Lab Chip* **2019**, *19*, 1436–1447. [CrossRef] [PubMed]
32. Pan, Y.X.; Jiang, D.M.; Gu, C.L.; Qiu, Y.; Wan, H.; Wang, P. 3D microgroove electrical impedance sensing to examine 3D cell cultures for antineoplastic drug assessment. *Microsyst. Nanoeng.* **2020**, *6*, 10. [CrossRef] [PubMed]
33. Hsiao, Y.P.; Mukundan, A.; Chen, W.C.; Wu, M.T.; Hsieh, S.C.; Wang, H.C. Design of a Lab-On-Chip for Cancer Cell Detection through Impedance and Photoelectrochemical Response Analysis. *Biosensors* **2022**, *12*, 405. [CrossRef] [PubMed]
34. Valekunja, R.B.; Kamakoti, V.; Peter, A.; Phadnis, S.; Prasad, S.; Nagaraj, V.J. The detection of papaya ringspot virus coat protein using an electrochemical immunosensor. *Anal. Methods* **2016**, *8*, 8534–8541. [CrossRef]

35. Jin, S.R.; Ye, Z.Z.; Wang, Y.X.; Ying, Y.B. A Novel Impedimetric Microfluidic Analysis System for Transgenic Protein Cry1Ab Detection. *Sci. Rep.* **2017**, *7*, 8. [CrossRef] [PubMed]
36. Chang, Y.J.; Ho, C.Y.; Zhou, X.M.; Yen, H.R. Determination of degree of RBC agglutination for blood typing using a small quantity of blood sample in a microfluidic system. *Biosens. Bioelectron.* **2018**, *102*, 234–241. [CrossRef]
37. Fuentes-Velez, S.; Fagoonee, S.; Sanginario, A.; Pizzi, M.; Altruda, F.; Demarchi, D. Electrical Impedance-Based Characterization of Hepatic Tissue with Early-Stage Fibrosis. *Biosensors* **2022**, *12*, 116. [CrossRef]
38. Zou, Z.W.; Kai, J.H.; Rust, M.J.; Han, J.; Ahn, C.H. Functionalized nano interdigitated electrodes arrays on polymer with integrated microfluidics for direct bio-affinity sensing using impedimetric measurement. *Sens. Actuator A Phys.* **2007**, *136*, 518–526. [CrossRef]
39. Sabounchi, P.; Morales, A.M.; Ponce, P.; Lee, L.P.; Simmons, B.A.; Davalos, R.V. Sample concentration and impedance detection on a microfluidic polymer chip. *Biomed. Microdevices* **2008**, *10*, 661–670. [CrossRef]
40. Dapra, J.; Lauridsen, L.H.; Nielsen, A.T.; Rozlosnik, N. Comparative study on aptamers as recognition elements for antibiotics in a label-free all-polymer biosensor. *Biosens. Bioelectron.* **2013**, *43*, 315–320. [CrossRef]
41. Pires, L.; Sachsenheimer, K.; Kleintschek, T.; Waldbaur, A.; Schwartz, T.; Rapp, B.E. Online monitoring of biofilm growth and activity using a combined multi-channel impedimetric and amperometric sensor. *Biosens. Bioelectron.* **2013**, *47*, 157–163. [CrossRef] [PubMed]
42. Rosati, G.; Dapra, J.; Cherre, S.; Rozlosnik, N. Performance Improvement by Layout Designs of Conductive Polymer Microelectrode Based Impedimetric Biosensors. *Electroanalysis* **2014**, *26*, 1400–1408. [CrossRef]
43. Shi, Z.Z.; Tian, Y.L.; Wu, X.S.; Li, C.M.; Yu, L. A one-piece lateral flow impedimetric test strip for label-free clenbuterol detection. *Anal. Methods* **2015**, *7*, 4957–4964. [CrossRef]
44. Sharif, S.; Wang, Y.X.; Ye, Z.Z.; Wang, Z.; Qiu, Q.M.; Ying, S.N.; Ying, Y.B. A novel impedimetric sensor for detecting LAMP amplicons of pathogenic DNA based on magnetic separation. *Sens. Actuator B Chem.* **2019**, *301*, 6. [CrossRef]
45. Lakey, A.; Ali, Z.; Scott, S.M.; Chebil, S.; Korri-Youssoufi, H.; Hunor, S.; Ohlander, A.; Kuphal, M.; Marti, J.S. Impedimetric array in polymer microfluidic cartridge for low cost point-of-care diagnostics. *Biosens. Bioelectron.* **2019**, *129*, 147–154. [CrossRef]
46. Ma, W.R.; Liu, L.L.; Xu, Y.; Wang, L.; Chen, L.; Yan, S.; Shui, L.L.; Wang, Z.J.; Li, S.B. A highly efficient preconcentration route for rapid and sensitive detection of endotoxin based on an electrochemical biosensor. *Analyst* **2020**, *145*, 4204–4211. [CrossRef]
47. Niaraki, A.; Shirsavar, M.A.; Aykar, S.S.; Taghavimehr, M.; Montazami, R.; Hashemi, N.N. Minute-sensitive real-time monitoring of neural cells through printed graphene microelectrodes. *Biosens. Bioelectron.* **2022**, *210*, 8. [CrossRef]
48. Chmayssem, A.; Tanase, C.E.; Verplanck, N.; Gougis, M.; Mourier, V.; Zebda, A.; Ghaemmaghami, A.M.; Mailley, P. New Microfluidic System for Electrochemical Impedance Spectroscopy Assessment of Cell Culture Performance: Design and Development of New Electrode Material. *Biosensors* **2022**, *12*, 452. [CrossRef]
49. Hantschke, M.; Triantis, I.F. Optimisation of an Electrical Impedance Sensor for Use in Microfluidic Chip Electrophoresis. *IEEE Sens. J.* **2022**, *22*, 16–24. [CrossRef]
50. Ruan, C.M.; Yang, L.J.; Li, Y.B. Immunobiosensor chips for detection of Escherichia coli O157:H7 using electrochemical impedance spectroscopy. *Anal. Chem.* **2002**, *74*, 4814–4820. [CrossRef]
51. Yang, L.J.; Li, Y.B.; Griffis, C.L.; Johnson, M.G. Interdigitated microelectrode (IME) impedance sensor for the detection of viable Salmonella typhimurium. *Biosens. Bioelectron.* **2004**, *19*, 1139–1147. [CrossRef]
52. Yang, L.J.; Li, Y.B.; Erf, G.F. Interdigitated array microelectrode-based electrochemical impedance immunosensor for detection of Escherichia coli O157:H7. *Anal. Chem.* **2004**, *76*, 1107–1113. [CrossRef] [PubMed]
53. Wang, R.H.; Ruan, C.M.; Kanayeva, D.; Lassiter, K.; Li, Y.B. TiO$_2$ nanowire bundle microelectrode based impedance immunosensor for rapid and sensitive detection of Listeria monocytogenes. *Nano Lett.* **2008**, *8*, 2625–2631. [CrossRef] [PubMed]
54. Tan, F.; Leung, P.H.M.; Liu, Z.B.; Zhang, Y.; Xiao, L.D.; Ye, W.W.; Zhang, X.; Yi, L.; Yang, M. A PDMS microfluidic impedance immunosensor for *E. coli* O157:H7 and *Staphylococcus aureus* detection via antibody-immobilized nanoporous membrane. *Sens. Actuator B Chem.* **2011**, *159*, 328–335. [CrossRef]
55. Lum, J.; Wang, R.H.; Lassiter, K.; Srinivasan, B.; Abi-Ghanem, D.; Berghman, L.; Hargis, B.; Tung, S.; Lu, H.G.; Li, Y.B. Rapid detection of avian influenza H5N1 virus using impedance measurement of immuno-reaction coupled with RBC amplification. *Biosens. Bioelectron.* **2012**, *38*, 67–73. [CrossRef]
56. Dastider, S.G.; Barizuddin, S.; Dweik, M.; Almasri, M. A micromachined impedance biosensor for accurate and rapid detection of *E. coli* O157:H7. *RSC Adv.* **2013**, *3*, 26297–26306. [CrossRef]
57. Couniot, N.; Vanzieleghem, T.; Rasson, J.; Van Overstraeten-Schlogel, N.; Poncelet, O.; Mahillon, J.; Francis, L.A.; Flandre, D. Lytic enzymes as selectivity means for label-free, microfluidic and impedimetric detection of whole-cell bacteria using ALD-Al2O3 passivated microelectrodes. *Biosens. Bioelectron.* **2015**, *67*, 154–161. [CrossRef]
58. Liu, J.Y.; Jasim, I.; Abdullah, A.; Shen, Z.Y.; Zhao, L.; El-Dweik, M.; Zhang, S.P.; Almasri, M. An integrated impedance biosensor platform for detection of pathogens in poultry products. *Sci. Rep.* **2018**, *8*, 10. [CrossRef] [PubMed]
59. Dastider, S.G.; Abdullah, A.; Jasim, I.; Yuksek, N.S.; Dweik, M.; Almasri, M. Low concentration *E. coli* O157:H7 bacteria sensing using microfluidic MEMS biosensor. *Rev. Sci. Instrum.* **2018**, *89*, 9. [CrossRef]
60. Liu, J.Y.; Jasim, I.; Shen, Z.Y.; Zhao, L.; Dweik, M.; Zhang, S.P.; Almasri, M. A microfluidic based biosensor for rapid detection of Salmonella in food products. *PLoS ONE* **2019**, *14*, e0216873. [CrossRef]

61. Mishra, N.N.; Retterer, S.; Zieziulewicz, T.J.; Isaacson, M.; Szarowski, D.; Mousseau, D.E.; Lawrence, D.A.; Turner, J.N. On-chip micro-biosensor for the detection of human CD4(+) cells based on AC impedance and optical analysis. *Biosens. Bioelectron.* **2005**, *21*, 696–704. [CrossRef] [PubMed]
62. Kuttel, C.; Nascimento, E.; Demierre, N.; Silva, T.; Braschler, T.; Renaud, P.; Oliva, A.G. Label-free detection of *Babesia bovis* infected red blood cells using impedance spectroscopy on a microfabricated flow cytometer. *Acta Trop.* **2007**, *102*, 63–68. [CrossRef] [PubMed]
63. Holmes, D.; Pettigrew, D.; Reccius, C.H.; Gwyer, J.D.; van Berkel, C.; Holloway, J.; Davies, D.E.; Morgan, H. Leukocyte analysis and differentiation using high speed microfluidic single cell impedance cytometry. *Lab Chip* **2009**, *9*, 2881–2889. [CrossRef] [PubMed]
64. Han, X.J.; van Berkel, C.; Gwyer, J.; Capretto, L.; Morgan, H. Microfluidic Lysis of Human Blood for Leukocyte Analysis Using Single Cell Impedance Cytometry. *Anal. Chem.* **2012**, *84*, 1070–1075. [CrossRef]
65. Lei, K.F.; Chen, K.H.; Tsui, P.H.; Tsang, N.M. Real-Time Electrical Impedimetric Monitoring of Blood Coagulation Process under Temperature and Hematocrit Variations Conducted in a Microfluidic Chip. *PLoS ONE* **2013**, *8*, e76243. [CrossRef]
66. Song, H.J.; Wang, Y.; Rosano, J.M.; Prabhakarpandian, B.; Garson, C.; Pant, K.; Lai, E. A microfluidic impedance flow cytometer for identification of differentiation state of stem cells. *Lab Chip* **2013**, *13*, 2300–2310. [CrossRef]
67. Du, E.; Ha, S.; Diez-Silva, M.; Dao, M.; Suresh, S.; Chandrakasan, A.P. Electric impedance microflow cytometry for characterization of cell disease states. *Lab Chip* **2013**, *13*, 3903–3909. [CrossRef] [PubMed]
68. Spencer, D.; Hollis, V.; Morgan, H. Microfluidic impedance cytometry of tumour cells in blood. *Biomicrofluidics* **2014**, *8*, 11. [CrossRef]
69. Liu, J.; Qiang, Y.H.; Alvarez, O.; Du, E. Electrical impedance microflow cytometry with oxygen control for detection of sickle cells. *Sens. Actuator B Chem.* **2018**, *255*, 2392–2398. [CrossRef]
70. Nguyen, T.A.; Yin, T.I.; Reyes, D.; Urban, G.A. Microfluidic Chip with Integrated Electrical Cell-Impedance Sensing for Monitoring Single Cancer Cell Migration in Three-Dimensional Matrixes. *Anal. Chem.* **2013**, *85*, 11068–11076. [CrossRef]
71. Liu, L.; Xiao, X.; Lei, K.F.; Huang, C.H. Quantitative impedimetric monitoring of cell migration under the stimulation of cytokine or anti-cancer drug in a microfluidic chip. *Biomicrofluidics* **2015**, *9*, 10. [CrossRef]
72. Huang, C.H.; Lei, K.F. Impedimetric quantification of migration speed of cancer cells migrating along a Matrigel-filled microchannel. *Anal. Chim. Acta* **2020**, *1121*, 67–73. [CrossRef] [PubMed]
73. Huang, C.H.; Lei, K.F. Quantitative study of tumor angiogenesis in three-dimensional matrigel barrier using electric impedance measurement technique. *Sens. Actuator B Chem.* **2022**, *370*, 8. [CrossRef]
74. Hong, J.L.; Lan, K.C.; Jang, L.S. Electrical characteristics analysis of various cancer cells using a microfluidic device based on single-cell impedance measurement. *Sens. Actuator B Chem.* **2012**, *173*, 927–934. [CrossRef]
75. Zhao, Y.; Zhao, X.T.; Chen, D.Y.; Luo, Y.N.; Jiang, M.; Wei, C.; Long, R.; Yue, W.T.; Wang, J.B.; Chen, J. Tumor cell characterization and classification based on cellular specific membrane capacitance and cytoplasm conductivity. *Biosens. Bioelectron.* **2014**, *57*, 245–253. [CrossRef] [PubMed]
76. Fan, W.H.; Chen, X.; Ge, Y.Q.; Jin, Y.; Jin, Q.H.; Zhao, J.L. Single-cell impedance analysis of osteogenic differentiation by droplet-based microfluidics. *Biosens. Bioelectron.* **2019**, *145*, 8. [CrossRef]
77. Lei, K.F.; Ho, Y.C.; Huang, C.H.; Huang, C.H.; Pai, P.C. Characterization of stem cell-like property in cancer cells based on single-cell impedance measurement in a microfluidic platform. *Talanta* **2021**, *229*, 8. [CrossRef] [PubMed]
78. Bieberich, E.; Guiseppi-Elie, A. Neuronal differentiation and synapse formation of PC12 and embryonic stem cells on interdigitated microelectrode arrays: Contact structures for neuron-to-electrode signal transmission (NEST). *Biosens. Bioelectron.* **2004**, *19*, 923–931. [CrossRef]
79. Jang, L.S.; Wang, M.H. Microfluidic device for cell capture and impedance measurement. *Biomed. Microdevices* **2007**, *9*, 737–743. [CrossRef]
80. Cho, Y.; Kim, H.S.; Frazier, A.B.; Chen, Z.G.; Shin, D.M.; Han, A. Whole-Cell Impedance Analysis for Highly and Poorly Metastatic Cancer Cells. *J. Microelectromech. Syst.* **2009**, *18*, 808–817. [CrossRef]
81. Hildebrandt, C.; Buth, H.; Cho, S.B.; Impidjati; Thielecke, H. Detection of the osteogenic differentiation of mesenchymal stem cells in 2D and 3D cultures by electrochemical impedance spectroscopy. *J. Biotechnol.* **2010**, *148*, 83–90. [CrossRef]
82. Houssin, T.; Folleta, J.; Follet, A.; Dei-Cas, E.; Senez, V. Label-free analysis of water-polluting parasite by electrochemical impedance spectroscopy. *Biosens. Bioelectron.* **2010**, *25*, 1122–1129. [CrossRef] [PubMed]
83. Dalmay, C.; Cheray, M.; Pothier, A.; Lalloue, F.; Jauberteau, M.O.; Blondy, P. Ultra sensitive biosensor based on impedance spectroscopy at microwave frequencies for cell scale analysis. *Sens. Actuator A Phys.* **2010**, *162*, 189–197. [CrossRef]
84. Bagnaninchi, P.O.; Drummond, N. Real-time label-free monitoring of adipose-derived stem cell differentiation with electric cell-substrate impedance sensing. *Proc. Natl. Acad. Sci. USA* **2011**, *108*, 6462–6467. [CrossRef]
85. Lei, K.F.; Wu, Z.M.; Huang, C.H. Impedimetric quantification of the formation process and the chemosensitivity of cancer cell colonies suspended in 3D environment. *Biosens. Bioelectron.* **2015**, *74*, 878–885. [CrossRef] [PubMed]
86. Chen, N.C.; Chen, C.H.; Chen, M.K.; Jang, L.S.; Wang, M.H. Single-cell trapping and impedance measurement utilizing dielectrophoresis in a parallel-plate microfluidic device. *Sens. Actuator B Chem.* **2014**, *190*, 570–577. [CrossRef]
87. Lei, K.F.; Kao, C.H.; Tsang, N.M. High throughput and automatic colony formation assay based on impedance measurement technique. *Anal. Bioanal. Chem.* **2017**, *409*, 3271–3277. [CrossRef]

88. Nguyen, N.V.; Jen, C.P. Impedance detection integrated with dielectrophoresis enrichment platform for lung circulating tumor cells in a microfluidic channel. *Biosens. Bioelectron.* **2018**, *121*, 10–18. [CrossRef]

89. Lei, K.F.; Lin, B.Y.; Tsang, N.M. Real-time and label-free impedimetric analysis of the formation and drug testing of tumor spheroids formed via the liquid overlay technique. *RSC Adv.* **2017**, *7*, 13939–13946. [CrossRef]

90. Lei, K.F.; Wu, M.H.; Hsu, C.W.; Chen, Y.D. Real-time and non-invasive impedimetric monitoring of cell proliferation and chemosensitivity in a perfusion 3D cell culture microfluidic chip. *Biosens. Bioelectron.* **2014**, *51*, 16–21. [CrossRef]

91. Sun, T.; Green, N.G.; Gawad, S.; Morgan, H. Analytical electric field and sensitivity analysis for two microfluidic impedance cytometer designs. *IET Nanobiotechnol.* **2007**, *1*, 69–79. [CrossRef] [PubMed]

92. Holmes, D.; She, J.K.; Roach, P.L.; Morgan, H. Bead-based immunoassays using a micro-chip flow cytometer. *Lab Chip* **2007**, *7*, 1048–1056. [CrossRef] [PubMed]

93. Sun, T.; Holmes, D.; Gawad, S.; Green, N.G.; Morgan, H. High speed multi-frequency impedance analysis of single particles in a microfluidic cytometer using maximum length sequences. *Lab Chip* **2007**, *7*, 1034–1040. [CrossRef]

94. Kummrow, A.; Theisen, J.; Frankowski, M.; Tuchscheerer, A.; Yildirim, H.; Brattke, K.; Schmidt, M.; Neukammer, J. Microfluidic structures for flow cytometric analysis of hydrodynamically focussed blood cells fabricated by ultraprecision micromachining. *Lab Chip* **2009**, *9*, 972–981. [CrossRef] [PubMed]

95. Spencer, D.; Morgan, H. Positional dependence of particles in microfludic impedance cytometry. *Lab Chip* **2011**, *11*, 1234–1239. [CrossRef]

96. Barat, D.; Spencer, D.; Benazzi, G.; Mowlem, M.C.; Morgan, H. Simultaneous high speed optical and impedance analysis of single particles with a microfluidic cytometer. *Lab Chip* **2012**, *12*, 118–126. [CrossRef]

97. Spencer, D.; Elliott, G.; Morgan, H. A sheath-less combined optical and impedance micro-cytometer. *Lab Chip* **2014**, *14*, 3064–3073. [CrossRef] [PubMed]

98. Haandbaek, N.; Burgel, S.C.; Heer, F.; Hierlemann, A. Characterization of subcellular morphology of single yeast cells using high frequency microfluidic impedance cytometer. *Lab Chip* **2014**, *14*, 369–377. [CrossRef]

99. David, F.; Hebeisen, M.; Schade, G.; Franco-Lara, E.; Di Berardino, M. Viability and membrane potential analysis of Bacillus megaterium cells by impedance flow cytometry. *Biotechnol. Bioeng.* **2012**, *109*, 483–492. [CrossRef]

100. Evander, M.; Ricco, A.J.; Morser, J.; Kovacs, G.T.A.; Leung, L.L.K.; Giovangrandi, L. Microfluidic impedance cytometer for platelet analysis. *Lab Chip* **2013**, *13*, 722–729. [CrossRef]

101. Lin, Z.; Cao, X.; Xie, P.; Liu, M.; Javanmard, M. PicoMolar level detection of protein biomarkers based on electronic sizing of bead aggregates: Theoretical and experimental considerations. *Biomed. Microdevices* **2015**, *17*, 7. [CrossRef] [PubMed]

102. Simon, P.; Frankowski, M.; Bock, N.; Neukammer, J. Label-free whole blood cell differentiation based on multiple frequency AC impedance and light scattering analysis in a micro flow cytometer. *Lab Chip* **2016**, *16*, 2326–2338. [CrossRef] [PubMed]

103. Xie, P.F.; Cao, X.N.; Lin, Z.T.; Javanmard, M. Top-down fabrication meets bottom-up synthesis for nanoelectronic barcoding of microparticles. *Lab Chip* **2017**, *17*, 1939–1947. [CrossRef] [PubMed]

104. McGrath, J.S.; Honrado, C.; Spencer, D.; Horton, B.; Bridle, H.L.; Morgan, H. Analysis of Parasitic Protozoa at the Single-cell Level using Microfluidic Impedance Cytometry. *Sci. Rep.* **2017**, *7*, 11. [CrossRef] [PubMed]

105. Tang, W.L.; Tang, D.Z.; Ni, Z.H.; Xiang, N.; Yi, H. Microfluidic Impedance Cytometer with Inertial Focusing and Liquid Electrodes for High-Throughput Cell Counting and Discrimination. *Anal. Chem.* **2017**, *89*, 3154–3161. [CrossRef]

106. Mansor, M.A.; Takeuchi, M.; Nakajima, M.; Hasegawa, Y.; Ahmad, M.R. Electrical Impedance Spectroscopy for Detection of Cells in Suspensions Using Microfluidic Device with Integrated Microneedles. *Appl. Sci.* **2017**, *7*, 170. [CrossRef]

107. Ahuja, K.; Rather, G.M.; Lin, Z.T.; Sui, J.Y.; Xie, P.F.; Le, T.; Bertino, J.R.; Javanmard, M. Toward point-of-care assessment of patient response: A portable tool for rapidly assessing cancer drug efficacy using multifrequency impedance cytometry and supervised machine learning. *Microsyst. Nanoeng.* **2019**, *5*, 11. [CrossRef]

108. Sui, J.Y.; Foflonker, F.; Bhattacharya, D.; Javanmard, M. Electrical impedance as an indicator of microalgal cell health. *Sci. Rep.* **2020**, *10*, 9. [CrossRef]

109. Mahesh, K.; Varma, M.; Sen, P. Double-peak signal features in microfluidic impedance flow cytometry enable sensitive measurement of cell membrane capacitance. *Lab Chip* **2020**, *20*, 4296–4309. [CrossRef]

110. Petchakup, C.; Hutchinson, P.E.; Tay, H.M.; Leong, S.Y.; Li, K.H.H.; Hou, H.W. Label-free quantitative lymphocyte activation profiling using microfluidic impedance cytometry. *Sens. Actuator B Chem.* **2021**, *339*, 9. [CrossRef]

111. Caselli, F.; Reale, R.; De Ninno, A.; Spencer, D.; Morgan, H.; Bisegna, P. Deciphering impedance cytometry signals with neural networks. *Lab Chip* **2022**, *22*, 1714–1722. [CrossRef] [PubMed]

112. Wang, R.H.; Wang, Y.; Lassiter, K.; Li, Y.B.; Hargis, B.; Tung, S.; Berghman, L.; Bottje, W. Interdigitated array microelectrode based impedance immunosensor for detection of avian influenza virus H5N1. *Talanta* **2009**, *79*, 159–164. [CrossRef] [PubMed]

113. Wang, R.H.; Lin, J.H.; Lassiter, K.; Srinivasan, B.; Lin, L.; Lu, H.G.; Tung, S.; Hargis, B.; Bottje, W.; Berghman, L.; et al. Evaluation study of a portable impedance biosensor for detection of avian influenza virus. *J. Virol. Methods* **2011**, *178*, 52–58. [CrossRef] [PubMed]

114. Shafiee, H.; Jahangir, M.; Inci, F.; Wang, S.Q.; Willenbrecht, R.B.M.; Giguel, F.F.; Tsibris, A.M.N.; Kuritzkes, D.R.; Demirci, U. Acute On-Chip HIV Detection Through Label-Free Electrical Sensing of Viral Nano-Lysate. *Small* **2013**, *9*, 2553–2563. [CrossRef]

115. Singh, R.; Hong, S.; Jang, J. Label-free Detection of Influenza Viruses using a Reduced Graphene Oxide-based Electrochemical Immunosensor Integrated with a Microfluidic Platform. *Sci. Rep.* **2017**, *7*, 11. [CrossRef]

116. Berdat, D.; Rodriguez, A.C.M.; Herrera, F.; Gijs, M.A.M. Label-free detection of DNA with interdigitated micro-electrodes in a fluidic cell. *Lab Chip* **2008**, *8*, 302–308. [CrossRef]

117. Javanmard, M.; Davis, R.W. A microfluidic platform for electrical detection of DNA hybridization. *Sens. Actuator B Chem.* **2011**, *154*, 22–27. [CrossRef]

118. Ben-Yoav, H.; Dykstra, P.H.; Bentley, W.E.; Ghodssi, R. A microfluidic-based electrochemical biochip for label-free diffusion-restricted DNA hybridization analysis. *Biosens. Bioelectron.* **2012**, *38*, 114–120. [CrossRef]

119. Chuang, C.H.; Du, Y.C.; Wu, T.F.; Chen, C.H.; Lee, D.H.; Chen, S.M.; Huang, T.C.; Wu, H.P.; Shaikh, M.O. Immunosensor for the ultrasensitive and quantitative detection of bladder cancer in point of care testing. *Biosens. Bioelectron.* **2016**, *84*, 126–132. [CrossRef]

120. Ondevilla, N.A.P.; Wong, T.W.; Lee, N.Y.; Chang, H.C. An AC electrokinetics-based electrochemical aptasensor for the rapid detection of microRNA-155. *Biosens. Bioelectron.* **2022**, *199*, 7. [CrossRef]

121. Wu, C.C.; Huang, W.C.; Hu, C.C. An ultrasensitive label-free electrochemical impedimetric DNA biosensing chip integrated with a DC-biased AC electroosmotic vortex. *Sens. Actuator B Chem.* **2015**, *209*, 61–68. [CrossRef]

122. Alsabbagh, K.; Hornung, T.; Voigt, A.; Sadir, S.; Rajabi, T.; Lange, K. Microfluidic Impedance Biosensor Chips Using Sensing Layers Based on DNA-Based Self-Assembled Monolayers for Label-Free Detection of Proteins. *Biosensors* **2021**, *11*, 80. [CrossRef] [PubMed]

123. Sheen, H.J.; Panigrahi, B.; Kuo, T.R.; Hsu, W.C.; Chung, P.S.; Xie, Q.Z.; Lin, C.Y.; Chang, Y.S.; Lin, C.T.; Fan, Y.J. Electrochemical biosensor with electrokinetics-assisted molecular trapping for enhancing C-reactive protein detection. *Biosens. Bioelectron.* **2022**, *210*, 9. [CrossRef]

124. Chinnadayyala, S.R.; Cho, S. Electrochemical Immunosensor for the Early Detection of Rheumatoid Arthritis Biomarker: Anti-Cyclic Citrullinated Peptide Antibody in Human Serum Based on Avidin-Biotin System. *Sensors* **2021**, *21*, 124. [CrossRef] [PubMed]

125. Selvam, S.P.; Chinnadayyala, S.R.; Cho, S. Electrochemical nanobiosensor for early detection of rheumatoid arthritis biomarker: Anti- cyclic citrullinated peptide antibodies based on polyaniline (PANI)/MoS2-modified screen-printed electrode with PANI-Au nanomatrix-based signal amplification. *Sens. Actuator B Chem.* **2021**, *333*, 12. [CrossRef]

126. Lin, C.Y.; Nguyen, U.T.N.; Hsieh, H.Y.; Tahara, H.; Chang, Y.S.; Wang, B.Y.; Gu, B.C.; Dai, Y.H.; Wu, C.C.; Tsai, I.J.; et al. Peptide-based electrochemical sensor with nanogold enhancement for detecting rheumatoid arthritis. *Talanta* **2022**, *236*, 9. [CrossRef]

127. Chiriaco, M.S.; Primiceri, E.; D'Amone, E.; Ionescu, R.E.; Rinaldi, R.; Maruccio, G. EIS microfluidic chips for flow immunoassay and ultrasensitive cholera toxin detection. *Lab Chip* **2011**, *11*, 658–663. [CrossRef]

128. Liu, F.; Nordin, A.N.; Li, F.; Voiculescu, I. A lab-on-chip cell-based biosensor for label-free sensing of water toxicants. *Lab Chip* **2014**, *14*, 1270–1280. [CrossRef]

129. Pradhan, R.; Kalkal, A.; Jindal, S.; Packirisamy, G.; Manhas, S. Four electrode-based impedimetric biosensors for evaluating cytotoxicity of tamoxifen on cervical cancer cells. *RSC Adv.* **2021**, *11*, 798–806. [CrossRef]

130. Guo, Y.M.; Liu, X.F.; Sun, X.; Cao, Y.Y.; Wang, X.Y. A PDMS Microfluidic Impedance Immunosensor for Sensitive Detection Of Pesticide Residues in Vegetable Real Samples. *Int. J. Electrochem. Sci.* **2015**, *10*, 4155–4164.

131. Zeng, L.; Wang, W.Q.; Rogers, F.; Zhang, H.P.; Zhang, X.M.; Yang, D.X. A High Sensitivity Micro Impedance Sensor Based on Magnetic Focusing for Oil Condition Monitoring. *IEEE Sens. J.* **2020**, *20*, 3813–3821. [CrossRef]

132. Lei, K.F.; Huang, C.H.; Tsang, N.M. Impedimetric quantification of cells encapsulated in hydrogel cultured in a paper-based microchamber. *Talanta* **2016**, *147*, 628–633. [CrossRef] [PubMed]

133. Teengam, P.; Siangproh, W.; Tuantranont, A.; Vilaivan, T.; Chailapakul, O.; Henry, C.S. Electrochemical impedance-based DNA sensor using pyrrolidinyl peptide nucleic acids for tuberculosis detection. *Anal. Chim. Acta* **2018**, *1044*, 102–109. [CrossRef]

134. Rengaraj, S.; Cruz-Izquierdo, A.; Scott, J.L.; Di Lorenzo, M. Impedimetric paper-based biosensor for the detection of bacterial contamination in water. *Sens. Actuator B Chem.* **2018**, *265*, 50–58. [CrossRef]

135. Moazeni, M.; Karimzadeh, F.; Kermanpur, A. Peptide modified paper based impedimetric immunoassay with nanocomposite electrodes as a point-of-care testing of Alpha-fetoprotein in human serum. *Biosens. Bioelectron.* **2018**, *117*, 748–757. [CrossRef] [PubMed]

136. Lei, K.F.; Liu, T.K.; Tsang, N.M. Towards a high throughput impedimetric screening of chemosensitivity of cancer cells suspended in hydrogel and cultured in a paper substrate. *Biosens. Bioelectron.* **2018**, *100*, 355–360. [CrossRef]

137. Congur, G.; Erdem, A. PAMAM dendrimer modified screen printed electrodes for impedimetric detection of miRNA-34a. *Microchem. J.* **2019**, *148*, 748–758. [CrossRef]

138. Vasantham, S.; Alhans, R.; Singhal, C.; Nagabooshanam, S.; Nissar, S.; Basu, T.; Ray, S.C.; Wadhwa, S.; Narang, J.; Mathur, A. Paper based point of care immunosensor for the impedimetric detection of cardiac troponin I biomarker. *Biomed. Microdevices* **2019**, *22*, 9. [CrossRef]

139. Li, X.; Qin, Z.; Fu, H.; Li, T.; Peng, R.; Li, Z.J.; Rini, J.M.; Liu, X.Y. Enhancing the performance of paper-based electrochemical impedance spectroscopy nanobiosensors: An experimental approach. *Biosens. Bioelectron.* **2021**, *177*, 8. [CrossRef]

140. Kare, S.P.O.; Das, D.; Chaudhury, K.; Das, S. Hand-drawn electrode based disposable paper chip for artificial sweat analysis using impedance spectroscopy. *Biomed. Microdevices* **2021**, *23*, 12. [CrossRef]

141. Eksin, E.; Torul, H.; Yarali, E.; Tamer, U.; Papakonstantinou, P.; Erdem, A. Paper-based electrode assemble for impedimetric detection of miRNA. *Talanta* **2021**, *225*, 6. [CrossRef] [PubMed]
142. Karuppiah, S.; Mishra, N.C.; Tsai, W.C.; Liao, W.S.; Chou, C.F. Ultrasensitive and Low-Cost Paper-Based Graphene Oxide Nanobiosensor for Monitoring Water-Borne Bacterial Contamination. *ACS Sens.* **2021**, *6*, 3214–3223. [CrossRef] [PubMed]
143. Yarali, E.; Eksin, E.; Torul, H.; Ganguly, A.; Tamer, U.; Papakonstantinou, P.; Erdem, A. Impedimetric detection of miRNA biomarkers using paper-based electrodes modified with bulk crystals or nanosheets of molybdenum disulfide. *Talanta* **2022**, *241*, 10. [CrossRef] [PubMed]
144. Mishra, A.; Pilloton, R.; Jain, S.; Roy, S.; Khanuja, M.; Mathur, A.; Narang, J. Paper-Based Electrodes Conjugated with Tungsten Disulfide Nanostructure and Aptamer for Impedimetric Detection of Listeria monocytogenes. *Biosensors* **2022**, *12*, 88. [CrossRef] [PubMed]
145. Furniturewalla, A.; Chan, M.; Sui, J.; Ahuja, K.; Javanmard, M. Fully integrated wearable impedance cytometry platform on flexible circuit board with online smartphone readout. *Microsyst. Nanoeng.* **2018**, *4*, 10. [CrossRef] [PubMed]
146. De Guzman, K.; Al-Kharusi, G.; Levingstone, T.; Morrin, A. Robust epidermal tattoo electrode platform for skin physiology monitoring. *Anal. Methods* **2019**, *11*, 1460–1468. [CrossRef]
147. Lee, H.B.; Meeseepong, M.; Trung, T.Q.; Kim, B.Y.; Lee, N.E. A wearable lab-on-a-patch platform with stretchable nanostructured biosensor for non-invasive immunodetection of biomarker in sweat. *Biosens. Bioelectron.* **2020**, *156*, 8. [CrossRef]
148. Huynh, V.L.; Trung, T.Q.; Meeseepong, M.; Lee, H.B.; Nguyen, T.D.; Lee, N.E. Hollow Microfibers of Elastomeric Nanocomposites for Fully Stretchable and Highly Sensitive Microfluidic Immunobiosensor Patch. *Adv. Funct. Mater.* **2020**, *30*, 12. [CrossRef]
149. Nah, J.S.; Barman, S.C.; Abu Zahed, M.; Sharifuzzaman, M.; Yoon, H.; Park, C.; Yoon, S.; Zhang, S.P.; Park, J.Y. A wearable microfluidics-integrated impedimetric immunosensor based on Ti3C2Tx MXene incorporated laser-burned graphene for noninvasive sweat cortisol detection. *Sens. Actuator B Chem.* **2021**, *329*, 9. [CrossRef]

Review

Capture-SELEX: Selection Strategy, Aptamer Identification, and Biosensing Application

Sin Yu Lam [1], Hill Lam Lau [1] and Chun Kit Kwok [1,2,*]

1   State Key Laboratory of Marine Pollution and Department of Chemistry, City University of Hong Kong, Hong Kong SAR 999077, China
2   Shenzhen Research Institute, City University of Hong Kong, Shenzhen 518057, China
*   Correspondence: ckkwok42@cityu.edu.hk

Abstract: Small-molecule contaminants, such as antibiotics, pesticides, and plasticizers, have emerged as one of the substances most detrimental to human health and the environment. Therefore, it is crucial to develop low-cost, user-friendly, and portable biosensors capable of rapidly detecting these contaminants. Antibodies have traditionally been used as biorecognition elements. However, aptamers have recently been applied as biorecognition elements in aptamer-based biosensors, also known as aptasensors. The systematic evolution of ligands by exponential enrichment (SELEX) is an in vitro technique used to generate aptamers that bind their targets with high affinity and specificity. Over the past decade, a modified SELEX method known as Capture-SELEX has been widely used to generate DNA or RNA aptamers that bind small molecules. In this review, we summarize the recent strategies used for Capture-SELEX, describe the methods commonly used for detecting and characterizing small-molecule–aptamer interactions, and discuss the development of aptamer-based biosensors for various applications. We also discuss the challenges of the Capture-SELEX platform and biosensor development and the possibilities for their future application.

Keywords: nucleic acid aptamer; DNA and RNA Capture-SELEX; small molecule contaminant; characterization; aptamer; biosensing applications

Citation: Lam, S.Y.; Lau, H.L.; Kwok, C.K. Capture-SELEX: Selection Strategy, Aptamer Identification, and Biosensing Application. *Biosensors* **2022**, *12*, 1142. https://doi.org/10.3390/bios12121142

Received: 5 November 2022
Accepted: 29 November 2022
Published: 7 December 2022

Publisher's Note: MDPI stays neutral with regard to jurisdictional claims in published maps and institutional affiliations.

1. Introduction

Accumulating evidence suggests that small-molecule contaminants, such as antibiotics, pesticides, and plasticizers, pose a severe threat to human health and the environment worldwide. Antibiotics have been extensively used for decades to treat bacterial infections, and most are recalcitrant to biodegradation and elimination [1]. Moreover, the global overuse of antibiotics has led to their long-term persistence in the marine environment. More than 700,000 human deaths per year are attributable to bacterial resistance to antimicrobials (including antibiotics), with this number projected to increase to 10 million by 2050 [2]. Pesticides are widely used in agriculture to control pests and disease carriers. According to the World Health Organization, 193,460 unintentional deaths in 2012 were due to pesticide poisoning [3]. Plasticizers are widely applied in plastic products, such as Di(2-ethylhexyl)phthalate (DEHP), contaminating the environment. Toxic plasticizers with their toxicological effects, including endocrine disruption, carcinogenicity, and bioaccumulation potential [4–6], may enter human body via inhalation, ingestion, and dermal contact, reflecting the high risk of harm to human health.

Some traditional techniques for detecting small-molecule contaminants with high accuracy include high-performance liquid chromatography [7,8], enzyme-linked immunosorbent assays [9,10], and liquid chromatography–mass spectrometry [11,12]. However, these techniques are costly and laborious, and can only be performed by professional technicians. Therefore, there is an urgent need for robust aptamer-based biosensors (also called aptasensors), for low-cost, user-friendly, and rapid detection of small-molecule contaminants.

Aptamers are short, single-stranded DNA or RNA oligonucleotides that are able to bind its target with high affinity and specificity. The development of an aptasensor for detecting small-molecule contaminants requires the generation of a novel aptamer. It has gradually replaced traditional antibodies in some applications as they have several advantages over antibodies, such as thermal stability, low cost, amenability to chemical modification, and ease of generation [13,14]. Furthermore, when designed appropriately, the chemical modification of aptamers often does not affect—and sometimes even strengthens—their binding affinity for their targets, whereas the chemical modification of antibodies may remove their affinity for their targets [13]. Moreover, antibodies can only bind to a narrow range of targets, such as immunogenic non-toxic molecules and other molecules that do not cause strong immune responses [13]. In contrast, aptamers can selectively and tightly bind to a broad range of targets, including proteins [15–17], cells [18,19], microorganisms [20], and small-molecule contaminants [21–24], which may be either immunogenic or non-immunogenic. In addition, the in vivo generation of antibodies is a lengthy process (~6 months or longer) and requires a living host, whereas aptamers can be generated through in vitro selection (~2–8 weeks) using a method known as systematic evolution of ligands by exponential enrichment (SELEX) [13].

Three research groups independently developed the first SELEX method and all reported their findings in 1990: Tuerk and Gold [25], Ellington and Szostak [26], and Joyce [27]. Since then, many derivative methods of SELEX have been developed, such as capillary electrophoresis (CE)-SELEX [28–30], graphene oxide (GO)-SELEX [31–33], cell-SELEX [34–36], in vivo SELEX [37,38], and Capture-SELEX [22,23,39]. Capture-SELEX selects DNA or RNA aptamers that bind small-molecule contaminants and is partly based on FluMag-SELEX. Stoltenburg et al. reported FluMag-SELEX in 2005 [40], and it is based on the magnetic bead (MB)-based SELEX, whereby the fluorescein-labeled oligonucleotides is to monitor the enrichment of target-specific aptamers with the quantification in each round, and biotinylated target molecules are immobilized onto the surface of streptavidin-coated magnetic beads (SA-MBs). However, it is more challenging to immobilize biotinylated small-molecule targets than biotinylated large-molecule targets such as proteins [41], as small-molecule targets have fewer functional groups. Thus, Capture-SELEX is an effective alternative to FluMag-SELEX, as Capture-SELEX involves the immobilization of a biotinylated hybridized library on a solid support, which is more feasible than the immobilization of biotinylated targets on a solid support (Figure 1). Due to the differences in the principle, Capture-SELEX also has some deficiencies compared to other SELEX methods [41]. The false-positive may happen in the immobilization process of Capture-SELEX. The complemented library may have a dynamic dissociation equilibrium due to the weak force, and some unbound sequences (non-biotinylated) are also dissociated from the solid support. Moreover, conformational changes of the target are essential for the success of aptamer screening and diversity in the selection process. This review discusses Capture-SELEX-based selection strategies for small-molecule contaminant targets, methods used to identify and characterize small molecule–aptamer interactions, and biosensing applications of aptamers generated using Capture-SELEX.

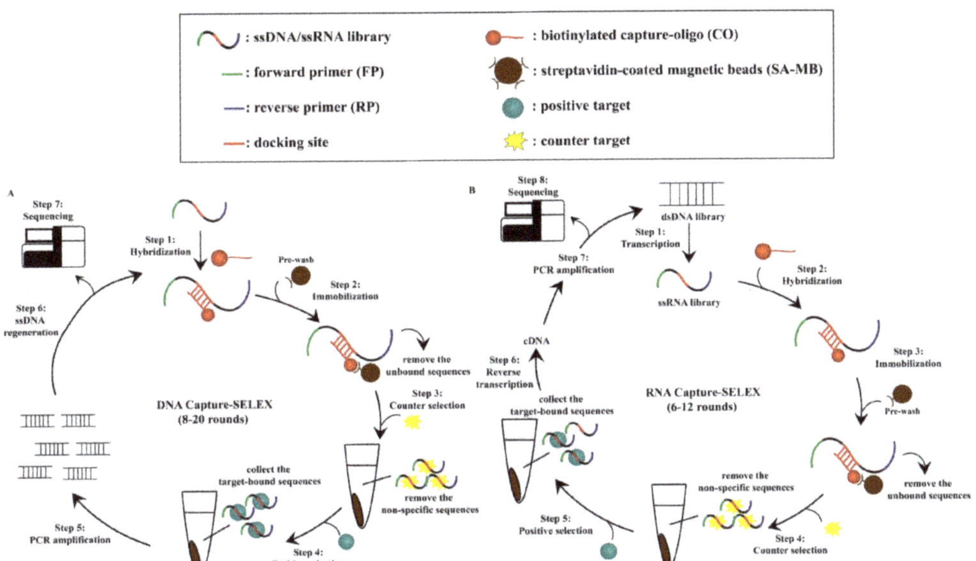

Figure 1. The experimental flowchart of DNA Capture-SELEX and RNA Capture-SELEX. (**A**) DNA Capture-SELEX. Each DNA Capture-SELEX selection round consists of 7 main steps: (1) The designed ssDNA library template is hybridized with biotinylated capture-oligonucleotide. (2) The hybridized product is immobilized on the washed streptavidin-coated magnetic beads. (3–4) Target-bound sequence elution: counter selection is followed by positive selection. (5) PCR amplification was performed on the eluted ssDNA to obtain the amplified dsDNA pool, (6) and regenerate ssDNA for the next DNA Capture-SELEX round (either gel purification or magnetic bead-based method). (7) After DNA Capture-SELEX, aptamer sequences were identified through sequencing. (**B**) RNA Capture-SELEX. The principles and procedures are similar to (**A**), except that RNA Capture-SELEX consists of additional transcription (step 1) to convert DNA to RNA and reverse transcription (step 6) to convert RNA to DNA steps.

2. Principle and General Procedure of Capture-SELEX for Aptamer Selection

2.1. Overview of DNA and RNA Capture-SELEX

The classical procedure of the DNA Capture-SELEX selection method involves hybridization, immobilization, target-bound sequence elution, polymerase chain reaction (PCR) amplification, regeneration of single-stranded DNA (ssDNA), and sequencing (Figure 1A). Briefly, in the first step, a random ssDNA library is designed with a docking site that is complementary to the biotinylated capture oligonucleotides, and these ssDNAs are hybridized with capture-oligonucleotides to form biotinylated hybridized library via Watson-Crick base pairing. Next, the biotinylated library is immobilized onto SA-MBs, which function as solid support, and the loaded SA-MBs are then washed several times to remove the residual and non-specific sequences using a magnetic rack and generate a washed biotinylated library on the MBs (which comprise superparamagnetic iron oxide). Subsequently, a magnetic rack also used to discard supernatant (unbound or non-specific sequences), after which the counter-selection process is performed, wherein a counter target with a similar chemical structure and concentration as the positive targets, enhances the binding affinity and specificity of aptamers towards it positive small-molecule target [42]. Counter selection can be performed either in the last few rounds [39,43] or between the selected rounds [44,45] of Capture-SELEX, and the concentration of the counter target should be the same as the positive target in step 3 (Figure 1A). The negative selection is followed by positive selection. Next, an elution is performed; if the binding affinity between the target and aptamer is higher than that between the aptamer and the capture oligonucleotide, the target-bound aptamer candidates are eluted. As the library sequences interact with the positive targets, only sequences with conformational changes due to the interaction can be eluted from the MBs [46].

In this review, we summarize two main strategies used to regenerate ssDNA from amplified dsDNA in step 6 (Figure 1A): the gel purification method [21,23,47] and the MB-based method [24]. A primer is designed and modified in order to identify the sense strand in the denaturing polyacrylamide gel, such as a reverse primer with poly-dA$_{20}$ extension or a forward primer labeled with either fluorescein or the cyanine dye Cy3. The gel purification method involves denaturing polyacrylamide gel electrophoresis (PAGE), followed by extraction and purification. The MB-based method begins with PCR products with biotinylated reverse primers being incubated to facilitate their deposition onto MBs. The supernatant is then removed, and the non-biotinylated forward strands of dsDNA PCR products are released from the MBs by treatment with sodium hydroxide and then neutralized by treatment with an appropriate amount of tris(hydroxymethyl)aminomethane hydrochloride or hydrochloric acid. The regenerated ssDNA in each round is collected for the next round of selection and then prepared for sequencing after several rounds of selection. Currently, Capture-SELEX is mostly applied for the selection of DNA aptamers rather than RNA aptamers. RNA Capture-SELEX adopts the same selection strategies as DNA Capture-SELEX, except for including two more steps: transcription and reverse transcription (Figure 1). In addition, more rounds need to be performed in DNA Capture-SELEX (8–20 rounds) than in RNA DNA Capture-SELEX (6–12 rounds) to enrich the target-bound pool sufficiently.

2.2. Sanger and Next-Generation Sequencing and Analysis

Sequencing is a crucial step performed after the SELEX-based selection of aptamer candidates and before aptamer–target characterization. Here, we review two of the main methods used for sequencing aptamers, those being Sanger sequencing and next-generation sequencing (NGS). Sanger sequencing is a DNA sequencing technology based on the chain termination method invented by Frederick Sanger and his colleagues [48] in 1997. The main workflow involves blocking the polymerase-mediated elongation of DNA through the incorporation of fluorophore-labeled dideoxynucleotides (dideoxyadenosine triphosphate, dideoxyguanine triphosphate, and dideoxythymine triphosphate) at the 3' ends of DNA sequences, resulting in various lengths of DNA fragments for size separation and fluorescent-based detection. This method has the advantages of high precision, high efficiency, and low radioactivity, and the disadvantages of being expensive and having low-quality primer binding in the first 15 to 40 base pairs [49]. Subsequently, high-throughput second- and third-generation sequencing were invented and superseded Sanger sequencing [50]. In 1998, Balasubramanian and Klenerman co-invented Solexa sequencing [51], also known as NGS. This method differs from Sanger sequencing mainly in its mode of chain termination: modified deoxynucleoside triphosphates (dNTPs) with a reversible terminator are used to terminate polymerization and are then removed to allow incorporation of the next modified dNTP [52]. NGS offers some advantages over Sanger sequencing, including enabling massively parallel sequencing in a short period and having a lower cost per base pair.

In order to analyze sequencing data, in silico technique towards SELEX facilitate the aptamer selection [53–55]. The initial step is primary sequence analysis using some bioinformatic tools, such as ClustalW [56] and Clustal Omega (https://www.ebi.ac.uk/Tools/msa/clustalo/ (accessed on 12 October 2022)) [57]. Multiple sequence alignment is performed to divide sequences into families or clusters. The conserved regions of the sequences are analyzed using Gblocks software. Furthermore, Multiple Expectation maximizations for Motif Elicitation (https://meme-suite.org/meme/tools/meme (accessed on 12 October 2022)) [58] is used to identify the sequence motif, which provides the basis for downstream analysis. Aptamers can form diverse secondary structures, such as stem-loop, triplex, G-quadruplex, and pseudoknot structures; thus, enriched sequences can also be selected based on secondary structure and Gibbs free energies (ΔG) prediction. Several nucleic acid structure-prediction Web servers are available and are often used to computationally predict secondary structures and Gibbs free energies (ΔG) of aptamer sequences. These include MFOLD (http://www.unafold.org/mfold/applications/dna-folding-form.php (accessed on 12 October 2022)) [59], KineFold (http://kinefold.curie.fr/ (accessed on 12 October 2022)) [60], and RNAstructure (https://rna.urmc.rochester.edu/RNAstructure.html (accessed on 12 October 2022)) [61]. The predicted secondary structures were then converted into unique three-dimensional (3D) structures using online web servers. These include fragment-based methods: RNAComposer (https://rnacomposer.cs.put.poznan.pl/ (accessed on 12 October 2022)), 3dRNA (http://biophy.hust.edu.cn/3dRNA (accessed on 12 October 2022)), and Vfold 3D (http://rna.physics.missouri.edu/vfold3D/ (accessed on 12 October 2022)); and energy-based method: simRNA (https://genesilico.pl/SimRNAweb (accessed on 12 October 2022)) [62]. Molecular docking tools include AutoDock, AutoDock Vina, and DOCK, which were used to predict the predominant binding modes and regions of the target molecule based on the generated binding scores for specific sequences [63–65]. After a few prediction steps, the aptamer candidates are shortlisted and subjected to binding tests and characterization.

3. Small-Molecule–Aptamer Interaction and Characterization

The secondary structures of aptamers often organize into complex three-dimensional structures to maintain global stability and form functional conformations. Therefore, a stable aptamer can strongly bind to small molecules and shape the aptamers-target complex via weak noncovalent interaction forces, including hydrogen bonding, π-π stacking, van der Waals forces, hydrophobic, and electrostatic interactions. The aptamer-target interactions may either depend on the availability of narrow binding pockets in the three-folded structures of aptamers or the chemical characteristics of targets and the aptamer candidate or the length of the oligonucleotides, for example, removing excess flanking nucleotides outside the binding site (truncated aptamer), which is also a strategy to improve binding affinity and specificity [66,67]. In order to assess the binding strength and specificity of the interaction between an aptamer and its target, binding assays are used to determine the dissociation constant (K_d) (Table 1), i.e., the binding affinity between an aptamer and its cognate target (and between an aptamer and its non-cognate targets), where the lower the K_d value, the stronger the interaction. Several factors may influence the K_d value, such as temperature, pH value, ionic concentrations, and hydrophobicity of the solution [66]. Below, we discuss the six standard assays used for characterizing the interactions between small-molecule contaminants and aptamers generated using DNA/RNA Capture-SELEX (Figure 2). In contrast with traditional methods, such as HPLC and LC-MS, these assays are highly sensitive, non-toxic, and inexpensive.

Table 1. A list of small molecule contaminant-specific aptamers developed by Capture-SELEX method as of to date.

Name	Class	Aptamer Sequence (5′–3′)	nt	Dissociation Constant (K_d)	Ref.
Kanamycin A CAS no.: 59-01-8	Aminoglycoside antibiotic	ATACCAGCTTATTCAATTAGCCCGGTATTGA GGTCGATCTCTTATCCTATGGCTT GTCCCCCAT GGCTCGGTTATATCCAGATAGTAAGTGCAATCT	97	3.9 μM	[22]

Table 1. *Cont.*

Name	Class	Aptamer Sequence (5′–3′)	nt	Dissociation Constant (K_d)	Ref.
Ofloxacin CAS no.: 82419-36-1	Antibiotic	ATACCAGCTTATTCAATTGCAGGGTAT CTGAGGCTTGATCTACTAAATGTCGT GGGGCATTGCTATTGGCGTTGATACGTA CAATCGTAATCAGTTAG	98	0.11 ± 0.06 nM	[39]
Furaneol CAS no.: 3658-77-3	Aroma compound	CGACCAGCTCATTCCTCACCACGA GAAAGGAGCTCGATGAACTGCGAGC CGGATTCGACCCTATGCGAGTAGGTG GTGGATCCGAGCTCACCAGTC	96	1.1 ± 0.4 μM	[68]
Vanillin CAS no.: 121-33-5	Flavoring	CGACCAGCTCATTCCTCAGGAGAAA CATGGAGTCTCGATGATGTAGGAGCGG CGGAACGTAGGAAGAGA GGATGACGGAGGATCCGAGCTCACCAGTC	98	0.9 ± 0.3 μM	[23]
Erythromycin CAS no.: 114-07-8	Antibiotic	AGGAATTCACGTCTCACTGGATTCACGCAC GCCAAGGACTGCACTTAAGGTTAGA TAGCCCCATGCAGTGAGTCAGGATATCG	83	20 ± 9 nM	[24]
Spermine CAS no.: 71-44-3	Biogenic amine	TATGAACGATTTACTCGTACAGA CGACACTTATCATTTGC	40	9.648 ± 0.896 nM	[44]
Tetrodotoxin CAS no.: 4368-28-9	Toxin	ATACCAGCTTATTCAATTTAATGCGGG GTGAGGCTCAATCAAGGAAAGATATAAGTAA GCAAAAAGGTCAAACAAGGGCGA GATAGTAAGTGCAATCT	98	7 ± 1 nM	[43]
Chlorpromazine CAS no.: 50-53-3	Phenothiazine	TCGGAGGGAAGTGCACCCATTCTTGGAAACAG GAGCTCCTGAACCGCCCACACGC	55	69.8 ± 9.81 nM	[69]
Roxithromycin CAS no.: 80214-83-1	Antibiotic	ATTGGCACTCCACGCATAGGCACACCCAC CGGCCTAGCCACACCATGCTGCTGTTGC CCACCTATGCGTGCTACCGTGAA	80	0.46 ± 0.08 μM	[70]
Di(2-ethylhexyl) phthalate (DEHP) CAS no.: 117-81-7	Plasticizer	ACGCATAGGGTGCGACCACAT ACGCCCCATGTATGTCCCTTGGTT GTGCCCTATGCGT	58	2.26 ± 0.06 nM	[71]

Table 1. *Cont.*

Name	Class	Aptamer Sequence (5′–3′)	nt	Dissociation Constant (K_d)	Ref.
Acyclovir (ACV) CAS no.: 59277-89-3	Aminoglycoside Antibiotics	TGAGCCCAAGCCCTGGTATGTGAAAACATACT AGACGTGGCTATGTATTTTTAAATCAA TGGCAGGTCTACTTTGGGATC	80	32.67 ± 4.127 nM	[72]
Glutamate CAS no.: 56-86-0	Excitatory neurotransmitter	GCATCAGTCCACTCGTGAGGTCGACTGA TGAGGCTCGATCAGGAGCGCCGCTCG ATCGCACTTTCACAGGATAGTAGTTG GTAGCGACCTCTGCTAGA	98	12 ± 6 µM	[73]
Famciclovir (FCV) CAS no.: 104227-87-4	Aminoglycoside Antibiotics	TGAGCCCAAGCCCTGGTATGTGAAAACA TACTAGACGTGGCTATGTATTTTTAAAT CAATGGCAGGTCTACTTTGGGATC	80	47.35 ± 10.42 nM	[72]
Ganciclovir (GCV) CAS no.: 82410-32-0	Aminoglycoside Antibiotics	TGAGCCCAAGCCCTGGTATGTGAAA ACATACTAGACGTGGCTATGTATTTTTAA ATCAATGGCAGGTCTACTTTGGGATC	80	47.91 ± 13.47 nM	[72]
Penciclovir (PCV) CAS no.: 39809-25-1	Aminoglycoside Antibiotics	TGAGCCCAAGCCCTGGTATGTGAAAA CATACTAGACGTGGCTATGTATTTTT AAATCAATGGCAGGTCTACTTTGGGATC	80	33.29 ± 5.851 nM	[72]
Levamisole (LEV) CAS no.: 14769-73-4	Veterinary drug	AATCAAACGCTAAGGTCAAGGGAGAG TGCACCCATTCTTGGGGCCCCGGGC CAGCCCCGACACGCCGCCGAAG CTTGGTACCCGTATCGT	90	66.15 ± 11.86 nM	[74]
Valaciclovir (VACV) CAS no.: 124832-26-4	Aminoglycoside Antibiotics	TGAGCCCAAGCCCTGGTATGTGAAA ACATACTAGACGTGGCTATGTATTTT TAAATCAATGGCAGGTCTACTTTGGGATC	80	44.26 ± 6.744 nM	[72]
Tetrodotoxin CAS no.: 4368-28-9	Toxin	ATACCAGCTTATTCAATTTAATGCGGGGTGAGG CTCAATCAAGGAAAGATATAAGTAAGCAAAAG GTCAAACAAGGGCGAGATAGTAAGTGCAATCT	98	7 ± 1 nM	[43]

Table 1. *Cont.*

Name	Class	Aptamer Sequence (5′–3′)	nt	Dissociation Constant (K_d)	Ref.
Ribavirin CAS no.: 36791-04-5	Antiviral agent	AAAGTAATGCCCGGTAGTTATTCAAAGATGA GTAGGAAAAGA	42	61.19 ± 21.48 nM	[45]
Gymnodimine-A CAS no.: 173792-58-0	Toxin	GCGACCGAAGTGAGGCTCGATCCAAGGTG GACGGGAGGTTGGATTGTGCGTG	52	34.5 ± 1.72 nM	[75]
λ-cyhalothrin CAS no.: 91456-08-6	Pesticide	AGGGGAAGCACGGGCGGGCG	20	10.27 ± 1.33 nM	[76]
Malachite green (MG) CAS no.: 569-64-2	Veterinary drug	(1) CGCAGCGCGGCAGACAGTCAGGCTC AGCACGTGGCA	36	102.46 μM	[77]
		(2) CACTCCACGCATAGGGACGCGAATTGCG GACCTATGTGTGGTGTG	45	2.3 ± 0.2 μM	[78]
Leucomalachite green (LMG) CAS no.: 129-73-7	Antimicrobial	CACTCCACGCATAGGGACGCGAATTGCGGA CCTATGTGTGGTGTG	45	8.2 ± 1.2 μM	[78]
Phorate CAS no.: 298-02-2	Organothiophosphate insecticide	AAGCTTGCTTTATAGCCTGCAGCGAT TCTTGATCGGAAAAGGCTGAGAGCTACGC	55	1.11 μM	[79]
Profenofos CAS no.: 41198-08-7	Organothiophosphate insecticide	AAGCTTGCTTTATAGCCTGCAGCGATTC TTGATCGGAAAAGGCTGAGAGCTACGC	55	1 μM	[79]
Isocarbophos CAS no.: 24353-61-5	Organothiophosphate insecticide	AAGCTTGCTTTATAGCCTGCAGCGATTCTTGATCG GAAAAGGCTGAGAGCTACGC	55	0.83 μM	[79]
Omethoate CAS no.: 1113-02-6	Organothiophosphate insecticide	AAGCTTTTTTGACTGACTGCAGCGATTC TTGATCGCCACGGTCTGGAAAAAGAG	54	2 μM	[79]

Table 1. *Cont.*

Name	Class	Aptamer Sequence (5′–3′)	nt	Dissociation Constant (K_d)	Ref.
Cortisol CAS no.: 50-23-7	Steroid hormone	GGAATGGATCCACATCCATGGATGGGCAATG CGGGGTGGAGAATGGTTGCCGCA CTTCGGCTTC ACTGCAGACTTGACGAAGCTT	85	6.9 ± 2.8 nM	[80]
Tobramycin CAS no.: 32986-56-4	Aminoglycoside antibiotic	CCATGATTCAACTTTACTGGTCTTGTCT TGGCTAGTCGTGTGTCATTCCCGTAAGGG	57	200 nM	[47]
Clenbuterol hydrochloride (CLB) CAS no.: 21898-19-1	beta-adrenergic agonist	AGCAGCACAGAGGTCAGATGTCATCTGAAG TGAATGAAGGTAAACATTATTTCATTAA CACCTATGCGTGCTACCGTGAA	80	76.61 ± 12.7 nM	[81]
Ractopamine CAS no.: 97825-25-7	Veterinary drug	AGCAGCACAGAGGTCAGATGGTCTCTAC TAAAAGTTTTGATCATACCGTTCACTAATTG ACCTATGCGTGCTACCGTGAA	80	54.22 ± 8.02 nM	[82]
Atrazine CAS no.: 1912-24-9	Herbicide	TGTACCGTCTGAGCGATTCGTACTTTATTCG GGAAGGGTATCAGCGGGGTTCAACAA GCCAGTCAGTCAGTGTTAAGGAGTGC	83	3.7 nM	[83]
Zearalenone CAS no.: 17924-92-4	Mycotoxin	ATACCAGCTTATTCAATTCTACCAGCTTT GAGGCTCGATCCAGCTTATTCAATTATA CCAGCTTATTCAATTATACCAGCACAATC GTAATCAGTTAG	98	15.2 ± 3.4 nM	[84]
Fenitrothion CAS no.: 122-14-5	Phosphorothioate insecticide	CTCTCGGGACGACGGGCCGAGTAGTCTCCAC GATTGATCGGAAGTCGTCCC	51	33.57 nM	[85]
Malathion CAS no.: 121-75-5	Organophosphate insecticide	GGGATACAGGTAGTATGGCATGTGCTAGCG GTTGCA	36	22.56 nM	[86]
Fipronil CAS no.: 120068-37-3	Insecticide	ACGACAGATAGTGTGTACATGAAGGGTCGT	30	15 nM	[87]

Table 1. *Cont.*

Name	Class	Aptamer Sequence (5′–3′)	nt	Dissociation Constant (K_d)	Ref.
Diazinon CAS no.: 333-41-5	Organophosphate insecticide	TTCCGATCAATCGTGGAGACTACTCGGCCC	30	4.571 ± 0.714 μM	[88]
Ethyl carbamate (EC) CAS no.: 51-79-6	Organic compound	GGGGGCACGGGAGGT	15	17.97 ± 0.98 nM	[89]
3,4-methylenedioxypyrov-alerone (MDPV) CAS no.: 687603-66-3	Synthetic cathinone	CTTACGACTCAGGCATTTTGCCGGGTAAC GAAGTTACTGTCGTAAG	46	6.1 ± 0.2 μM	[90]
Dichlorodiphenyltrichloroethane (o,p′—DDT) CAS no.: 789-02-6	Insecticide	TCCAGCACTCCACGCATAACGAATTGTGC TCAATGCGCCCCTGCAGTGAATGTGGAATT TGTTATGCGTGCGACGGTGAA	80	412.3 ± 124.6 nM	[91]
Penicillin G CAS no.: 61-33-6	β-lactam antibiotic	GGGAGGACGAAGCGGAACGAGATGTAGAT GAGGCTCGATCCGAATGCGTGACGTCTATCGGA ATACTCGTTTTTACGCCTCAGAAGACACGCCCGACA	98	–	[21]

3.1. SYBR Green I (SGI) Assay

The binding affinity of Capture-SELEX-generated aptamers for small molecules can be identified by a novel label-free fluorescence SGI assay [24,68,70,73,83]. SGI is a fluorescent nucleic-acid-intercalating dye that was introduced in the 1990s and is widely applied in real-time PCR [92–94], fluorescent gel imaging [95,96], and flow cytometry [97,98]. It intercalates into the minor groove of DNA base pairs or adenine–thymine-rich stem-loop sites to form a fluorescent complex with DNA [99]. The fluorescence intensity of SGI–DNA complexes is measured in the range of 505 to 650 nm, with an excitation wavelength of 495 nm and an emission wavelength of 525 nm, using a spectrophotometer or a microplate reader; thus, an SGI assay identifies the binding affinity of aptamers for small-molecule targets by detecting the change in the fluorescent signal induced by the dissociation of SGI from the aptamers upon their binding with these small-molecule targets, and therefore the fluorescence intensity decreases (Figure 2A). The fluorescence intensity will be recovered as the concentration of the aptamers' small-molecule targets increases.

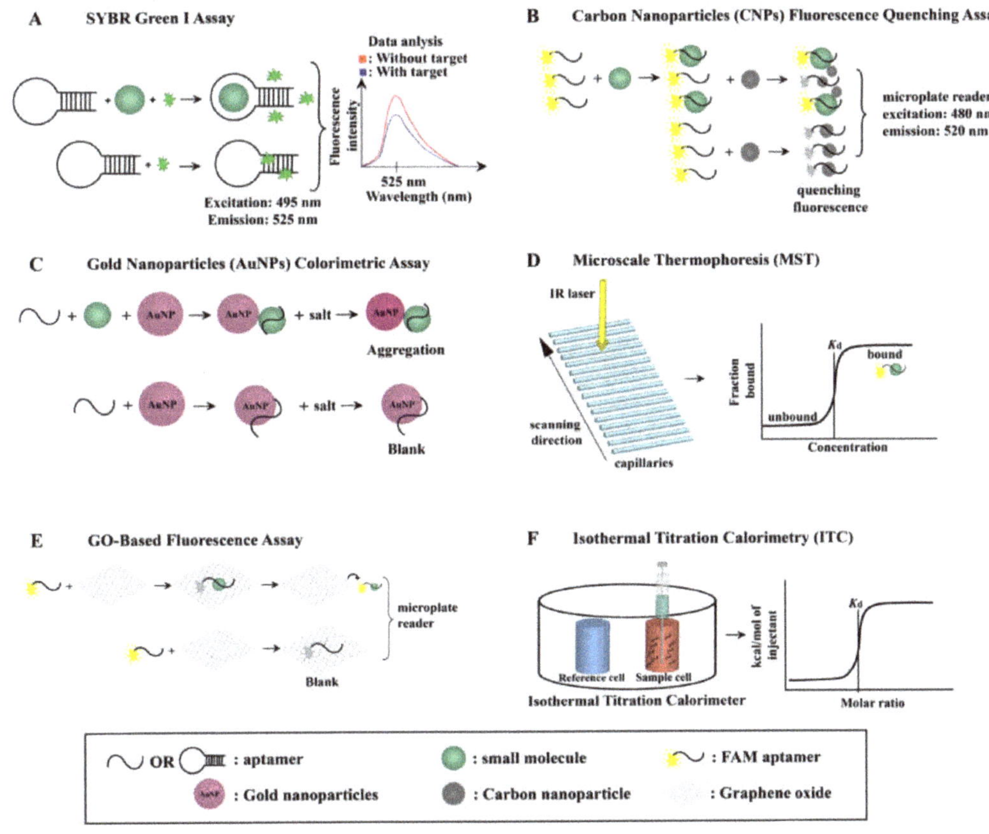

Figure 2. A schematic illustration of six commonly used methods for detecting interactions between small molecule and aptamer. (**A**) SYBR green I (SGI) assay. (**B**) Carbon nanoparticles (CNPs) fluorescence quenching assay. (**C**) Gold nanoparticles (AuNPs) colorimetric assay. (**D**) Microscale thermophoresis (MST) assay. (**E**) Graphene oxide (GO)-based fluorescent assay. (**F**) Isothermal titration calorimetry (ITC) assay.

3.2. Carbon Nanoparticle (CNP) Fluorescence Quenching Assay

CNPs can be readily synthesized [45] in research laboratories, have low toxicity, and are biocompatible [100]. Due to the strong fluorescence-quenching ability of CNPs' sp^2- and π-rich structures, they are useful for characterizing the binding affinity of aptamers for small molecules. This means that when an aptamer is adsorbed onto the surface of CNPs, π-π stacking interactions that form between the nucleobases and nucleosides of the aptamer and the CNPs result in fluorescence quenching. Therefore, in the absence of aptamers' small-molecule targets, aptamers are bound to CNPs, resulting in fluorescence quenching, whereas in the presence of aptamers' small-molecule targets, aptamers are bound to these targets rather than CNPs, and thus fluorescence is not quenched (Figure 2B) [101].

3.3. Gold Nanoparticle (AuNP) Colorimetric Assay

The AuNP colorimetric assay is widely used for characterizing the binding affinity of an aptamer for its target [24,45,70,71,76,80,83,84]. It is a label-free assay as AuNPs allow the identification of a target reaction based on colorimetric signals. AuNPs can be synthesized by reducing chloroauric acid with citric acid and ascorbic acid (the Turkevich method) [102] or with sodium citrate [103]. The Turkevich method affords AuNPs coated with negatively charged citrate ions, and thus the AuNPs are treated with sodium chloride to neutralize their surface charge and thereby induce their aggregation (Figure 2C) [104]. Owing to the electrostatic screening effect, the surface plasma resonance absorption peak of AuNPs undergoes a red shift from 520 nm to 650 nm, which is measured using a microplate reader and reflects the binding affinity of aptamers to their small-molecule targets. In the absence of their small-molecule target, aptamers are adsorbed onto the surface of AuNPs, increasing their colloidal stability of AuNPs and reducing their salt-induced aggregation [105]. However, in the presence of their small-molecule targets, aptamers binds to their targets rather than AuNPs, which decreases the colloidal stability of AuNPs and thereby increases their salt-induced aggregation. This results in a solution containing AuNPs, aptamers, and their small-molecule targets changing color from red to pink to purple or even to blue or gray [70].

3.4. Microscale Thermophoresis (MST) Assay

The MST assay is a powerful tool for testing the binding affinity of aptamers for their small-molecule targets as it only requires a small amount of sample (up to 4 µL), involves a simple preparation process, and rapidly provides accurate results (within 15 min) [75,76,80]. It is a fluorophore-labeled and immobilization-free assay that is carried out using a Monolith NT.115 instrument. As shown in Figure 2D, the capillary tray of this instrument can accommodate up to 16 thin glass capillaries for operation. An infrared laser heats a specific capillaries area, creating a microscopic temperature gradient on them. Thus, when an aptamer binds to its small-molecule target, there is a significant change in fluorescence intensity due to thermophoresis. An MST binding curve (dose-response curve) is plotted using Affinity Analysis Software to represent specific binding that is specific to its small-molecule target, which automatically calculates the K_d of an aptamer for its small-molecule target.

3.5. GO-Based Fluorescent Assay

GO can be synthesized from graphite via the Hummers and Offeman method [106], which uses potassium permanganate (as the oxidizing agent) and sodium nitrate in a solution of sulfuric acid [107]. GO has an sp^2 structure and numerous oxygen-containing functional groups [108], such as carbonyl groups, hydroxyl groups, and carboxylic acid groups, and thus can strongly quench fluorescence. This quenching occurs when a $5'$-fluorescein-labeled aptamer is adsorbed onto the surface of GO through π-π stacking interactions and hydrogen bonding (Figure 2E). When $5'$-fluorescein labeled aptamer is absorbed onto GO, GO-mediated fluorescence quenching occurs. The fluorescence intensity will be recovered when the labeled aptamer is desorbed from the surface of GO in the presence of small-molecule target. The fluorescence intensity change are measured using a microplate reader, and eventually the K_d value of the aptamer for its small-molecule target is calculated [72,77,109].

3.6. Isothermal Titration Calorimetry (ITC) Assay

The ITC assay is a label-free and immobilization-free assay for measuring the binding affinity of an aptamer for its small-molecule target [44,83] in an instrument comprising a reference cell and a sample cell (Figure 2F). The reference cell is filled with a buffer and water (without the aptamer or its small-molecule target), while the sample cell is filled with the buffer and the aptamer and is titrated via syringe with up to 20 volumes of the aptamer's small-molecule target [77]. The temperature in the two cell units is maintained at a constant value, and when the small molecule is titrated against the aptamer in the sample cell, an exothermic binding reaction occurs. The feedback system of the ITC instrument then reduces the power supply to the sample cell to prevent an increase in temperature, leading to a difference in the amount of energy supplied to the reference cell and the sample cell. Therefore, during this titration, the power supply to the sample cell continuously decreases until the small-molecule–aptamer binding has been completed. The binding affinity of the aptamer can then be determined from the energy curve generated [110].

The ITC instrument measures the exothermic reaction (heat released) and endothermic reaction (heat absorbed) when the molecules interact. The peak goes downward when the exothermic reaction occurs, whereas an upward peak is measured when an endothermic reaction occurs. The downward peak between the interaction of aptamer and small-molecule target are measured from the ITC instrument using software such as MicroCal ITC200 and MicroCal PEAQ-ITC.

4. Development of Aptamer-Based Biosensors

To date, Capture-SELEX has been used for screening and generating a series of aptamers against small-molecule contaminants (Table 1). As aptasensors can be used for low-cost, rapid, and real-time detection of small-molecule contaminants, they have increasingly supplanted conventional antibody-based biosensors. Fluorescent aptasensors are the most popular (Table 2), and their working principle is similar to that of fluorescent-based characterization methods. Therefore, two other types of aptasensor are discussed here (Figure 3).

Table 2. Representative biosensors using aptamers against small molecule contaminants that are generated from Capture-SELEX.

Types	Aptamer Targets	Limit of Detection (LOD)	Detection Range	Ref.
Ion-sensitive field-effect transistor (ISFET)	Vanillin	0.155 µM	0.155–1.0 µM	[23]
	Furaneol	–	0.1–10 µM	[68]
Electrochemical impedance spectroscopy (EIS)	Penicillin G	0.00051 µM	1.2 nM–2.99 µM	[21]
	Di(2-ethylhexyl) phthalate (DEHP)	0.264 pM	–	[71]
	Glutamate	0.0013 pM	0.01 pM–1 nM	[73]
	Spermine	0.052 nM	0.1–20 nM	[44]
	Tyramine (TYR)	2.48 nM	3.64–728.97 nM	[109]
	β-phenethylamine	3.22 nM	4.13–825.22 nM	[109]
	Acyclovir	2.13 nM	8.88–444.03 nM	[72]
	Famciclovir	1.74 nM	6.22–311.2 nM	[72]
	Ganciclovir	2.08 nM	7.84–391.80 nM	[72]
	Penciclovir	1.97 nM	7.9–394.85 nM	[72]
	Valaciclovir	1.17 nM	6.17–308.32 nM	[72]
	Ribavirin	2.74 nM	4.09–204.75 nM	[45]
Fluorescent	Malachite Green	5.84 nM	4.69 nM–2.35 µM	[77]
	Cadmium ions (Cd(II))	40 nM	0–1000 nM	[111]
	Clenbuterol hydrochloride (CLB)	0.22 nM	0.32–159.44 nM	[81]
	Ractopamine	0.13 nM	0.33–331.79 nM	[82]
	Chlorpromazine	0.67 nM	1–100 nM	[69]
	Fenitrothion	14 nM	0–80 nM	[85]
	Malathion	6.08 nM	–	[86]
	Fipronil	3.4 nM	0–70 nM	[87]
	Diazinon	148 nM	0.1–25 µM	[88]
Lateral flow aptasensor (LFA)	Erythromycin	3 pM	1 pM–10 nM	[24]
	Ethyl carbamate	0.024 µM	0.11–0.67 µM	[89]
	Roxithromycin	0.077 µM	0–4.44 µM	[70]
Colorimetric	λ-cyhalothrin	41 nM	0.22–1.11 µM	[76]
	Zearalenone	12.5 nM	12.5–402.1 nM	[84]
	Levamisole	1.12 nM	1–200 nM	[74]
Biolayer interferometry (BLI)	Gymnodimine-A	6.21 nM	55–1400 nM	[75]

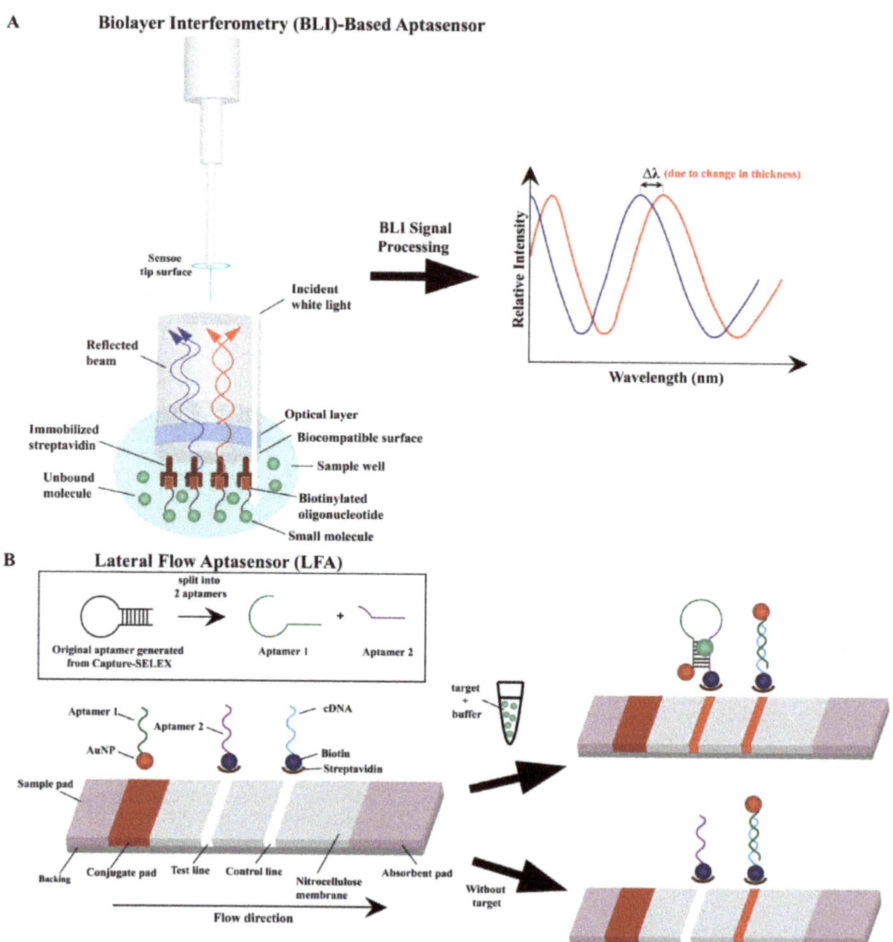

Figure 3. The schematic illustration of aptasensors for the detection of small molecules using novel aptamer generated from Capture-SELEX. (**A**) Biolayer interferometry (BLI)-based aptasensor. (**B**) A lateral flow aptasensor (LFA) is generally constructed by 4 sections: sample pad, conjugate pad, nitrocellulose membrane with test line and control line, and absorbent pad. The sample flow from left to right laterally (left). When the target is loaded onto the sample pad, positive result (red band for both control and test line) is observed after 15 min (right).

4.1. Biolayer Interferometry (BLI)-Based Aptasensor

Optical-fiber-based BLI biosensors have been developed for real-time, sensitive, and rapid measurement [112] of interactions between biomolecules, including DNA–protein [113,114], antibody–antigen [115], and DNA–small-molecule-contaminant interactions [116,117]. The sensor tip of an optical fiber consists of two reflective surfaces: a streptavidin-coated biocompatible surface, which is immersed in the sample solution, and an optical layer, which functions as an internal surface (Figure 3A). The biocompatible surface is functionalized to immobilize a biotinylated aptamer and minimize non-specific binding. Inside the sensor, these two surfaces generate an interference pattern by reflecting incident white light. When an aptamer binds to its small-molecule target, a spectral red-shift ($\Delta\lambda$) occurs as the bound compound immobilizes on the tip surface and the surface thickness increases. The false-positive results from this sensor can be minimized as the non-specific and unbound molecules can be differentiated from the molecules with high binding affinity, resulting in high detection accuracy [116].

4.2. Lateral Flow Aptasensors (LFAs)

LFAs have been developed as portable detection devices with cost-effective, rapid ($\leq$15 min), and easy operation and broad applications, such as in the detection of pregnancy, severe acute respiratory syndrome coronavirus 2 [118,119], and infectious diseases [120]. Owing to the drawbacks of antibodies, LFAs are gradually replacing antibody-based lateral flow biosensors, although the latter remains predominantly used. In addition, LFAs have been used to detect antibody-inaccessible small-molecule contaminants [24,89,121]. Figure 3B depicts the LFA concept [121–123], in which AuNPs are typically used for colorimetric identification of the test line and control line on the strip [124]. The high affinity aptamer generated from Capture-SELEX against the specific small-molecule target that splits into two fragmented aptamers (aptamer 1 and 2; Figure 3B) [123]. A sample solution (a solution of an aptamer's small-molecule target in buffer) is loaded onto the sample pad and flows through the conjugate pad (containing AuNP–aptamer 1 conjugates) by capillary action. When aptamer–small-molecule-target binding occurs, two fragmented aptamers could rejoin the three-dimensional structure without affecting the binding affinity, and thus a red band appears in the test line. A complementary DNA (cDNA) is designed to hybridize with aptamer 1 in the control line and generate a visual signal, which is used to verify whether the aptasensor is working properly based on the appearance of red band in both presence or absence of the aptamer target. The excess sample reagent flows from the test line to the control line and to the absorbent pad eventually. Two slightly different LFAs have also been developed that use aptamers generated using Capture-SELEX. The LFA devised by Du et al. [24] combines lateral flow strips and the recombinase polymerase amplification technique for detecting erythromycin in tap water, whereas that devised by Xia et al. [89] uses aptamer EC1-34 as a recognition probe for detecting ethyl carbamate. The latter LFA differs from the former LFA only in terms of its use of a cationic polymer (such as poly(dimethyldiallylammonium chloride)) instead of streptavidin in the test line.

5. Biosensor Applications

5.1. Food Safety Analysis

5.1.1. Veterinary Drug Residues and Pesticides

Ractopamine (RAC) is a β-adrenergic agonist used illegally as an animal feed additive for increasing skeletal muscle mass, reducing fat deposition, and increasing protein accretion in livestock [125,126]. The accumulation of RAC in animals may increase the risk of food poisoning in humans and have other adverse effects on human health, such as causing headache, tachycardia, and muscle tremors [127]. Therefore, the development of a rapid and cost-effective biosensor for RAC contamination in food is warranted. Duan et al. [82] obtained nine aptamer candidates for RAC through 16 selection rounds of Capture-SELEX. The aptamer RAC-6 showed the highest binding affinity for RAC in a GO-based fluorescent assay, with weak binding affinity for other off-target species (<22%). The researchers further developed a fluorescent aptasensor based on RAC-6 that exhibited a linear detection range of 0.33 to 331.79 nM, a low limit of detection (LOD; 0.13 nM; Table 2), and high recovery rates (82.57–104.65%). They found that RAC-6 was especially useful for detecting RAC contamination in pork samples.

λ-Cyhalothrin is a broad-spectrum pyrethroid insecticide used to control agricultural insect pests, such as Coleoptera, Lepidoptera, and mites, and thus increase agricultural productivity [128]. Compared with older-generation pesticides, the insecticidal effect of λ-cyhalothrin is 10–100 times stronger, and overuse of this insecticide may lead to food contamination [129]. Due to its toxicity, the ingestion of λ-cyhalothrin residues in food may cause serious adverse effects, including mouth ulcers, nausea, abdominal pain, and vomiting [130]. Yang et al. [76] used Capture-SELEX to obtain several candidate aptamers against λ-cyhalothrin. The aptamer LCT-1 showed the strongest affinity and specificity for λ-cyhalothrin in a colorimetric assay and its binding affinity was further optimized by truncation. The dissociation constant of the truncated aptamer, named LCT-1-39, was improved by approximately 40 nM relative to LCT-1, and similar results were obtained using the MST assay. LCT-1 and LCT-1-39 were thus used to establish colorimetric aptasensors for detecting λ-cyhalothrin. These aptasensors demonstrated low LODs for LCT-1 and LCT-1-39 (43.8 nM and 41.35 nM, respectively; Table 2) and mean λ-cyhalothrin recovery rates of 82.93–95.50%. Compared with traditional quantification methods, these colorimetric aptasensors demonstrated more rapid detection of λ-cyhalothrin in cucumber and pear samples.

5.1.2. Food Additives and Flavoring Agents

Vanillin is the second most popular flavoring agent worldwide and is used as a food additive in sweet foods and beverages, and a masking agent in numerous pharmaceutical formulations [131]. It is a phenolic aldehyde that has demonstrated antioxidant, antimicrobial, and antifungal activities in various food products [132,133]. Hence, a rapid detection biosensor is required to monitor vanillin concentrations during processed food production. Through Capture-SELEX, Kuznetsov et al. [23] obtained six aptamer candidates against vanillin and found that Van_74 had the highest binding affinity by nondenaturing PAGE. Its specificity for vanillin was also confirmed in the presence of interferents, such as benzaldehyde, guaiacol, furaneol, ethyl guaiacol, and ethyl vanillin. The authors also found that Van_74 was sensitive to the composition of the selection buffer. Van_74 was then used in the development of an ion-sensitive field-effect transistor (ISFET)-based biosensor that demonstrated a low LOD (0.155 μM) and a dynamic detection range of 0.155–1 μM (Table 2). This novel aptasensor can be applied for the rapid on-site detection of vanillin contamination in coffee extracts and mixtures.

Another aroma compound, furaneol, is extensively used as an artificial flavoring agent as it imparts fruit flavor to food [134], and thus a rapid detection biosensor is required to monitor furaneol concentrations during processed food production. Komarova et al. [68] obtained eight aptamer candidates against furaneol through 13 selection rounds of Capture-SELEX. These aptamers' binding affinity for furaneol was analyzed by three methods: an exonuclease protection assay, an SGI assay, and an MB-associated elution assay. The results revealed that the aptamer Fur_14 had the highest binding affinity for furaneol; therefore, Fur_14 was used to develop an ISFET-based aptasensor. Fur_14 was further modified with an alkyne label at its 5′-end, and this Fur_14 derivative exhibited a furaneol detection range of 0.1–10 μM (Table 2).

Spermine, tyramine (TYR), and β-phenethylamine (PHE) are biogenic amines (BAs) that are typically present in foodstuffs. As the consumption of foods containing high concentrations of BAs may cause toxic effects, biosensors are needed for BA detection in foods [135,136]. Tian et al. [44] obtained aptamer candidates against spermine by Capture-SELEX selection and tested them using ITC and fluorescence assays. The aptamer APJ-6 showed the highest affinity and specificity for spermine and was, thus, used to develop a fluorescent aptasensor for spermine detection in pork samples. This aptasensor demonstrated a linear detection range of 0.1–20 nM and a low LOD (0.052 nM). For detecting TYR and PHE, Kuznetsov et al. selected and isolated several aptamers using Capture-SELEX. The selection process was monitored by the melting temperature (T_m) in the screening process, and T_m peaked during the 14th round for both TYR and PHE. The aptamers TYR-2 and PHE-2 were identified to have the strongest binding affinity and specificity for TYR and PHE, respectively, based on a GO-based fluorescent assay. TYR-2 and PHE-2 were then used to develop fluorescent aptasensors for the detection of TYR and PHE in pork and bear meat samples. These aptasensors demonstrated LODs for TYR and PHE of 2.48 and 3.22 nM, respectively (Table 2), with target recovery rates in the range of 95.6–104.2%, suggesting their efficacy in detecting TYR and PHE in foods.

5.2. Aquatic Environment

5.2.1. Veterinary Drug Residues and Pesticides

Erythromycin is a broad-spectrum macrolide antibiotic used to treat diseases such as diphtheria, pertussis, and bacillary angiomatosis [137]. The natural degradation of erythromycin is prolonged due to its stable structure, leading to increased erythromycin resistance among bacteria [138]. As erythromycin diffuses rapidly into most tissues of the human body, erythromycin pollution of environmental media poses a serious threat to human health, in addition to the ecosystem. Du et al. [24] obtained 10 aptamer candidates against erythromycin through 20 selection rounds of Capture-SELEX. The binding affinity and specificity of the candidates were determined using an SGI fluorometric assay, an AuNP-based colorimetric assay, a quartz crystal microbalance with dissipation assay, and an agarose chasing diffusion assay, resulting in the selection of the aptamer Ery_06 for the development of a novel LFA. This LFA demonstrated an erythromycin-detection range of 250–500 pM in water samples, with a low LOD (3 pM; Table 2) and rapid detection (within 15 min), suggesting its efficacy for erythromycin detection in water.

Roxithromycin is a macrolide antibiotic that poses a similar risk to the ecosystem and human health as erythromycin, indicating the need to establish a rapid and effective detection device for monitoring roxithromycin residues in environmental media. Jiang et al. selected aptamer candidates against roxithromycin after 16 selection rounds of Capture-SELEX. The aptamer Ap01 demonstrated the highest affinity and specificity for roxithromycin, as indicated by the results of an SGI assay, and was therefore selected for the development of a colorimetric aptasensor for roxithromycin. The developed aptasensor demonstrated a low LOD (0.077 μM) for roxithromycin in water samples (Table 2) and high recovery rates in the range of 90.48–109.39%.

5.2.2. Toxins and Plasticizers

Gymnodimines (GYMs) are fast-acting cyclic imine toxins that are biosynthesized by dinoflagellates and have deleterious effects on the aquatic environment with the accumulation. The contaminated environment can have serious toxic effects on filter feeding shellfish and thereby pose a threat to human health [139]. Zhang et al. [75] used Capture-SELEX to screen and obtain six aptamer candidates against gymnodimine-A (GYM-A). G48 exhibited the highest binding affinity (K_d: 288 nM) and was therefore chosen for further optimization and investigation. The truncated aptamer G48nop demonstrated an improved K_d value of 34.5 ± 1.72 nM and high specificity for GYM-A. A novel BLI-based aptasensor was established using this aptamer that detected GYM-A in the range of 55–1400 nM (linear range of 55–875 nM) and had a low LOD (6.21 nM; Table 2). This BLI-based aptasensor also demonstrated high recovery rates in the range of 96.65%–109.67%, indicating that is reliable and efficient in detecting and monitoring GYM-A in water samples.

Di(2-ethylhexyl) phthalate (DEHP) is a plasticizer that is widely used as an additive in packaging materials, and its residues are known to accumulate and dissolve in water [140]. DEHP is also a well-known endocrine disruptor that can enter the human body through ingestion of food or water and inhalation with contaminated air that disrupts the immune system. Lu et al. [71] selected aptamer candidates against DEHP through eight rounds of Capture-SELEX. Upon high-throughput sequencing and characterizing the candidate aptamers using an AuNP colorimetric assay and localized surface plasmon resonance, aptamer 31 was revealed to have high affinity and specificity. Aptamer 31 was thus used to develop an ultrasensitive electrochemical impedance spectroscopy aptasensor to detect DEHP in real water samples; this aptasensor demonstrated a low LOD (0.264 pM; Table 2) and a mean recovery rate ranging from 76.07% to 141.32%.

5.3. Other Potential Applications

Synthetic riboswitches can have several biotechnological applications, such as regulating gene expression, e.g., the construction of genetic circuits [141,142]. Natural riboswitches are mainly found in bacteria, while synthetic riboswitches are artificially generated by combining aptamer domains (using in vitro SELEX method) with expression platforms to regulate gene expression via small-molecule-RNA interactions [143,144]. However, using the conventional SELEX method, only a ciprofloxacin riboswitch aptamer has been developed, as most aptamers have limitations, such as excellent binding affinity and conformation switching, and require cellular screening after in vitro selection [145]. Subsequently, Boussebayle et al. [146] identified a paromomycin riboswitch aptamer using Capture-SELEX and found it had a high affinity (K_d: 21 nM) using an ITC assay. Through further in vivo selection, this aptamer was revealed to have riboswitching properties. This work has introduced an efficient protocol for developing synthetic riboswitches and boosted the development of real-time intracellular biosensors for monitoring metabolic flows in living cells.

Zearalenone (ZEN) is a nonsteroidal estrogenic mycotoxin produced by fungi and is known to contaminate cereal grains and other crops [147]. Due to its high estrogenic activity, long-term intake of ZEN residues adversely affects human health by causing cervical cancer or hyperestrogenic syndrome [147,148]. Zhang et al. [84] obtained aptamer Z100 against ZEN after eight rounds of Capture-SELEX. Z100 was shown to have high affinity and specificity for ZEN in a fluorescence assay, and hence was selected to develop a rapid and on-site AuNP-based label-free aptasensor for detecting ZEN in agricultural produce. The developed aptasensor had a low LOD (12.5 nM), a linear detection range of 12.5–402.1 nM (Table 2), and high recovery rates in corn powder and feed (96.42–99.78% and 95.99–103.73%, respectively). This study revealed the great potential for developing aptamer-based inhibitors for ZEN to enhance animal feed safety.

Fenitrothion (FEN) is a broad-spectrum organophosphorus insecticide mainly used to control insect pests in agriculture [149]. As FEN is available at a low cost, large amounts of FEN are frequently applied in agriculture. High concentrations of FEN residues have been found in foods, which has become a great concern for human health and the environment. Trinh et al. [85] screened aptamer candidates against FEN through Capture-SELEX. In the thioflavin T (ThT) displacement assay, the aptamer FenA2 was identified to exhibit high-affinity FEN binding, as indicated by the loss of fluorescence. FenA2 was further optimized and used to develop a label-free ThT sensor. The developed aptasensor has a G4-quadruplex-like structure and a low LOD (14 nM; Table 2). This aptamer may be further optimized to develop a real-time FEN-detecting aptasensor.

6. Conclusions

Small-molecule contaminants are ubiquitous in the aquatic environment, agriculture produce, and animal feed due to the overuse of small molecules such as antibiotics and pesticides, and these contaminants pose a serious threat to human health and the environment. Owing to the few functional groups on these small molecules, screening aptamers against them is more challenging than doing so against large-molecule targets. Compared with other SELEX approaches, which involve the immobilization of small-molecule targets, the Capture-SELEX approach is more feasible, as they involve the immobilization of a biotinylated ssDNA/RNA library against which the binding affinity and specificity of small-molecule targets can be screened. To date, fewer than 50 studies have reported using Capture-SELEX to identify novel aptamers against specific small-molecule contaminants, suggesting that this process remains challenging, such as false-positive and lack of diversity. However, this research field has recently been receiving increasing attention from scientists. We hope that this review will encourage further research into the use of Capture-SELEX in generating aptamers against small-molecule contaminants. Six small-molecule aptamer characterization methods are introduced in this review. The high affinity and specificity aptamer work as a biorecognition element in aptasensor to detect specific small-molecule contaminants in environmental media and agricultural produce. This will help to improve food safety, aquatic environments, and agricultural crop production.

Author Contributions: Conceptualization, S.Y.L. and C.K.K.; writing—original draft preparation, S.Y.L. and C.K.K.; writing—review and editing, S.Y.L., H.L.L. and C.K.K.; visualization, S.Y.L. and C.K.K.; supervision, C.K.K.; funding acquisition, C.K.K. All authors have read and agreed to the published version of the manuscript.

Funding: This work was supported by the Shenzhen Basic Research Project [JCYJ20180507181642811]; National Natural Science Foundation of China Project [32222089]; Research Grants Council of the Hong Kong SAR, China Projects [CityU 11100222, CityU 11100421, CityU 11101519, CityU 11100218, N_CityU110/17]; Croucher Foundation Project [9509003]; State Key Laboratory of Marine Pollution Director Discretionary Fund; City University of Hong Kong projects [7005503, 9667222, 9680261] to C.K.K.

Institutional Review Board Statement: Not applicable.

Informed Consent Statement: Not applicable.

Data Availability Statement: Not applicable.

Acknowledgments: Not applicable.

Conflicts of Interest: The authors declare no conflict of interest.

References

1. Minh, T.B.; Leung, H.W.; Loi, I.H.; Chan, W.H.; So, M.K.; Mao, J.Q.; Choi, D.; Lam, J.C.; Martin, M.; Lee, J.H.W.; et al. Antibiotics in the Hong Kong metropolitan area: Ubiquitous distribution and fate in Victoria Harbour. *Mar. Pollut. Bull.* **2009**, *58*, 1052–1062. [CrossRef] [PubMed]
2. Gothwal, R.; Shashidhar, T. Antibiotic Pollution in the Environment: A Review. CLEAN–Soil. *Air Water* **2015**, *43*, 479–489. [CrossRef]
3. Parra-Arroyo, L.; Gonzalez-Gonzalez, R.B.; Castillo-Zacarias, C.; Melchor Martinez, E.M.; Sosa-Hernandez, J.E.; Bilal, M.; Iqbal, H.M.N.; Barcelo, D.; Parra-Saldivar, R. Highly hazardous pesticides and related pollutants: Toxicological, regulatory, and analytical aspects. *Sci. Total Environ.* **2022**, *807*, 151879. [CrossRef] [PubMed]
4. Pan, W.; Zeng, D.; Ding, N.; Luo, K.; Man, Y.B.; Zeng, L.; Zhang, Q.; Luo, J.; Kang, Y. Percutaneous Penetration and Metabolism of Plasticizers by Skin Cells and Its Implication in Dermal Exposure to Plasticizers by Skin Wipes. *Environ. Sci. Technol.* **2020**, *54*, 10181–10190. [CrossRef] [PubMed]
5. Eales, J.; Bethel, A.; Galloway, T.; Hopkinson, P.; Morrissey, K.; Short, R.E.; Garside, R. Human health impacts of exposure to phthalate plasticizers: An overview of reviews. *Environ. Int.* **2022**, *158*, 106903. [CrossRef]
6. Qadeer, A.; Kirsten, K.L.; Ajmal, Z.; Jiang, X.; Zhao, X. Alternative Plasticizers As Emerging Global Environmental and Health Threat: Another Regrettable Substitution? *Environ. Sci. Technol.* **2022**, *56*, 1482–1488. [CrossRef]
7. Schuster, C.; Sterz, S.; Teupser, D.; Brugel, M.; Vogeser, M.; Paal, M. Multiplex Therapeutic Drug Monitoring by Isotope-dilution HPLC-MS/MS of Antibiotics in Critical Illnesses. *J. Vis. Exp.* **2018**. [CrossRef]
8. Li, S.; Wnag, L.; Wu, J.; Inoue, H. Syntheses and spectra of Mn(III)-chlorophyll-a and Mn(II)-chlorophyll-a. *Guang Pu Xue Yu Guang Pu Fen Xi* **1997**, *17*, 55–59.
9. Sheng, W.; Xia, X.; Wei, K.; Li, J.; Li, Q.X.; Xu, T. Determination of marbofloxacin residues in beef and pork with an enzyme-linked immunosorbent assay. *J. Agric. Food Chem.* **2009**, *57*, 5971–5975. [CrossRef]
10. Wang, G.; Zhang, H.C.; Liu, J.; Wang, J.P. A receptor-based chemiluminescence enzyme linked immunosorbent assay for determination of tetracyclines in milk. *Anal. Biochem.* **2019**, *564-565*, 40–46. [CrossRef]
11. Dahal, U.P.; Jones, J.P.; Davis, J.A.; Rock, D.A. Small molecule quantification by liquid chromatography-mass spectrometry for metabolites of drugs and drug candidates. *Drug Metab. Dispos.* **2011**, *39*, 2355–2360. [CrossRef]
12. Li, B.; Van Schepdael, A.; Hoogmartens, J.; Adams, E. Characterization of impurities in tobramycin by liquid chromatography-mass spectrometry. *J. Chromatogr. A* **2009**, *1216*, 3941–3945. [CrossRef]
13. Zhou, J.; Rossi, J. Aptamers as targeted therapeutics: Current potential and challenges. *Nat. Rev. Drug Discov.* **2017**, *16*, 440. [CrossRef]
14. Thiviyanathan, V.; Gorenstein, D.G. Aptamers and the next generation of diagnostic reagents. *Proteom. Clin. Appl.* **2012**, *6*, 563–573. [CrossRef]
15. He, Y.; Zhou, L.; Deng, L.; Feng, Z.; Cao, Z.; Yin, Y. An electrochemical impedimetric sensing platform based on a peptide aptamer identified by high-throughput molecular docking for sensitive l-arginine detection. *Bioelectrochemistry* **2021**, *137*, 107634. [CrossRef]
16. Lin, N.; Wu, L.; Xu, X.; Wu, Q.; Wang, Y.; Shen, H.; Song, Y.; Wang, H.; Zhu, Z.; Kang, D.; et al. Aptamer Generated by Cell-SELEX for Specific Targeting of Human Glioma Cells. *ACS Appl. Mater. Interfaces* **2021**, *13*, 9306–9315. [CrossRef]
17. Chen, X.; Qiu, L.; Cai, R.; Cui, C.; Li, L.; Jiang, J.H.; Tan, W. Aptamer-Directed Protein-Specific Multiple Modifications of Membrane Glycoproteins on Living Cells. *ACS Appl. Mater. Interfaces* **2020**, *12*, 37845–37850. [CrossRef]

18. Lohlamoh, W.; Soontornworajit, B.; Rotkrua, P. Anti-Proliferative Effect of Doxorubicin-Loaded AS1411 Aptamer on Colorectal Cancer Cell. *Asian Pac. J. Cancer Prev.* **2021**, *22*, 2209–2219. [CrossRef]
19. Amraee, M.; Oloomi, M.; Yavari, A.; Bouzari, S. DNA aptamer identification and characterization for E. coli O157 detection using cell based SELEX method. *Anal. Biochem.* **2017**, *536*, 36–44. [CrossRef]
20. Chinnappan, R.; Eissa, S.; Alotaibi, A.; Siddiqua, A.; Alsager, O.A.; Zourob, M. In vitro selection of DNA aptamers and their integration in a competitive voltammetric biosensor for azlocillin determination in waste water. *Anal. Chim. Acta* **2020**, *1101*, 149–156. [CrossRef]
21. Paniel, N.; Istamboulie, G.; Triki, A.; Lozano, C.; Barthelmebs, L.; Noguer, T. Selection of DNA aptamers against penicillin G using Capture-SELEX for the development of an impedimetric sensor. *Talanta* **2017**, *162*, 232–240. [CrossRef] [PubMed]
22. Stoltenburg, R.; Nikolaus, N.; Strehlitz, B. Capture-SELEX: Selection of DNA Aptamers for Aminoglycoside Antibiotics. *J. Anal. Methods Chem.* **2012**, *2012*, 415697. [CrossRef] [PubMed]
23. Kuznetsov, A.; Komarova, N.; Andrianova, M.; Grudtsov, V.; Kuznetsov, E. Aptamer based vanillin sensor using an ion-sensitive field-effect transistor. *Mikrochim. Acta* **2017**, *185*, 3. [CrossRef] [PubMed]
24. Du, Y.; Liu, D.; Wang, M.; Guo, F.; Lin, J.S. Preparation of DNA aptamer and development of lateral flow aptasensor combining recombinase polymerase amplification for detection of erythromycin. *Biosens. Bioelectron.* **2021**, *181*, 113157. [CrossRef] [PubMed]
25. Tuerk, C.; Gold, L. Systematic evolution of ligands by exponential enrichment: RNA ligands to bacteriophage T4 DNA polymerase. *Science* **1990**, *249*, 505–510. [CrossRef]
26. Ellington, A.D.; Szostak, J.W. In vitro selection of RNA molecules that bind specific ligands. *Nature* **1990**, *346*, 818–822. [CrossRef]
27. Robertson, D.L.; Joyce, G.F. Selection in vitro of an RNA enzyme that specifically cleaves single-stranded DNA. *Nature* **1990**, *344*, 467–468. [CrossRef]
28. Yang, J.; Bowser, M.T. Capillary electrophoresis-SELEX selection of catalytic DNA aptamers for a small-molecule porphyrin target. *Anal. Chem.* **2013**, *85*, 1525–1530. [CrossRef]
29. Eaton, R.M.; Shallcross, J.A.; Mael, L.E.; Mears, K.S.; Minkoff, L.; Scoville, D.J.; Whelan, R.J. Selection of DNA aptamers for ovarian cancer biomarker HE4 using CE-SELEX and high-throughput sequencing. *Anal. Bioanal. Chem.* **2015**, *407*, 6965–6973. [CrossRef]
30. Zhu, C.; Li, L.; Fang, S.; Zhao, Y.; Zhao, L.; Yang, G.; Qu, F. Selection and characterization of an ssDNA aptamer against thyroglobulin. *Talanta* **2021**, *223*, 121690. [CrossRef]
31. Kou, Q.; Wu, P.; Sun, Q.; Li, C.; Zhang, L.; Shi, H.; Wu, J.; Wang, Y.; Yan, X.; Le, T. Selection and truncation of aptamers for ultrasensitive detection of sulfamethazine using a fluorescent biosensor based on graphene oxide. *Anal. Bioanal. Chem.* **2021**, *413*, 901–909. [CrossRef]
32. Ozyurt, C.; Canbay, Z.C.; Dinckaya, E.; Evran, S. A highly sensitive DNA aptamer-based fluorescence assay for sarcosine detection down to picomolar levels. *Int. J. Biol. Macromol.* **2019**, *129*, 91–97. [CrossRef]
33. Kim, S.H.; Lee, J.; Lee, B.H.; Song, C.S.; Gu, M.B. Specific detection of avian influenza H5N2 whole virus particles on lateral flow strips using a pair of sandwich-type aptamers. *Biosens. Bioelectron.* **2019**, *134*, 123–129. [CrossRef]
34. Wang, G.; Liu, J.; Chen, K.; Xu, Y.; Liu, B.; Liao, J.; Zhu, L.; Hu, X.; Li, J.; Pu, Y.; et al. Selection and characterization of DNA aptamer against glucagon receptor by cell-SELEX. *Sci. Rep.* **2017**, *7*, 7179. [CrossRef]
35. Yilmaz, D.; Muslu, T.; Parlar, A.; Kurt, H.; Yuce, M. SELEX against whole-cell bacteria resulted in lipopolysaccharide binding aptamers. *J. Biotechnol.* **2022**, *354*, 10–20. [CrossRef]
36. Wan, J.; Ye, L.; Yang, X.; Guo, Q.; Wang, K.; Huang, Z.; Tan, Y.; Yuan, B.; Xie, Q. Cell-SELEX based selection and optimization of DNA aptamers for specific recognition of human cholangiocarcinoma QBC-939 cells. *Analyst* **2015**, *140*, 5992–5997. [CrossRef]
37. Wang, H.; Zhang, Y.; Yang, H.; Qin, M.; Ding, X.; Liu, R.; Jiang, Y. In Vivo SELEX of an Inhibitory NSCLC-Specific RNA Aptamer from PEGylated RNA Library. *Mol. Ther. Nucleic Acids* **2018**, *10*, 187–198. [CrossRef]
38. Huang, X.; Zhong, J.; Ren, J.; Wen, D.; Zhao, W.; Huan, Y. A DNA aptamer recognizing MMP14 for in vivo and in vitro imaging identified by cell-SELEX. *Oncol. Lett.* **2019**, *18*, 265–274. [CrossRef]
39. Reinemann, C.; Freiin von Fritsch, U.; Rudolph, S.; Strehlitz, B. Generation and characterization of quinolone-specific DNA aptamers suitable for water monitoring. *Biosens. Bioelectron.* **2016**, *77*, 1039–1047. [CrossRef]
40. Stoltenburg, R.; Reinemann, C.; Strehlitz, B. FluMag-SELEX as an advantageous method for DNA aptamer selection. *Anal. Bioanal. Chem.* **2005**, *383*, 83–91. [CrossRef]
41. Lyu, C.; Khan, I.M.; Wang, Z. Capture-SELEX for aptamer selection: A short review. *Talanta* **2021**, *229*, 122274. [CrossRef] [PubMed]
42. Jenison, R.D.; Gill, S.C.; Pardi, A.; Polisky, B. High-resolution molecular discrimination by RNA. *Science* **1994**, *263*, 1425–1429. [CrossRef]
43. Shkembi, X.; Skouridou, V.; Svobodova, M.; Leonardo, S.; Bashammakh, A.S.; Alyoubi, A.O.; Campas, M.; CK, O.S. Hybrid Antibody-Aptamer Assay for Detection of Tetrodotoxin in Pufferfish. *Anal. Chem.* **2021**. [CrossRef] [PubMed]
44. Tian, H.; Duan, N.; Wu, S.; Wang, Z. Selection and application of ssDNA aptamers against spermine based on Capture-SELEX. *Anal. Chim. Acta* **2019**, *1081*, 168–175. [CrossRef] [PubMed]
45. Song, M.; Lyu, C.; Duan, N.; Wu, S.; Khan, I.M.; Wang, Z. The isolation of high-affinity ssDNA aptamer for the detection of ribavirin in chicken. *Anal. Methods* **2021**, *13*, 3110–3117. [CrossRef]

46. Wang, T.; Chen, C.; Larcher, L.M.; Barrero, R.A.; Veedu, R.N. Three decades of nucleic acid aptamer technologies: Lessons learned, progress and opportunities on aptamer development. *Biotechnol. Adv.* **2019**, *37*, 28–50. [CrossRef]

47. Spiga, F.M.; Maietta, P.; Guiducci, C. More DNA-Aptamers for Small Drugs: A Capture-SELEX Coupled with Surface Plasmon Resonance and High-Throughput Sequencing. *ACS Comb. Sci.* **2015**, *17*, 326–333. [CrossRef]

48. Sanger, F.; Nicklen, S.; Coulson, A.R. DNA sequencing with chain-terminating inhibitors. *Proc. Natl. Acad. Sci. USA* **1977**, *74*, 5463–5467. [CrossRef]

49. Crossley, B.M.; Bai, J.; Glaser, A.; Maes, R.; Porter, E.; Killian, M.L.; Clement, T.; Toohey-Kurth, K. Guidelines for Sanger sequencing and molecular assay monitoring. *J. Vet. Diagn. Investig.* **2020**, *32*, 767–775. [CrossRef]

50. Levy, S.E.; Myers, R.M. Advancements in Next-Generation Sequencing. *Annu. Rev. Genom. Hum. Genet.* **2016**, *17*, 95–115. [CrossRef]

51. Balasubramanian, S. Solexa sequencing: Decoding genomes on a population scale. *Clin. Chem.* **2015**, *61*, 21–24. [CrossRef]

52. Slatko, B.E.; Gardner, A.F.; Ausubel, F.M. Overview of Next-Generation Sequencing Technologies. *Curr. Protoc. Mol. Biol.* **2018**, *122*, e59. [CrossRef]

53. Eisold, A.; Labudde, D. Detailed Analysis of 17beta-Estradiol-Aptamer Interactions: A Molecular Dynamics Simulation Study. *Molecules* **2018**, *23*, 1690. [CrossRef]

54. Bavi, R.; Liu, Z.; Han, Z.; Zhang, H.; Gu, Y. In silico designed RNA aptamer against epithelial cell adhesion molecule for cancer cell imaging. *Biochem. Biophys. Res. Commun.* **2019**, *509*, 937–942. [CrossRef]

55. Ahmad, N.A.; Mohamed Zulkifli, R.; Hussin, H.; Nadri, M.H. In silico approach for Post-SELEX DNA aptamers: A mini-review. *J. Mol. Graph Model* **2021**, *105*, 107872. [CrossRef]

56. Larkin, M.A.; Blackshields, G.; Brown, N.P.; Chenna, R.; McGettigan, P.A.; McWilliam, H.; Valentin, F.; Wallace, I.M.; Wilm, A.; Lopez, R.; et al. Clustal W and Clustal X version 2.0. *Bioinformatics* **2007**, *23*, 2947–2948. [CrossRef]

57. Sievers, F.; Higgins, D.G. Clustal omega. *Curr. Protoc. Bioinform.* **2014**, *48*, 3–13. [CrossRef]

58. Bailey, T.L.; Boden, M.; Buske, F.A.; Frith, M.; Grant, C.E.; Clementi, L.; Ren, J.; Li, W.W.; Noble, W.S. MEME SUITE: Tools for motif discovery and searching. *Nucleic Acids Res.* **2009**, *37*, W202–W208. [CrossRef]

59. Zuker, M. Mfold web server for nucleic acid folding and hybridization prediction. *Nucleic Acids Res.* **2003**, *31*, 3406–3415. [CrossRef]

60. Xayaphoummine, A.; Bucher, T.; Isambert, H. Kinefold web server for RNA/DNA folding path and structure prediction including pseudoknots and knots. *Nucleic Acids Res.* **2005**, *33*, W605–W610. [CrossRef]

61. Xu, Z.Z.; Mathews, D.H. Experiment-Assisted Secondary Structure Prediction with RNAstructure. *Methods Mol. Biol.* **2016**, *1490*, 163–176. [CrossRef] [PubMed]

62. Chen, Z.; Hu, L.; Zhang, B.T.; Lu, A.; Wang, Y.; Yu, Y.; Zhang, G. Artificial Intelligence in Aptamer-Target Binding Prediction. *Int. J. Mol. Sci.* **2021**, *22*, 3605. [CrossRef] [PubMed]

63. Soon, S.; Nordin, N.A. In silico predictions and optimization of aptamers against streptococcus agalactiae surface protein using computational docking. *Mater. Today-Proc.* **2019**, *16*, 2096–2100. [CrossRef]

64. Douaki, A.; Garoli, D.; Inam, A.; Angeli, M.A.C.; Cantarella, G.; Rocchia, W.; Wang, J.; Petti, L.; Lugli, P. Smart Approach for the Design of Highly Selective Aptamer-Based Biosensors. *Biosensors* **2022**, *12*, 574. [CrossRef] [PubMed]

65. Mousivand, M.; Anfossi, L.; Bagherzadeh, K.; Barbero, N.; Mirzadi-Gohari, A.; Javan-Nikkhah, M. In silico maturation of affinity and selectivity of DNA aptamers against aflatoxin B1 for biosensor development. *Anal. Chim. Acta* **2020**, *1105*, 178–186. [CrossRef]

66. Kadam, U.S.; Hong, J.C. Advances in aptameric biosensors designed to detect toxic contaminants from food, water, human fluids, and the environment. *Trends Environ. Anal. Chem.* **2022**, *36*, e00184. [CrossRef]

67. Tian, Y.; Wang, Y.; Sheng, Z.; Li, T.; Li, X. A colorimetric detection method of pesticide acetamiprid by fine-tuning aptamer length. *Anal. Biochem.* **2016**, *513*, 87–92. [CrossRef]

68. Komarova, N.; Andrianova, M.; Glukhov, S.; Kuznetsov, A. Selection, Characterization, and Application of ssDNA Aptamer against Furaneol. *Molecules* **2018**, *23*, 3159. [CrossRef]

69. Song, M.; Li, C.; Wu, S.; Duan, N. Screening of specific aptamers against chlorpromazine and construction of novel ratiometric fluorescent aptasensor based on metal-organic framework. *Talanta* **2022**, *252*, 123850. [CrossRef]

70. Jiang, L.; Wang, M.; Zhang, Y.; Chen, H.; Su, Y.; Wang, Y.; Lin, J.S. Preparation and characterization of DNA aptamers against roxithromycin. *Anal. Chim. Acta* **2021**, *1164*, 338509. [CrossRef]

71. Lu, Q.; Liu, X.; Hou, J.; Yuan, Q.; Li, Y.; Chen, S. Selection of Aptamers Specific for DEHP Based on ssDNA Library Immobilized SELEX and Development of Electrochemical Impedance Spectroscopy Aptasensor. *Molecules* **2020**, *25*, 747. [CrossRef]

72. Ren, L.; Qi, S.; Khan, I.M.; Wu, S.; Duan, N.; Wang, Z. Screening and application of a broad-spectrum aptamer for acyclic guanosine analogues. *Anal. Bioanal. Chem.* **2021**, *413*, 4855–4863. [CrossRef]

73. Wu, C.; Barkova, D.; Komarova, N.; Offenhausser, A.; Andrianova, M.; Hu, Z.; Kuznetsov, A.; Mayer, D. Highly selective and sensitive detection of glutamate by an electrochemical aptasensor. *Anal. Bioanal. Chem.* **2021**, *414*, 1609–1622. [CrossRef]

74. Li, C.; Song, M.; Wu, S.; Wang, Z.; Duan, N. Selection of aptamer targeting levamisole and development of a colorimetric and SERS dual-mode aptasensor based on AuNPs/Cu-TCPP(Fe) nanosheets. *Talanta* **2023**, *251*, 123739. [CrossRef]

75. Zhang, X.; Gao, Y.; Deng, B.; Hu, B.; Zhao, L.; Guo, H.; Yang, C.; Ma, Z.; Sun, M.; Jiao, B.; et al. Selection, Characterization, and Optimization of DNA Aptamers against Challenging Marine Biotoxin Gymnodimine-A for Biosensing Application. *Toxins* **2022**, *14*, 195. [CrossRef]

76. Yang, Y.; Tang, Y.; Wang, C.; Liu, B.; Wu, Y. Selection and identification of a DNA aptamer for ultrasensitive and selective detection of lambda-cyhalothrin residue in food. *Anal. Chim. Acta* **2021**, *1179*, 338837. [CrossRef]

77. Xie, M.; Chen, Z.; Zhao, F.; Lin, Y.; Zheng, S.; Han, S. Selection and Application of ssDNA Aptamers for Fluorescence Biosensing Detection of Malachite Green. *Foods* **2022**, *11*, 801. [CrossRef]

78. Wu, W.; Sun, Q.; Li, T.; Liu, K.; Jiang, Y.; Wang, Y.; Yang, Y. Selection and characterization of bispecific aptamers against malachite green and leucomalachite green. *Anal. Biochem.* **2022**, *658*, 114849. [CrossRef]

79. Wang, L.; Liu, X.; Zhang, Q.; Zhang, C.; Liu, Y.; Tu, K.; Tu, J. Selection of DNA aptamers that bind to four organophosphorus pesticides. *Biotechnol. Lett.* **2012**, *34*, 869–874. [CrossRef]

80. Martin, J.A.; Chavez, J.L.; Chushak, Y.; Chapleau, R.R.; Hagen, J.; Kelley-Loughnane, N. Tunable stringency aptamer selection and gold nanoparticle assay for detection of cortisol. *Anal. Bioanal. Chem.* **2014**, *406*, 4637–4647. [CrossRef]

81. Duan, N.; Gong, W.; Wu, S.; Wang, Z. Selection and Application of ssDNA Aptamers against Clenbuterol Hydrochloride Based on ssDNA Library Immobilized SELEX. *J. Agric. Food Chem.* **2017**, *65*, 1771–1777. [CrossRef] [PubMed]

82. Duan, N.; Gong, W.; Wu, S.; Wang, Z. An ssDNA library immobilized SELEX technique for selection of an aptamer against ractopamine. *Anal. Chim. Acta* **2017**, *961*, 100–105. [CrossRef] [PubMed]

83. Abraham, K.M.; Roueinfar, M.; Ponce, A.T.; Lussier, M.E.; Benson, D.B.; Hong, K.L. In Vitro Selection and Characterization of a Single-Stranded DNA Aptamer Against the Herbicide Atrazine. *ACS Omega* **2018**, *3*, 13576–13583. [CrossRef]

84. Zhang, Y.; Lu, T.; Wang, Y.; Diao, C.; Zhou, Y.; Zhao, L.; Chen, H. Selection of a DNA Aptamer against Zearalenone and Docking Analysis for Highly Sensitive Rapid Visual Detection with Label-Free Aptasensor. *J. Agric. Food Chem.* **2018**, *66*, 12102–12110. [CrossRef] [PubMed]

85. Trinh, K.H.; Kadam, U.S.; Song, J.; Cho, Y.; Kang, C.H.; Lee, K.O.; Lim, C.O.; Chung, W.S.; Hong, J.C. Novel DNA Aptameric Sensors to Detect the Toxic Insecticide Fenitrothion. *Int. J. Mol. Sci.* **2021**, *22*, 846. [CrossRef] [PubMed]

86. Kadam, U.S.; Trinh, K.H.; Kumar, V.; Lee, K.W.; Cho, Y.; Can, M.T.; Lee, H.; Kim, Y.; Kim, S.; Kang, J.; et al. Identification and structural analysis of novel malathion-specific DNA aptameric sensors designed for food testing. *Biomaterials* **2022**, *287*, 121617. [CrossRef]

87. Trinh, K.H.; Kadam, U.S.; Rampogu, S.; Cho, Y.; Yang, K.A.; Kang, C.H.; Lee, K.W.; Lee, K.O.; Chung, W.S.; Hong, J.C. Development of novel fluorescence-based and label-free noncanonical G4-quadruplex-like DNA biosensor for facile, specific, and ultrasensitive detection of fipronil. *J. Hazard Mater.* **2022**, *427*, 127939. [CrossRef]

88. Can, M.T.; Kadam, U.S.; Trinh, K.H.; Cho, Y.; Lee, H.; Kim, Y.; Kim, S.; Kang, C.H.; Kim, S.H.; Chung, W.S.; et al. Engineering Novel Aptameric Fluorescent Biosensors for Analysis of the Neurotoxic Environmental Contaminant Insecticide Diazinon from Real Vegetable and Fruit Samples. *Front Biosci. (Landmark Ed.)* **2022**, *27*, 92. [CrossRef]

89. Xia, L.; Yang, Y.; Yang, H.; Tang, Y.; Zhou, J.; Wu, Y. Screening and identification of an aptamer as novel recognition molecule in the test strip and its application for visual detection of ethyl carbamate in liquor. *Anal. Chim. Acta* **2022**, *1226*, 340289. [CrossRef]

90. Yu, H.; Yang, W.; Alkhamis, O.; Canoura, J.; Yang, K.A.; Xiao, Y. In vitro isolation of small-molecule-binding aptamers with intrinsic dye-displacement functionality. *Nucleic Acids Res.* **2018**, *46*, e43. [CrossRef]

91. Zhang, W.; Li, D.; Zhang, J.; Jiang, L.; Li, Z.; Lin, J.S. Preparation and Characterization of Aptamers Against O,p'-DDT. *Int. J. Mol. Sci.* **2020**, *21*, 2211. [CrossRef]

92. Matsenko, N.U.; Rijikova, V.S.; Kovalenko, S.P. Comparison of SYBR Green I and TaqMan real-time PCR formats for the analysis of her2 gene dose in human breast tumors. *Bull. Exp. Biol. Med.* **2008**, *145*, 240–244. [CrossRef]

93. Marinowic, D.R.; Zanirati, G.; Rodrigues, F.V.F.; Grahl, M.V.C.; Alcara, A.M.; Machado, D.C.; Da Costa, J.C. A new SYBR Green real-time PCR to detect SARS-CoV-2. *Sci. Rep.* **2021**, *11*, 2224. [CrossRef]

94. Schnerr, H.; Niessen, L.; Vogel, R.F. Real time detection of the tri5 gene in Fusarium species by lightcycler-PCR using SYBR Green I for continuous fluorescence monitoring. *Int. J. Food Microbiol.* **2001**, *71*, 53–61. [CrossRef]

95. Vitzthum, F.; Geiger, G.; Bisswanger, H.; Brunner, H.; Bernhagen, J. A quantitative fluorescence-based microplate assay for the determination of double-stranded DNA using SYBR Green I and a standard ultraviolet transilluminator gel imaging system. *Anal. Biochem.* **1999**, *276*, 59–64. [CrossRef]

96. Iida, R.; Yasuda, T.; Tsubota, E.; Nakashima, Y.; Sawazaki, K.; Aoyama, M.; Matsuki, T.; Kishi, K. Detection of isozymes of deoxyribonucleases I and II on electrophoresed gels with picogram sensitivity using SYBR Green I. *Electrophoresis* **1998**, *19*, 2416–2418. [CrossRef]

97. Izumiyama, S.; Omura, M.; Takasaki, T.; Ohmae, H.; Asahi, H. Plasmodium falciparum: Development and validation of a measure of intraerythrocytic growth using SYBR Green I in a flow cytometer. *Exp. Parasitol.* **2009**, *121*, 144–150. [CrossRef]

98. Assuncao, P.; Rosales, R.S.; Rifatbegovic, M.; Antunes, N.T.; de la Fe, C.; Ruiz de Galarreta, C.M.; Poveda, J.B. Quantification of mycoplasmas in broth medium with sybr green-I and flow cytometry. *Front. Biosci.* **2006**, *11*, 492–497. [CrossRef]

99. Dragan, A.I.; Pavlovic, R.; McGivney, J.B.; Casas-Finet, J.R.; Bishop, E.S.; Strouse, R.J.; Schenerman, M.A.; Geddes, C.D. SYBR Green I: Fluorescence properties and interaction with DNA. *J. Fluoresc.* **2012**, *22*, 1189–1199. [CrossRef]

100. Wang, Y.; Bao, L.; Liu, Z.; Pang, D.W. Aptamer biosensor based on fluorescence resonance energy transfer from upconverting phosphors to carbon nanoparticles for thrombin detection in human plasma. *Anal. Chem.* **2011**, *83*, 8130–8137. [CrossRef]

101. Li, H.L.; Zhang, Y.W.; Wang, L.; Tian, J.Q.; Sun, X.P. Nucleic acid detection using carbon nanoparticles as a fluorescent sensing platform. *Chem. Commun.* **2011**, *47*, 961–963. [CrossRef]

102. Kimling, J.; Maier, M.; Okenve, B.; Kotaidis, V.; Ballot, H.; Plech, A. Turkevich method for gold nanoparticle synthesis revisited. *J. Phys. Chem. B* **2006**, *110*, 15700–15707. [CrossRef] [PubMed]
103. Wu, Y.; Zhan, S.; Wang, L.; Zhou, P. Selection of a DNA aptamer for cadmium detection based on cationic polymer mediated aggregation of gold nanoparticles. *Analyst* **2014**, *139*, 1550–1561. [CrossRef] [PubMed]
104. Alba-Molina, D.; Martin-Romero, M.T.; Camacho, L.; Giner-Casares, J.J. Ion-Mediated Aggregation of Gold Nanoparticles for Light-Induced Heating. *Appl. Sci.* **2017**, *7*, 916. [CrossRef]
105. Zhao, W.; Chiuman, W.; Lam, J.C.; McManus, S.A.; Chen, W.; Cui, Y.; Pelton, R.; Brook, M.A.; Li, Y. DNA aptamer folding on gold nanoparticles: From colloid chemistry to biosensors. *J. Am. Chem. Soc.* **2008**, *130*, 3610–3618. [CrossRef] [PubMed]
106. Smith, A.T.; LaChance, A.M.; Zeng, S.; Liu, B.; Sun, L. Synthesis, properties, and applications of graphene oxide/reduced graphene oxide and their nanocomposites. *Nano Mater. Sci.* **2019**, *1*, 31–47. [CrossRef]
107. Offeman, W.S.H.J.a.R.E. Preparation of Graphitic Oxide. *J. Am. Chem. Soc.* **1958**, *80*, 1339. [CrossRef]
108. Lakowicz, J.R.; Weber, G. Quenching of fluorescence by oxygen. A probe for structural fluctuations in macromolecules. *Biochemistry* **1973**, *12*, 4161–4170. [CrossRef]
109. Duan, N.; Song, M.; Mi, W.; Wang, Z.; Wu, S. Effectively Selecting Aptamers for Targeting Aromatic Biogenic Amines and Their Application in Aptasensing Establishment. *J. Agric. Food Chem.* **2021**. [CrossRef]
110. Lin, K.; Wu, G. Isothermal Titration Calorimetry Assays to Measure Binding Affinities In Vitro. *Methods Mol. Biol.* **2019**, *1893*, 257–272. [CrossRef]
111. Wang, H.; Cheng, H.; Wang, J.; Xu, L.; Chen, H.; Pei, R. Selection and characterization of DNA aptamers for the development of light-up biosensor to detect Cd(II). *Talanta* **2016**, *154*, 498–503. [CrossRef]
112. Overacker, R.D.; Plitzko, B.; Loesgen, S. Biolayer interferometry provides a robust method for detecting DNA binding small molecules in microbial extracts. *Anal. Bioanal. Chem.* **2021**, *413*, 1159–1171. [CrossRef]
113. Ciesielski, G.L.; Hytonen, V.P.; Kaguni, L.S. Biolayer Interferometry: A Novel Method to Elucidate Protein-Protein and Protein-DNA Interactions in the Mitochondrial DNA Replisome. *Methods Mol. Biol.* **2016**, *1351*, 223–231. [CrossRef]
114. Teague, J.L.; Barrows, J.K.; Baafi, C.A.; Van Dyke, M.W. Discovering the DNA-Binding Consensus of the Thermus thermophilus HB8 Transcriptional Regulator TTHA1359. *Int. J. Mol. Sci.* **2021**, *22*, 42. [CrossRef]
115. Kamat, V.; Rafique, A. Designing binding kinetic assay on the bio-layer interferometry (BLI) biosensor to characterize antibody-antigen interactions. *Anal. Biochem.* **2017**, *536*, 16–31. [CrossRef]
116. Wartchow, C.A.; Podlaski, F.; Li, S.; Rowan, K.; Zhang, X.; Mark, D.; Huang, K.S. Biosensor-based small molecule fragment screening with biolayer interferometry. *J. Comput. Aided Mol. Des.* **2011**, *25*, 669–676. [CrossRef]
117. Cusano, A.M.; Aliberti, A.; Cusano, A.; Ruvo, M. Detection of small DNA fragments by biolayer interferometry. *Anal. Biochem.* **2020**, *607*, 113898. [CrossRef]
118. Mahmoudinobar, F.; Britton, D.; Montclare, J.K. Protein-based lateral flow assays for COVID-19 detection. *Protein Eng. Des. Sel.* **2021**, *34*, gzab010. [CrossRef]
119. Sachdeva, S.; Davis, R.W.; Saha, A.K. Microfluidic Point-of-Care Testing: Commercial Landscape and Future Directions. *Front. Bioeng. Biotechnol.* **2020**, *8*, 602659. [CrossRef]
120. Boehringer, H.R.; O'Farrell, B.J. Lateral Flow Assays in Infectious Disease Diagnosis. *Clin. Chem.* **2021**, *68*, 52–58. [CrossRef]
121. Wang, L.; Wang, R.; Wei, H.; Li, Y. Selection of aptamers against pathogenic bacteria and their diagnostics application. *World J. Microbiol. Biotechnol.* **2018**, *34*, 149. [CrossRef] [PubMed]
122. Zhu, C.; Zhao, Y.; Yan, M.; Huang, Y.; Yan, J.; Bai, W.; Chen, A. A sandwich dipstick assay for ATP detection based on split aptamer fragments. *Anal. Bioanal. Chem.* **2016**, *408*, 4151–4158. [CrossRef] [PubMed]
123. Wang, T.; Chen, L.; Chikkanna, A.; Chen, S.; Brusius, I.; Sbuh, N.; Veedu, R.N. Development of nucleic acid aptamer-based lateral flow assays: A robust platform for cost-effective point-of-care diagnosis. *Theranostics* **2021**, *11*, 5174–5196. [CrossRef] [PubMed]
124. Petrakova, A.V.; Urusov, A.E.; Zherdev, A.V.; Dzantiev, B.B. Gold nanoparticles of different shape for bicolor lateral flow test. *Anal. Biochem.* **2019**, *568*, 7–13. [CrossRef] [PubMed]
125. Mohammadi, M.; Abazari, M.; Nourozi, M. Effects of two beta-adrenergic agonists on adipose tissue, plasma hormones and metabolites of Moghani ewes. *Small Ruminant. Res.* **2006**, *63*, 84–90. [CrossRef]
126. Fan, F.S. Assessing the Possible Influence of Residues of Ractopamine, a Livestock Feed Additive, in Meat on Alzheimer Disease. *Dement. Geriatr. Cogn. Dis. Extra* **2021**, *11*, 110–113. [CrossRef]
127. Wang, M.Y.; Zhu, W.; Ma, L.; Ma, J.J.; Zhang, D.E.; Tong, Z.W.; Chen, J. Enhanced simultaneous detection of ractopamine and salbutamol—Via electrochemical-facial deposition of MnO2 nanoflowers onto 3D RGO/Ni foam templates. *Biosens. Bioelectron.* **2016**, *78*, 259–266. [CrossRef]
128. Bownik, A.; Kowalczyk, M.; Banczerowski, J. Lambda-cyhalothrin affects swimming activity and physiological responses of Daphnia magna. *Chemosphere* **2019**, *216*, 805–811. [CrossRef]
129. Djouaka, R.; Soglo, M.F.; Kusimo, M.O.; Adeoti, R.; Talom, A.; Zeukeng, F.; Paraiso, A.; Afari-Sefa, V.; Saethre, M.G.; Manyong, V.; et al. The Rapid Degradation of Lambda-Cyhalothrin Makes Treated Vegetables Relatively Safe for Consumption. *Int. J. Environ. Res. Public Health* **2018**, *15*, 1536. [CrossRef]
130. Bradberry, S.M.; Cage, S.A.; Proudfoot, A.T.; Vale, J.A. Poisoning due to pyrethroids. *Toxicol. Rev.* **2005**, *24*, 93–106. [CrossRef]
131. Banerjee, G.; Chattopadhyay, P. Vanillin biotechnology: The perspectives and future. *J. Sci. Food Agric.* **2019**, *99*, 499–506. [CrossRef]

132. Mourtzinos, I.; Konteles, S.; Kalogeropoulos, N.; Karathanos, V.T. Thermal oxidation of vanillin affects its antioxidant and antimicrobial properties. *Food Chem.* **2009**, *114*, 791–797. [CrossRef]

133. Yadav, M.; Pandey, R.; Chauhan, N.S. Catabolic Machinery of the Human Gut Microbes Bestow Resilience Against Vanillin Antimicrobial Nature. *Front. Microbiol.* **2020**, *11*, 588545. [CrossRef]

134. Schwab, W. Natural 4-hydroxy-2,5-dimethyl-3(2H)-furanone (Furaneol(R)). *Molecules* **2013**, *18*, 6936–6951. [CrossRef]

135. Prester, L. Biogenic amines in fish, fish products and shellfish: A review. *Food Addit. Contam. Part A Chem. Anal. Control. Expo. Risk Assess* **2011**, *28*, 1547–1560. [CrossRef]

136. Bogdanovic, T.; Petricevic, S.; Brkljaca, M.; Listes, I.; Pleadin, J. Biogenic amines in selected foods of animal origin obtained from the Croatian retail market. *Food Addit. Contam. Part A Chem. Anal. Control Expo. Risk Assess* **2020**, *37*, 815–830. [CrossRef]

137. Alvarez-Elcoro, S.; Enzler, M.J. The macrolides: Erythromycin, clarithromycin, and azithromycin. *Mayo. Clin. Proc.* **1999**, *74*, 613–634. [CrossRef]

138. Ashraf, A.; Liu, G.; Yousaf, B.; Arif, M.; Ahmed, R.; Irshad, S.; Cheema, A.I.; Rashid, A.; Gulzaman, H. Recent trends in advanced oxidation process-based degradation of erythromycin: Pollution status, eco-toxicity and degradation mechanism in aquatic ecosystems. *Sci. Total Environ.* **2021**, *772*, 145389. [CrossRef]

139. Harju, K.; Koskela, H.; Kremp, A.; Suikkanen, S.; de la Iglesia, P.; Miles, C.O.; Krock, B.; Vanninen, P. Identification of gymnodimine D and presence of gymnodimine variants in the dinoflagellate Alexandrium ostenfeldii from the Baltic Sea. *Toxicon* **2016**, *112*, 68–76. [CrossRef]

140. Rowdhwal, S.S.S.; Chen, J. Toxic Effects of Di-2-ethylhexyl Phthalate: An Overview. *Biomed. Res. Int.* **2018**, *2018*, 1750368. [CrossRef]

141. Deigan, K.E.; Ferre-D'Amare, A.R. Riboswitches: Discovery of drugs that target bacterial gene-regulatory RNAs. *Acc. Chem. Res.* **2011**, *44*, 1329–1338. [CrossRef] [PubMed]

142. Born, J.; Weitzel, K.; Suess, B.; Pfeifer, F. A Synthetic Riboswitch to Regulate Haloarchaeal Gene Expression. *Front Microbiol.* **2021**, *12*, 696181. [CrossRef] [PubMed]

143. Topp, S.; Reynoso, C.M.; Seeliger, J.C.; Goldlust, I.S.; Desai, S.K.; Murat, D.; Shen, A.; Puri, A.W.; Komeili, A.; Bertozzi, C.R.; et al. Synthetic riboswitches that induce gene expression in diverse bacterial species. *Appl. Environ. Microbiol.* **2010**, *76*, 7881–7884. [CrossRef] [PubMed]

144. Lotz, T.S.; Suess, B. Small-Molecule-Binding Riboswitches. *Microbiol. Spectr.* **2018**, *6*, RWR-0025-2018. [CrossRef]

145. Groher, F.; Bofill-Bosch, C.; Schneider, C.; Braun, J.; Jager, S.; Geissler, K.; Hamacher, K.; Suess, B. Riboswitching with ciprofloxacin-development and characterization of a novel RNA regulator. *Nucleic Acids Res.* **2018**, *46*, 2121–2132. [CrossRef]

146. Boussebayle, A.; Torka, D.; Ollivaud, S.; Braun, J.; Bofill-Bosch, C.; Dombrowski, M.; Groher, F.; Hamacher, K.; Suess, B. Next-level riboswitch development-implementation of Capture-SELEX facilitates identification of a new synthetic riboswitch. *Nucleic Acids Res.* **2019**, *47*, 4883–4895. [CrossRef]

147. Mally, A.; Solfrizzo, M.; Degen, G.H. Biomonitoring of the mycotoxin Zearalenone: Current state-of-the art and application to human exposure assessment. *Arch. Toxicol.* **2016**, *90*, 1281–1292. [CrossRef]

148. Liew, W.P.; Mohd-Redzwan, S. Mycotoxin: Its Impact on Gut Health and Microbiota. *Front. Cell. Infect. Microbiol.* **2018**, *8*, 60. [CrossRef]

149. Saber, T.M.; Abd El-Aziz, R.M.; Ali, H.A. Quercetin mitigates fenitrothion-induced testicular toxicity in rats. *Andrologia* **2016**, *48*, 491–500. [CrossRef]

MDPI
St. Alban-Anlage 66
4052 Basel
Switzerland
www.mdpi.com

Biosensors Editorial Office
E-mail: biosensors@mdpi.com
www.mdpi.com/journal/biosensors